M. Mehring

High Resolution NMR Spectroscopy in Solids

With 104 Figures

Springer-Verlag
Berlin Heidelberg New York 1976

Professor Dr. Michael Mehring

Universität Dortmund, Institut für Physik
Baroper Straße, D-4600 Dortmund 50

ISBN 3-540-07704-9 Springer-Verlag Berlin Heidelberg New York
ISBN 0-387-07704-9 Springer-Verlag Berlin Heidelberg New York

Library of Congress Cataloging in Publication Data. Mehring, Michael, 1937 – High resolution NMR in solids. (NMR, basic principles and progress; v. 11) Bibliography: p. 1. Nuclear magnetic resonance spectroscopy. I. Title. II. Series. QC490.N2 vol. 11 [QC762] 538'.3 76-10680

Typesetting, printing and bookbinding: Brühlsche Universitätsdruckerei, Gießen.

NMR

Basic Principles and Progress
Grundlagen und Fortschritte

11

Editors: P. Diehl E. Fluck R. Kosfeld

Editorial

Since the Series "NMR—Basic Principles and Progress" was founded in 1969 it has dealt primarily with the theoretical and physical aspects of the methods. Today nuclear-magnetic resonancespectroscopy has become one of the principal techniques of the chemist and is finding increasing use in the fields of Biology, Pharmacy, Medicine, and Criminology. The growing significance of applied spectroscopy has earned it a correspondingly important place for the future in this Series. With the aim of achieving a balanced representation of theoretical and practical problems and results, the present Editors have asked several world-renowned scientists in the field of NMR-spectroscopy to join an International Editorial Board.

The international nature of this Board will facilitate closer contact among research groups and authors throughout the world, making it possible to follow comprehensively the developments in pure and applied NMR-spectroscopy. On this basis, the readers of the Series will be assured of up-to-date contributions not only of current significance, but of long-term value as well.

Prof. E. Fluck   Prof. P. Diehl   Prof. R. Kosfeld 1976

Preface

Manipulation and Dilution
Tools for Ruling Abundant Species

"NMR is dead" was the slogan heard in the late 1960s at least among physicists, until John S. Waugh and his co-workers initiated a series of new NMR experiments, which employed the coherent modulation of interactions by strong radiofrequency fields. A wealth of new phenomena was observed, which are summarized in the introduction for the convenience of the unbiased reader, whereas Section 2 collects the basic spin interactions observed in solids.

Line-narrowing effects in dipolar coupled solids by the application of multiple-pulse experiments are extensively discussed in Section 3. Numerous extensions of the basic Waugh, Huber, and Haeberlen experiment have been developed by different groups and have been applied to the nuclei ^{1}H, ^{9}Be, ^{19}F, ^{27}Al, ^{31}P, ^{63}Cu in solids. Application of this technique to a variety of systems is still in progress and should reveal interesting insights into weak spin interactions in solids.

It was soon realized that rare spins could be used as monitors for molecular fields in the solid state; however, rare spin observation is difficult because of the small signal-to-noise ratio. Pines, Gibby, and Waugh introduced a new concept of cross-polarization, based on ideas of Hahn and co-workers, which allows the detection of rare spins with increased sensitivity. The dynamics involved are treated in detail. Other sections merely list results obtained by the techniques described and demonstrate their usefulness in the investigation of dynamical problems in molecular and solid state physics.

Prerequisite to reading this monograph is some familiarity with the fundamental book by A. Abragam and M. Goldman. Additional reading of the review article written by U. Haeberlen is highly recommended.

I have tried very hard to cover the whole current literature in this growing field; however, I am aware that I have certainly missed important contributions. Of those who suffer from this, I herewith beg pardon.

Among my friends and colleagues I am particularly indebted, for their patient criticisms, discussions and comments to O. Kanert, A. Pines and J. S. Waugh. Among these my friend A. Pines has encouraged and excited me continuously and I have benefited tremendously from this ingenious restlessness. I gratefully acknowledge the kind hospitality of J. S. Waugh during my stay at the Massachusetts Institute of Technology during 1969–1971, where I learned about these fascinating experiments. This monograph would certainly not have been written without the help of my friend and colleague O. Kanert, who took most of the administrative burden off my shoulders during the time of writing.

I am very much obliged to my co-workers, who have supported me tremendously with the preparation of the manuscript. Several drawings had to be specially computed and material had to be prepared. Special thanks are due to J. Becker, H. Raber, G. Sinning, D. Suwelack and E. Wolff. There are numerous scientists from whose discussion I have benefited greatly in the past, among these I am particularly indebted to R. G. Griffin, R. W. Vaughan and especially to U. Haeberlen, who kindly let me have the manuscript of his review article prior to publication.

I also gratefully acknowledge the patience and endurance of Miss A. Gassner, who typed the manuscript in several versions, and of Mrs. A. Schröder and Mrs. L. Sinning who have drawn and photographed most of the figures.

Finally I want to apologize to my wife Sabine and the children for spoiling many sunny weekends by working on this monograph. Their patience and understanding are gratefully acknowledged here.

The editors of this series have contributed much to the final layout of this volume and their continuous interest and support are highly appreciated.

Dortmund, June 21, 1976 M. Mehring

Table of Contents

List of Editors

Managing Editors

Professor Dr. Peter Diehl, Physikalisches Institut der Universität Basel, Klingelbergstr. 82, CH-4056 Basel

Professor Dr. Ekkehard Fluck, Institut für Anorganische Chemie der Universität Stuttgart, Pfaffenwaldring 55, D-7000 Stuttgart 80

Professor Dr. Robert Kosfeld, Institut für Physikalische Chemie der Rhein.-Westf. Technischen Hochschule Aachen, Tempelgraben 59, D-5100 Aachen

Editorial Board

Professor Sture Forsén, Department of Physical Chemistry, Chemical Centre, University of Lund, P.O.B. 740, S-22007 Lund

Professor Dr. Shizuo Fujiwara, Department of Chemistry, Faculty of Science, The University of Tokyo, Bunkyo-Ku, Tokyo, Japan

Dr. R. K. Harris, School of Chemical Sciences, The University of East Anglia, Norwich NR4 7TJ, Great Britain

Professor C. L. Khetrapal, Raman Research Institute, Bangalore - 560006, India

Professor E. Lippmaa, Department of Physics, Institute of Cybernetics Academy of Sciences of the Estonian SSR, Lenini puiestee 10, Tallinn 200001, USSR

Professor G. J. Martin, Chimie Organique Physique, Université de Nantes, UER de Chimie, 38, Bd. Michelet, F-44 Nantes, B.P. 1044

Professor A. Pines, Department of Chemistry, University of California, Berkeley, CA 94720, USA

Professor Franz H. A. Rummens, Department of Chemistry, University of Regina, Regina, Saskatschewan S4S OA2, Canada

Professor Dr. Bernard L. Shapiro, Department of Chemistry, Texas A and M University, College Station, TX 77843, USA

1. Introduction

Spin engineering has brought about a wealth of techniques to overcome the natural line broadening mechanisms in solids, such as dipole-dipole and quadrupole interactions. We are going to review in this monograph the different techniques involved and we shall discuss the results obtained. For the convenience of the unbiased reader let us first take a look at some representative results.

As is well known to the chemist, the NMR spectrum of a liquid consists of numerous sharp lines typically with less than 1 Hz linewidth, due to magnetic field inhomogeneities or spin relaxation [*1*]. In order to supply a reference to this concept of "High Resolution NMR", Fig. 1.1 displays as a representative example the spectrum of ethyl alcohol [*2*]. Neither *manipulation* nor *dilution* is indicated in order to obtain the NMR spectrum of this compound in the liquid state. It may be obtained in a rather standard fashion by taking simply the NMR spectrum of the liquid sample. However, also high resolution NMR spectroscopists like to manipulate on their spectra as is demonstrated

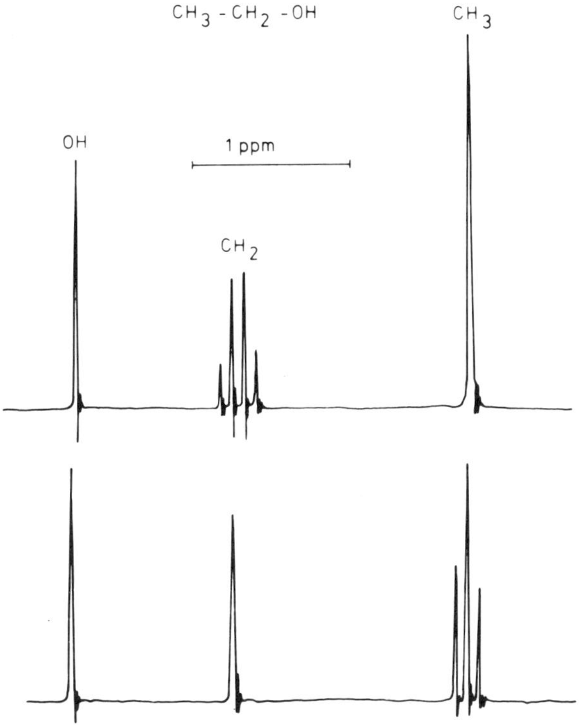

Fig. 1.1. Highly resolved proton spectrum of ethanol using spin decoupling. Top: While recording the methyl group line, irradiation was performed on the methylene resonance. Bottom: While recording the methylene group line, irradiation was performed on the methyl group resonance [*2*]

in Fig. 1.1. We note in advance that in general "High Resolution NMR in Solids" has not lived up to this state of the art. The interaction Hamiltonian in a liquid sample is represented by isotropic chemical shift and scalar spin-spin interactions. All possible anisotropic interactions, namely chemical shift anisotropy, dipole-dipole interaction, quadrupole interaction etc. are averaged to zero due to the rapid isotropic molecular motion i.e., nature performs some manipulations in this case.

In the solid state, however, all these anisotropic interactions are retained and may be used to monitor the symmetry properties and the electronic state of the solid [*1*]. Unfortunately in many cases (like ^{19}F and ^{1}H) the dipole-dipole interaction is overwhelming at ordinary magnetic field strength (1–6 Tesla). This results in a more or less bell shaped, structureless line shape, from which very little information can be extracted about the local symmetry and electronic configuration. In this sense the goal of high resolution NMR in solids can be formulated as designing methods to repress the "unwanted" dipolar interaction considerably, leaving chemical shift anisotropies, scalar interactions etc. more or less unaffected.

The natural way of achieving this goal would be by *dilution* of the spins, since the dipolar interaction is proportional to r^{-3}, where r is the distance between the spins. This, however, leads to "High Resolution" only in that case, where no other heteronuclear spins are present. In favourable cases such as in a dilute spin system with small gyromagnetic ratio and large chemical shift anisotropy, already the ordinary NMR spectrum yields a "High Resolution Spectrum" in the sense that anisotropic shift interactions are observable. If in addition considerable molecular motion is present, highly resolved spectral lines can be observed in a solid, as was demonstrated by Andrew and co-workers [*3*] in the case of solid P_4S_3 whose spectrum is shown in Fig. 1.2.

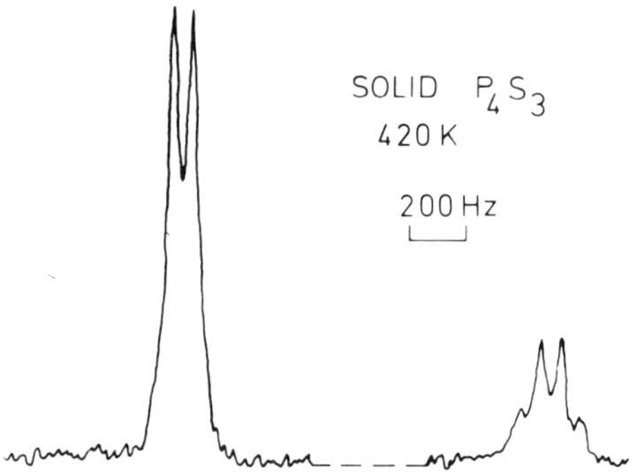

Fig. 1.2. ^{31}P NMR spectrum in solid P_4S_3 at 420 K (melting point 446 K) by Andrew, Hinshaw and Jasinski [*3*]. The spectrum is strongly narrowed by molecular motion in the solid. The AB_3 type fine structure is represented by a chemical shift separation between the doublet (basal nuclei) and the quartet (apical nuclei) of 185 ± 2 ppm. The coupling constant is J = 70 ± 3 Hz

In this monograph, however, we are dealing with the case where the natural line broadening due to dipolar and quadrupolar interactions masks other weak interactions, such as shift interactions and scalar couplings. The first attempt to overcome this obstacle was made indepently by E. R. Andrew *et al.* [*4*] and by I. Lowe [*5*] by using a specimen rotation method. The whole sample is rapidly rotated, in this method, about an axis tilted by the "magic angle" $\vartheta_m = 54°44'\ 8''\ 12'''$ ($\tan\vartheta_m = \sqrt{2}$) with respect to the static magnetic field H_0. It can be shown that the average dipolar interaction vanishes in this case. A representative example is shown in Fig. 1.3. The reader will realize that shielding anisotropies are also averaged to

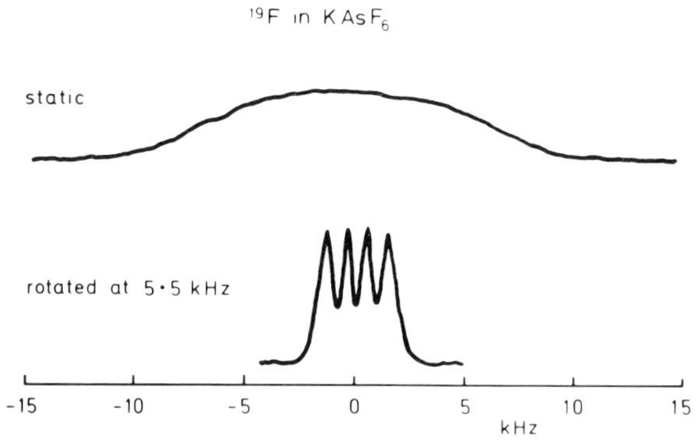

Fig. 1.3. Application of the magic angle specimen rotation method to ^{19}F in polycrystalline $KAsF_6$ by Andrew, Farnell and Gledhill [*46*]. The upper spectrum shows the ordinary NMR line without sample rotation. Dipolar coupling among the spins governs the line width. The lower spectrum corresponds to a magic angle rotation of the sample with 5.5 KHz, displaying a quartet structure due to electron-coupled interaction between ^{19}F and ^{75}As nuclei ($I = 3/2$). The coupling constant is $J = 905$ Hz

zero and only the isotropic shift is retained. On the other hand, homonuclear and heteronuclear dipolar interactions vanish, since this technique is not spin specific. Excellent review articles have been written on this subject by E. R. Andrew [*6*]. Therefore we are not treating this subject in detail; however, we shall touch upon it occasionally. One of our main purposes is to review the recently developed techniques of *manipulation*, which operate in spin space and which are capable of reducing dipolar as well as quadrupolar interactions considerably. The first useful multiple pulse cycle which has succussfully been applied in this sense, was the four-pulse cycle of J. W. Waugh, L. M. Huber and U. Haeberlen [*7*], often referred to as WAHUHA experiment. Patent holders: J. S. Waugh and U. Haeberlen, U.S. patent No. 3,530, 374. Figure 1.4 represents the "High Resolution" spectrum of ^{19}F in C_6F_{12} obtained with such a four-pulse experiment in comparison with the ordinary NMR spectrum [*8*]. The great potential of these techniques for the resolution of weak spin interactions is demonstrated clearly. The reason why no such anisotropy is observed in Fig. 1.4 is because nature supplies enough motion at the applied temperature, to average anisotropic shielding, leaving only the scalar interactions. Since this method operates in spin space, all anisotropic spin interactions which are linear in the spin varaible are retained and have been extensively studied in powder samples as well as in single crystals [*9–12*]. Modifications of this basic multiple-pulse line narrowing

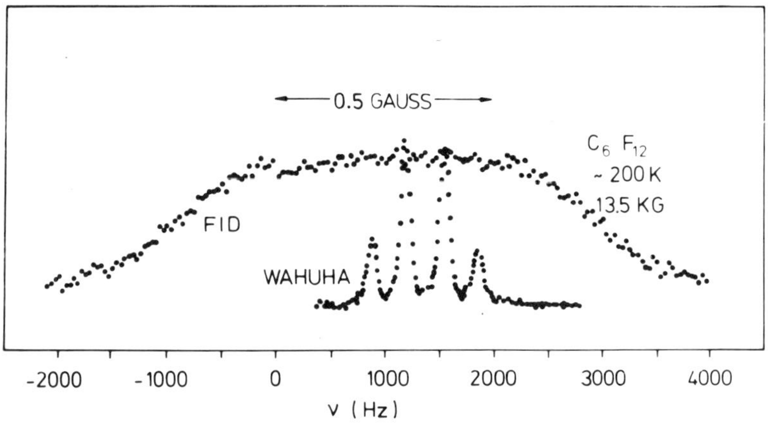

Fig. 1.4. Application of the multiple pulse method to ^{19}F in polycrystalline C_6F_{12} at 200 K by Ellett, Haeberlen and Waugh [*8*]. Upper curve: Normal NMR spectrum governed by dipolar coupling among the ^{19}F nuclei. Lower curve AB type spectrum (J = 310 Hz, δ = 17.5 ppm) after removal of the dipolar interaction by applying a multiple-pulse sequence (WAHUHA). Chemical shift anisotropy is removed by isotropic molecular motion in the solid, but is observed at a lower temperature

experiment have been developed which in some cases are capable of extremely high resolution [*13*].

Dilution of spin systems is often done on purpose, however, nature supplies a wealth of diluted spin systems with very low natural abundance, such as ^{13}C, ^{15}N, ^{43}Ca etc. in a surrounding of abundant spins like ^{1}H and ^{19}F. In order to obtain a high signal to noise, high resolution spectrum, the rare spins are first polarized by the virtue of spin order transfer from the abundant spins. During the subsequent observation of the rare spins, the abundant spins are decoupled, in order to repress the broadening due to heteronuclear dipolar interaction. This technique was first applied by A. Pines, M. Gibby, and J. S. Waugh and is referred to as "Proton Enhanced Induction Spectroscopy: PENIS" [*14*]. Patentholders: A. Pines, M. Gibby and J. S. Waugh, U.S. Pat. No. 3,792,346 (1974). Figure 1.5 shows the ^{13}C spectrum of natural abundant (1.1%) ^{13}C in adamantane, obtained by the PENIS technique. There is considerable motion present in adamantane at room temperature to average out chemical shift anisotropies, leaving only the isotropic shift [*15*].

In Chapter 2 we remind the reader of the tensorial character of all spin interactions which manifest itself by second rank tensors in ordinary solids. There may be interactions of higher rank in principle, but no valid experimental verification has been given of this to my knowledge. There has been a discussion on the transformation properties of these interactions. We restrict ourselves, however, in this chapter to rotations in real space. In this context also the specimen rotation method is discussed as well as molecular reorientation.

Chapter 3 deals with multiple-pulse experiments, discussing the application of pulse cycles to the spin system, mainly serving the goal of *repressing* the dipolar interaction. Coherent averaging, second averaging and the influence of pulse imperfections are treated in detail.

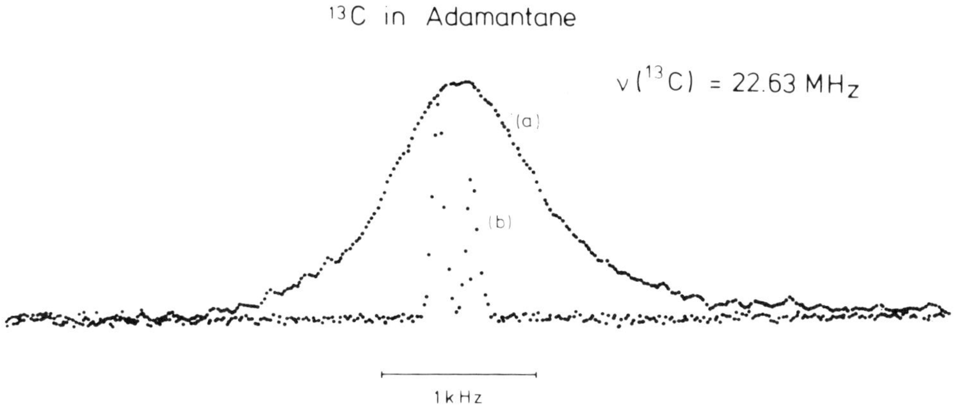

Fig. 1.5 a and b. Application of the cross-polarization method of Pines, Gibby and Waugh [*14*] to ^{13}C in polycrystalline adamantane at room temperature. (a) Undecoupled ^{13}C spectrum governed by ^{13}C-^{1}H dipolar coupling, (b) Decoupled ^{13}C spectrum displaying two chemically inequivalent ^{13}C nuclei. Chemical shift anisotropy is not observed due to the rapid isotropic motion of the molecules. (Courtesy of G. Sinning)

Double resonance experiments on rare spins with the purpose of obtaining high resolution spectra in solids are analyzed in Chapter 4.

After reviewing the principles of double resonance we turn to cross-polarization experiments, which recently have supplied a wealth of ^{13}C spectra in solids. Cross-polarization dynamics and spin decoupling dynamics are discussed also.

The main application of the techniques of *manipulation* and *dilution* has been the determination of magnetic shielding tensors, which we deal with in Chapter 5. After a brief introduction to the concepts we summarize the shielding tensors, thus far determined by these techniques.

Spin lattice relaxation has not played a dominant role in high resolution NMR in solids so far. However, as we see in Chapter 6 there are some useful applications of the techniques which have been described in the preceeding chapters about the investigation of spin lattice relaxation processes. The *appendices* summarize for the convenience of the reader some useful aspects to which we refer in the text. Reviews on the subject have been written by P. Mansfield [*16*], R. W. Vaughan [*17*] and E. R. Andrew [*6*]. A forthcoming review article by U. Haeberlen [*18*] covers part of this subject taking a slightly different approach and is highly recommended to the reader.

2. Nuclear Spin Interactions in Solids

For the convenience of the reader we summarize in this section the basic nuclear spin interactions which occur in solids. The notation will be in tensorial form throughout to emphasize the anisotropy of these interactions. In order to keep the presentation as compact as possible we shall avoid detailed derivations and refer the reader to the outstanding book by A. Abragam [*1*]. Also Pool and Farach [*2*] have given a detailed description of spin interactions in tensorial form.

2.1 Basic Nuclear Spin Interactions in Solids

Figure 2.1 can serve as a guide to the different basic interactions occurring in solids. It represents the sevenfold way a nuclear spin can communicate with its surrounding.

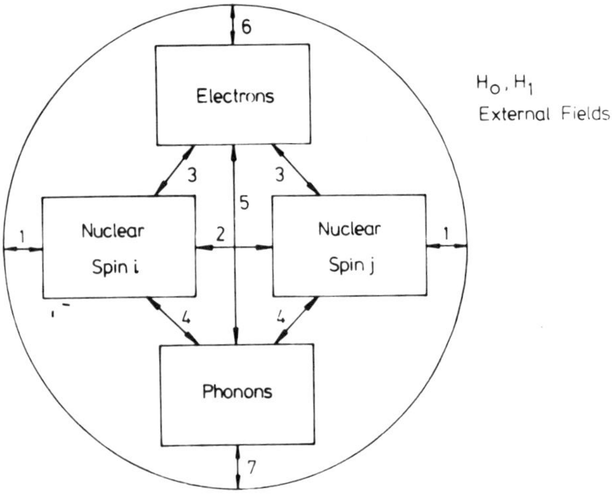

Fig. 2.1. The seven-fold way a nuclear spin system can interact with its surrounding. 1: Zeeman interaction of spins. 2: Direct spin interaction. 3: Nuclear spin-electron interaction and indirect spin interaction. 4: Direct spin-lattice interaction. 3–5: Indirect spin lattice interaction via electrons. 3–6: Shielding and polarization of nuclear spins by electrons. 4–7: Coupling of nuclear spins to sound fields etc.

We shall refer to this figure later in the text. We first distinguish nuclear spin interactions between external fields (H_0,H_1) and internal fields:

$$\mathcal{H} = \mathcal{H}_{\text{ext.}} + \mathcal{H}_{\text{int.}} \tag{2.1}$$

where

$$\mathcal{H}_{\text{ext.}} = \mathcal{H}_0 + \mathcal{H}_1.$$

$\mathcal{H}_0$ and $\mathcal{H}_1$ are the Zeeman interactions with the external fields H_0 and H_1, respectively. This distinction is appropriate in fact, because we are going to work in a regime where the "size" of $\mathcal{H}_0$ and $\mathcal{H}_1$ is much larger than the "size" of $\mathcal{H}_{\text{int}}$. By the "size" of an Hamiltonian we mean unless otherwise stated:

$$\|\mathcal{H}\| = [\text{Tr}\{\mathcal{H}^2\}]^{1/2}. \tag{2.2}$$

The spin interactions of two different types of spins (I, S) (gyromagnetic ratios γ_I and γ_S) with internal fields may be written as

$$\mathcal{H}_{\text{int}} = \mathcal{H}_{II} + \mathcal{H}_{SS} + \mathcal{H}_{IS} + \mathcal{H}_Q + \mathcal{H}_S + \mathcal{H}_L \tag{2.3}$$

where $\mathcal{H}_{II}$ and $\mathcal{H}_{SS}$ represent the direct (dipolar) as well as the indirect interactions among I spins and S spins respectively (path 2 and 3 in Fig. 2.1). The same paths are involved in the I-S interaction as expressed by $\mathcal{H}_{IS}$, which covers the direct as well as the indirect interactions between I spins and S spins. $\mathcal{H}_Q$ is the quadrupole Hamiltonian of the I and S spins respectively. $\mathcal{H}_S$ contains all shielding Hamiltonians (chemical shift and Knight shift) of the I and S spins (path 3,6 in Fig. 2.1).

$\mathcal{H}_L$ describes the spin lattice interaction (path 4 and 3,5 in Fig. 2.1). We find it convenient to express all Hamiltonian in frequency units (ω-units) throughout this monograph. In the following we assume that the symmetry of the solid is such that all the spin interactions can be represented by second rank Cartesian tensors in the following way [*1,2*]:

$$\mathcal{H} = \mathbf{X} \cdot \mathbf{A} \cdot \mathbf{Y} = \sum_{i,j} X_i \cdot A_{ij} \cdot Y_j \tag{2.4}$$

where $\mathbf{X}$ and $\mathbf{Y}$ are vectors and $\mathbf{A}$ is a second rank Cartesian tensor (3×3 matrix).

Coupling of the spin I to external fields is now expressed as:

$$\begin{aligned} \mathcal{H}_{0I} &= \mathbf{H}_0 \cdot \mathbf{Z} \cdot \mathbf{I} \\ \mathcal{H}_{1I} &= \mathbf{H}_1 \cdot \mathbf{Z} \cdot \mathbf{I} \end{aligned} \tag{2.5}$$

where $\mathbf{H}_0 = (H_x, H_y, H_z)$,
$\mathbf{H}_1(t) = 2(H_{1x}(t), H_{1y}(t), H_{1z}(t)) \cos\omega t$ and $\mathbf{Z} = -\gamma_I \mathbf{E}$

with the unit matrix

$$\mathbf{E} = \begin{pmatrix} 1 & 0 & 0 \\ 0 & 1 & 0 \\ 0 & 0 & 1 \end{pmatrix}$$

being the coupling matrix between the spin I and the magnetic field H_0. Similarly we express the coupling between two spins $I^{(i)}$ and $I^{(j)}$ as

$$\mathcal{H}_{II} = \mathbf{I}^{(i)} \cdot \mathbf{D}_{II} \cdot \mathbf{I}^{(j)} \tag{2.6}$$

or between I and S spins as

$$\mathcal{H}_{IS} = \mathbf{I} \cdot \mathbf{D}_{IS} \cdot \mathbf{S}. \tag{2.7}$$

Note that the tensor $\mathbf{D}$ in Eqs. (2.6) and (2.7) covers both, direct spin-spin interactions (dipolar interactions) as well as indirect spin-spin interactions (scalar coupling, pseudo dipolar interaction and its antisymmetric counterpart) i.e. in general

$$\mathrm{Tr}\{\mathbf{D}\} \neq 0$$

and $\mathbf{D}$ is not necessarily symmetric. In the case of *dipolar interaction*, however, $\mathbf{D}$ is axially symmetric and additionally

$$\mathrm{Tr}\{\mathbf{D}\} = 0.$$

In this case $\mathcal{H}_{II}$ and $\mathcal{H}_{IS}$ can be expanded into the dipolar alphabet (see A. Abragam [*1*]) and D_{ij}, $(i, j = x, y, z)$ may be expressed as

$$D_{ij} = \frac{\gamma_I \gamma_S \hbar}{r^3} (\delta_{ij} - 3e_i e_j) \quad (i, j) = x, y, z \tag{2.8}$$

where r is the distance between the spins, δ_{ij} is the Kronecker delta and $e_i (i = x, y, z)$ are x, y and z-components of a unit vector pointing from one spin to the other.

We consider the quadrupole Hamiltonian as being a degenerate case of Eq. (2.6) with

$$\mathcal{H}_{QI} = \mathbf{I} \cdot \mathbf{Q} \cdot \mathbf{I} \tag{2.9}$$

where

$$\mathbf{Q} = \frac{eQ}{2I(2I-1)\hbar} \mathbf{V}$$

with the nuclear quadrupole moment Q and the electric field gradient tensor $\mathbf{V}$. The trace of $\mathbf{V}$ vanishes due to the Laplace equation.

The shielding Hamiltonian of the spin I can be visualized as the coupling of the spin I with the magnetic field H_0 via the shielding tensor $\mathbf{S}$:

$$\mathcal{H}_S = \mathbf{H}_0 \cdot \mathbf{S} \cdot \mathbf{I}. \tag{2.10}$$

The tensor $\mathbf{S}$, however, is no longer the unit matrix as in the case of the Zeeman interaction [Eq. (2.5)], since the coupling is established via the electronic surrounding (path 3,6 in Fig. 2.1) rather than being direct (path 1 in Fig. 2.1). This is why the investigation of shielding interactions can give valuable information about electronic states in solids and molecules. The main purpose of the techniques to be described in this book serves the goal of measuring just this quantity. Here we would like to point out that the corresponding tensor interaction in esr is the g-tensor. The shielding tensor $\mathbf{S}$ in Eq. (2.10) may be expressed in the case of the chemical shift tensor as

$$\mathbf{S} = \gamma_I \begin{pmatrix} \sigma_{xx} & \sigma_{xy} & \sigma_{xz} \\ \sigma_{yx} & \sigma_{yy} & \sigma_{yz} \\ \sigma_{zx} & \sigma_{zy} & \sigma_{zz} \end{pmatrix} \tag{2.11}$$

and similarly in the case of the Knight shift tensor as

$$\mathbf{K} = -\gamma_I (K_{ij}).$$

In the following we do not distinguish between $\mathbf{S}$ and $\mathbf{K}$ and assume all shielding interactions which are linear in the spin variable to be represented by $\mathbf{S}$. In general $\mathrm{Tr}\{\mathbf{S}\} \neq 0$ and $\mathbf{S}$ is not necessarily symmetric. However, as shown in Appendix C antisymmetric components of $\mathbf{S}$ contribute to the resonance shift only in second order and can usually be ignored.

The spin rotation interaction $\mathcal{H}_J$, which represents the coupling of the nuclear spin I with the magnetic moment produced by the angular momentum $\mathbf{J}$ of the molecule can be expressed as

$$\mathcal{H}_J = \mathbf{J} \cdot \mathbf{C} \cdot \mathbf{I}. \tag{2.12}$$

The spin rotation interaction tensor $\mathbf{C}$ has some common features with the shielding tensor $\mathbf{S}$, since both are associated with the "dequenching" of orbital angular momentum, whether by H_0 or by $\mathbf{J}$.

The corresponding Hamiltonians for the S spins may be easily obtained by exchanging I with S in Eqs. (2.5–2.7, 2.9, 2.10, 2.12). The spin lattice interaction $\mathcal{H}_L$ involves the scattering of a phonon with the spin. In NMR however, this interaction is usually treated semiclassically by rendering the spin interactions time dependent.

In Appendix A we have derived the irreducible tensor representation of spin interactions where we express the interactions by spherical, rather than Cartesian tensors. This is particularly convenient when rotations are involved. We remark, however, that the clear physical apprehension of direction cosines leaves sufficient validity to the representation in Cartesian coordinates, and we shall use both representations in the following sections. Instead of Eq. (2.4) we may express the interaction Hamiltonian by irreducible spherical tensors as [*3, 4*] (see Appendix A):

$$\mathcal{H} = \sum_{k=0}^{2} \sum_{q=-k}^{k} (-1)^q A_{kq} T_{k-q} \tag{2.13}$$

where A_{kq} refers to spatial coordinates and T_{kq} to spin variables. The spherical tensors A_{kq} and T_{kq} can be readily expressed by their Cartesian counterparts by using Eq. (A.12).

Under rapid isotropic molecular reorientation, which is the general case in liquids, but which can occur in solids also, the second rank tensor interaction is averaged and only the scalar invariant of the tensor $\{T_{ij}\}$ i.e.

$$\frac{1}{3}\mathrm{Tr}\{T_{ij}\} = \frac{1}{3}\sum_{k} T_{kk}$$

is retained. Notice, that $\mathcal{H}_{ext}$ is not changed under rotation of the sample. The other interaction Hamiltonians can be written in the isotropic average as follows:
Indirect spin spin interaction:

$$\mathcal{H}_{IS} = J\mathbf{I}\cdot\mathbf{S} \tag{2.14}$$

where

$$J = \frac{1}{3}\mathrm{Tr}\{\mathbf{D}_{IS}\}. \tag{2.15}$$

Chemical shift interaction:

$$\mathcal{H}_{S} = \gamma_I \sigma \mathbf{I}\cdot\mathbf{H}_0 \tag{2.16}$$

where

$$\sigma = \frac{1}{3}\mathrm{Tr}\{\sigma_{ij}\}$$

and similar in the case of Knight shift interaction.
In the case of spin rotation interaction:

$$\mathcal{H}_{J} = c\mathbf{I}\cdot\mathbf{J} \tag{2.17}$$

where

$$c = \frac{1}{3}\mathrm{Tr}\,\{C_{ij}\}.$$

With $\mathbf{H}_0 = (0,0,\mathrm{H}_0)$ the total Hamiltonian of the spin interactions may be expressed in the isotropic average as:

$$\mathcal{H}_{\mathrm{int}} = -\gamma_I H_0(1-\sigma_I+K_I)\,\bar{I}_z - \gamma_S H_0(1-\sigma_S+K_S)S_z + J\mathbf{I}\cdot\mathbf{S}. \tag{2.18}$$

This is the usual Hamiltonian in a liquid. "High Resolution NMR" is supplied by nature in this case without any extra manipulations on the spin system. We will not be concerned with this degenerate case in this book. Our ultimate goal, however, is to approach a related situation by *manipulation* and/or *dilution* in a solid sample, while retaining some tensorial character of the spin interactions.

2.2 Spin Interactions in High Magnetic Fields

If we apply a static magnetic field H_0 to the spin system, forces (angular momentum) are exerted on the spin *via* the Zeeman interaction. As a consequence, the spin frame is accelerated with respect to the laboratory frame, i.e. the spin interactions become time dependent [5]. The Hamiltonian is no longer a scalar quantity, but becomes a "tensor operator". If H_0 is large in the sense $\|\mathcal{H}_0\| \gg \|\mathcal{H}_{\text{int}}\|$ only those parts of $\mathcal{H}_{\text{int}}$ which commute with $\mathcal{H}_0$ contribute in first order to the spectrum, i.e. we can split $\mathcal{H}_{\text{int}}$ into two parts

$$\mathcal{H}_{\text{int}} = \mathcal{H}'_{\text{int}} + \mathcal{H}''_{\text{int}} \tag{2.19}$$

with

$$[\mathcal{H}_0, \mathcal{H}'_{\text{int}}] = 0 \text{ and } [\mathcal{H}_0, \mathcal{H}''_{\text{int}}] \neq 0$$

where $\mathcal{H}'_{\text{int}}$ is the so called secular part and $\mathcal{H}''_{\text{int}}$ is the non-secular part of $\mathcal{H}_{\text{int}}$. The NMR spectrum is "generated" by $\mathcal{H}'_{\text{int}}$ in first order.
With

$$\mathcal{H}_0 = -\omega_{0S} S_z - \omega_{0I} I_z \tag{2.20}$$

where

$$\omega_{0I} = \gamma_I H_0 \text{ and } \omega_{0S} = \gamma_S H_0$$

an interaction representation is imposed, which renders the interaction Hamiltonian $\mathcal{H}_{\text{int}}$ time dependent as follows:

$$\tilde{\mathcal{H}}_{\text{int}}(t) = e^{i(\omega_0 I_z + \omega_{0S} S_z)t} \mathcal{H}_{\text{int}} e^{-i(\omega_0 I_z + \omega_{0S} S_z)t}. \tag{2.21}$$

We note, that the application of the static magnetic field H_0 already imposes a manipulation of the spin system which separates certain parts of the interaction Hamiltonian. In fact, only the time average of $\tilde{\mathcal{H}}_{\text{int}}(t)$ survives in first order

$$\mathcal{H}'_{\text{int}} = \frac{1}{t_c} \int_0^{t_c} dt \tilde{\mathcal{H}}_{\text{int}}(t) \tag{2.22}$$

where t_c is a suitable chosen cycle time, with the condition

$$\omega_{0I} t_c = m 2\pi \text{ and } \omega_{0S} t_c = n 2\pi.$$

For the interaction between a spin I and a spin S we obtain in the case

(i) $\omega_{0I} \neq \omega_{0S}$

$$\mathcal{H}'_{IS} = D_{zz} I_z S_z \tag{2.23}$$

and in the case

(ii) $\omega_{0I} = \omega_{0S}$

$$\mathcal{H}'_{IS} = D_{zz} I_z S_z + \frac{1}{2}(D_{xx} + D_{yy})(I_x S_x + I_y S_y)$$

$$+ \frac{1}{2}(D_{xy} - D_{yx})(I_x S_y - I_y S_x). \tag{2.24}$$

Equation (2.24) may be rewritten as:

$$\mathcal{H}'_{IS} = J\mathbf{I} \cdot \mathbf{S} + \frac{1}{2}(D_{zz} - J)(3I_z S_z - \mathbf{I} \cdot \mathbf{S})$$

$$+ \frac{1}{2}(D_{xy} - D_{yx})(I_x S_y - I_y S_x) \tag{2.25}$$

where $J = \frac{1}{3}\mathrm{Tr}\{D_{ij}\}$.

Note that antisymmetric components of the interaction appear in the secular Hamiltonian. However, these parts are not involved in line shifts in first order.

Often $\mathbf{D}$ is a symmetric tensor i.e. $D_{ij} = D_{ji}$ and the last term in Eq. (2.25) vanishes. In the case of dipolar interaction in addition $\mathrm{Tr}\{D_{ij}\} = 0$ and the familiar truncated or secular dipolar Hamiltonian results ($\omega_{0I} = \omega_{0S}$):

$$\mathcal{H}'_D = \frac{\gamma_I \gamma_S \hbar}{r^3} \frac{1}{2}(1 - 3\cos^2\vartheta_{IS})(3I_z S_z - \mathbf{I} \cdot \mathbf{S}) \tag{2.26}$$

where ϑ_{IS} is the angle between the magnetic field H_0 and the vector connecting spin I and S.

The secular part of the shielding Hamiltonian is immediately obtained as

$$\mathcal{H}'_s = S_{zz} H_0 I_z. \tag{2.27}$$

It contains only components of the symmetric part of the tensor i.e. the eigenvalues of this symmetric part of the shielding tensor exist and can be determined by different orientations of the magnetic field H_0 with respect to the principal axes of the tensor. Antisymmetric and other non-secular constituents of the shielding tensor contribute only in second order (see Appendix C).

Suppose the interaction Hamiltonian is represented in the spherical tensor notation as in Eq. (2.13), the secular part of the interactions is given by [*3, 4*]

$$\mathcal{H}'_{\mathrm{int}} = A_{00} T_{00} + A_{10} T_{10} + A_{20} T_{20}$$

since

$$[I_z, T_{kq}] = q T_{kq}$$

which vanishes besides trivial cases only for $q = 0$.

Inserting the Cartesian counterparts in A_{k0} and T_{k0} leads to the same expressions for the secular spin interactions as before.

The magnetic shielding Hamiltonian according to Eq. (2.10) may be also visualized as the interaction of the nuclear spin I with the magnetic field H_S as (see Appendix A)

$$\mathbf{H}_S = -\overset{\rightrightarrows}{\sigma} \cdot \mathbf{H}_0. \tag{2.28}$$

The vector $\mathbf{H}_S$ is no longer parallel to the vector $\mathbf{H}_0$, because of the tensor $\overset{\rightrightarrows}{\sigma}$ (see Fig. 2.2). The nucleus under consideration now "sees" an effective field

$$\mathbf{H}_{\text{eff}} = \mathbf{H}_0 + \mathbf{H}_S \tag{2.29}$$

where

$$\mathbf{H}_0 = (0, 0, H_0)$$

and

$$\mathbf{H}_S = -(\sigma_{xz}, \sigma_{yz}, \sigma_{zz})H_0.$$

If we simply ignore the deviation of the effective field $\mathbf{H}_{\text{eff}}$ at the nuclear site from the magnetic field direction $\mathbf{H}_0$ we could use the projection of $\mathbf{H}_S$ onto $\mathbf{H}_0$, which results in

$$H_{\text{eff}} = (1 - \sigma_{zz})H_0.$$

This was called the secular contribution before. If we want to calculate the size of H_{eff}, however exactly, we have to write

$$H_{\text{eff}}^2 = H_0^2 + H_S^2 + 2\mathbf{H}_0 \cdot \mathbf{H}_S \tag{2.30}$$

with

$$H_{\text{eff}} = H_0(1 + \sigma_{xz}^2 + \sigma_{yz}^2 + \sigma_{zz}^2 - 2\sigma_{zz})^{1/2} \tag{2.31}$$

or

$$H_{\text{eff}} = H_0(1 - \sigma_{zz})[1 + \frac{\sigma_{xz}^2 + \sigma_{yz}^2}{(1 - \sigma_{zz})^2}]^{1/2}. \tag{2.32}$$

Because the σ values are very small numbers we expand the square root in Eq. (2.32) to obtain

$$H_{\text{eff}} = H_0(1 - \sigma_{zz})[1 + \frac{1}{2}\frac{\sigma_{xz}^2 + \sigma_{yz}^2}{(1 - \sigma_{zz})^2}]. \tag{2.33a}$$

Using Eq. (2.29) we may write

$$H_S = -H_0\sigma_{zz}\left(1 - \frac{\sigma_{xz}^2 + \sigma_{yz}^2}{2\sigma_{zz}(1 - \sigma_{zz})}\right) = H_{\text{eff}} - H_0 \tag{2.33b}$$

A similar expression is obtained in Appendix C [Eq. (C 14)], when the so called non-secular terms are taken into account. Since only the symmetric part of the shielding tensor contributes to σ_{zz} and the σ values are usually very small numbers, the contribution of antisymmetric and other non-secular parts is usually neglected.

The symmetric part of the shielding tensor can always be diagonalized and σ_{zz} changes with the orientation of the static magnetic field H_0 with respect to the principal axis system (1, 2, 3) of this tensor.

Let us now turn to the pictorial representation of the shielding tensor. In Fig. 2.2 we have plotted the variation of $\mathbf{S} = \mathbf{H}_S/H_0$, which we refer to as the "shielding vector" with respect to the magnetic field. The shielding vector $\mathbf{S} = (S_1, S_2, S_3)$ may be represented by

$$\frac{S_1^2}{\sigma_{11}^2} + \frac{S_2^2}{\sigma_{22}^2} + \frac{S_3^2}{\sigma_{33}^2} = 1. \tag{2.34}$$

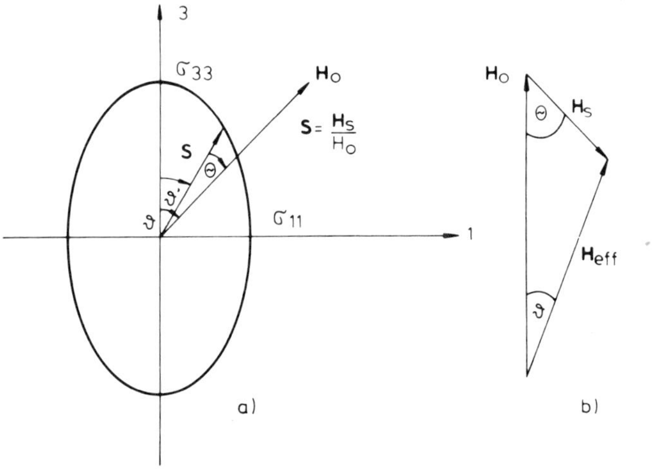

Fig. 2.2. (a) Graphical representation of the "shielding vector" $\mathbf{S} = (S_1, S_2, S_3)$ according to Eq. (2.34), (b) Vectorial representation of the shielding field $\mathbf{H}_S$ due to nuclear spin–electron coupling. Notice, that $\mathbf{S}$ and $\mathbf{H}_S$ are not parallel to H_0 due to the tensorial character of the interaction (see text)

This is the representation of an ellipsoid for the vector $\mathbf{S}$, which gives a direct measure of the σ_{ii} ($i = 1, 2, 3$) values, when the magnetic field H_0 points along the i-axis. In between of course, the $\mathbf{S}$ vector is no longer parallel to H_0 as outlined above. We prefer this representation with respect to the so called "quartic" representation, where $1/\sqrt{\sigma_{ii}}$ appears at the i-axis.

There are still different possible choices for obtaining a pictorial representation of the shielding tensor, as far as the reference is concerned to which the measured σ_{ii} values are related. Since no absolute shielding data are available we are going to define a suitable reference for the pictorial representation as follows:

(i) we always use the convention: $\sigma_{11} \leqslant \sigma_{22} \leqslant \sigma_{33}$
(ii) choose reference (zero point of σ-scale) such that $\sigma_{11} = (\sigma_{33} - \sigma_{11})/n$

where a suitable $n = 1, 2, 3, 4 \ldots$ may be chosen for aesthetical reasons. We find $n = 3$ most appealing. Condition (ii) has to be applied to each shielding tensor in a molecule individually. This assures that the anisotropy of the shielding is clearly visible. Most shielded, intermediate and least shielded axes are clearly distinguishable. Furthermore, the size of the anisotropy is directly related to the size of the ellipsoid. Examples of such an "aestheticized" tensor representation are given in Figs. 5.5 and 5.7. A circle of constant radius may be added to each nucleus, if this is found to be convenient for representing e.g. more or less isotropic tensors.

2.3 Transformation Properties of Spin Interactions in Real Space

Since we have expressed the spin interactions by second rank tensors, we can resort to the transformation properties of those tensors in order to express the interactions in different reference frames (see Appendix B). Suppose $\mathbf{A}$ is a second rank tensor expressed in the coordinate system (x, y, z) and $\mathbf{A}'$ is the same tensor expressed in the system (x', y', z') which is related to (x, y, z) by a unitary transformation $R(\alpha, \beta, \gamma)$ with the Euler angles (α, β, γ) i.e.

$$\mathbf{r}' = R\,\mathbf{r} \tag{2.35}$$

where $\mathbf{r}$ is a vector in (x, y, z) and $\mathbf{r}'$ the same vector in (x', y', z'). Now $\mathbf{A}'$ and $\mathbf{A}$ are related by the unitary transformation

$$\mathbf{A}' = R\,\mathbf{A}\,R^{-1}. \tag{2.36}$$

$\mathbf{A}$ represents here any of the spatial coupling tensors of the spin interactions as discussed in Section 2.1.

Instead of expressing the matrix R by the Euler angles, R may be represented also by the direction cosines [*3, 4*].

$$R = \{r_{ij}\} = \{\cos(\mathbf{r}', \mathbf{r})\} \tag{2.37}$$

with $\{r_{ij}\} = \{r_{ji}\}^{-1}$.

The matrix elements r_{ij} are given in Appendix B Eq. (B.2).

Equation (2.36) can be readily expressed in Cartesian components as

$$A'_{ij} = \sum_{k,l} A_{kl} r_{ik} r_{jl}. \tag{2.38}$$

The tensor **A** is expressed in the principal axis system as

$$A_{kl} = A_{kk}\delta_{kl} \text{ with the Kronecker } \delta_{kl} = \begin{Bmatrix} 1 \text{ if } k = l \\ 0 \text{ if } k \neq l \end{Bmatrix}.$$

If only the z component of the tensor is wanted, which is the case for the secular part of the shielding interaction for example, we obtain from Eq. (2.38)

$$A'_{zz} = \sum_k A_{kk} r_{3k}^2 = A_{11} r_{31}^2 + A_{22} r_{32}^2 + A_{33} r_{33}^2. \tag{2.39}$$

This can be written in terms of Euler angles by using the r_{3k} according to Appendix B as

$$A'_{zz} = A_{11}\cos^2\alpha\sin^2\beta + A_{22}\sin^2\alpha\sin^2\beta + A_{33}\cos^2\beta \tag{2.40}$$

which is the well-known formula for the variation of the secular part of a second rank tensor with the Euler angles (α, β). Equation (2.40) may be applied to the shift tensor **S** as well as to the field gradient tensor **V**, for example. A complete determination of the tensor elements A_{11}, A_{22}, and A_{33}, as well as the determination of the principal axis system with respect to some reference frame, can be achieved by measuring the secular part A'_{zz} of the spin interaction at different orientations of a single crystal.

In the following we always assume that dipolar couplings have been eliminated by the techniques of manipulation and dilution, mentioned in the introduction (Chapter 1).

Suppose a single crystal is mounted on a goniometer in a well defined fashion, i.e. the crystal axes have a known relation to the goniometer axes (x_g, y_g, z_g). The goniometer rotates the crystal by an angle φ about its rotation axis (z_g), which may have an angle ϑ with respect to the magnetic field (see Fig. 2.3). How does the secular component of the spin interaction change with the angle φ?

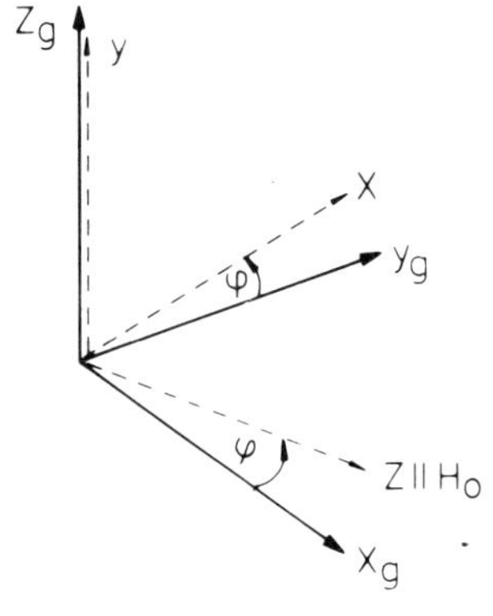

Fig. 2.3. Relation between the goniometer frame (x_g, y_g, z_g) and the laboratory frame (x, y, z). The angle φ represents the "rotation angle" by which the goniometer is rotated away from the starting position: $x \| y_g; y \| z_g; z \| x_g$

The transformation from the goniometer frame (x_g, y_g, z_g) to the laboratory frame (x, y, z) may be described by $R(\varphi, \vartheta, \psi)$:

$$\begin{array}{ccc} \mathbf{A} & & \mathbf{A}' \\ (x_g, y_g, z_g) & \xrightarrow{R(\varphi, \vartheta, \psi)} & (x, y, z) \\ z_g \parallel \text{rot. axis} & & z \parallel H_0 \end{array}$$

where

$$R(\varphi, \vartheta, \psi) = R_{zg}(\psi) R_{yg}(\vartheta) R_{zg}(\varphi). \tag{2.41}$$

It immediately follows from Eq. (2.37) that the rotation in Fig. 2.3 can be represented by

$$R(\varphi, 90°, 90°) = \begin{pmatrix} -\sin\varphi & \cos\varphi & 0 \\ 0 & 0 & 1 \\ \cos\varphi & \sin\varphi & 0 \end{pmatrix} \tag{2.42}$$

with

$$A'_{zz} = \sum_{i,j} A_{ij} r_{3i} r_{3j} \tag{2.43}$$

where the A_{ij} are the tensor elements in the goniometer frame (not necessarily the principal axis or even crystal frame).

With

$$r_{31} = \cos\varphi; r_{32} = \sin\varphi; r_{33} = 0.$$

we obtain

$$A'_{zz} = \frac{1}{2}(A_{11} + A_{22}) + \frac{1}{2}(A_{11} - A_{22})\cos 2\varphi + \frac{1}{2}(A_{12} + A_{21})\sin 2\varphi. \tag{2.44}$$

This means that the observed line shift which is proportional to A'_{zz} varies harmonically with 2φ. The three distinct tensor elements (A_{11}, A_{22}, A_{12}) of the symmetric part $(A_{12} = A_{21})$ of the spin interaction tensor with respect to the goniometer axis system can be determined from the corresponding rotation pattern. Examples of this will be shown in later sections.

In general, the goniometer may have an arbitrary orientation with respect to the laboratory frame, describable by the three Euler angles $(\varphi', \vartheta, \psi)$.

If φ is still the angle by which H_0 is rotated with respect to the goniometer starting position, the corresponding transformation has to be performed in two steps:

$$R_{\text{total}} = R'(\varphi', \vartheta, \psi) R(\varphi). \tag{2.45}$$

The second rank tensor representing the spin interactions in the lab-frame (x, y, z) can be expressed by

$$\mathbf{A}' = R' R \mathbf{A} R^{-1} R'^{-1} \tag{2.46}$$

or by using the unitary nature of R and R'

$$A'_{ij} = \sum_{k,l,m,n} A_{lm} r'_{ik} r'_{jn} r_{kl} r_{nm} \tag{2.47}$$

where r'_{ik} and r_{kl} are the corresponding matrix elements of R' and R respectively (see Appendix B).

As noted before, in most cases only the z component ($i = j = 3$) of the tensor $\mathbf{A}'$ is measured in an experiment leading to

$$A'_{zz} = A + D\cos\varphi + E\sin\varphi + B\cos 2\varphi + C\sin 2\varphi \tag{2.48}$$

where it can be easily realized that

$$\begin{aligned}
A &= \frac{1}{2}(A_{11} + A_{22})(r'^2_{31} + r'^2_{32}) + A_{33}r'^2_{33} \\
D &= (A_{13} + A_{31})\, r'_{31}r'_{33} + (A_{23} + A_{32})r'_{32}r'_{33} \\
E &= (A_{23} + A_{32})\, r'_{31}\, r'_{33} - (A_{13} + A_{31})\, r'_{32}\, r'_{33} \\
B &= (A_{12} + A_{21})r'_{31}r'_{32} + \frac{1}{2}(A_{11} - A_{22})(r'^2_{31} - r'^2_{32}) \\
C &= \frac{1}{2}(A_{12} + A_{21})(r'^2_{31} - r'^2_{32}) + (A_{22} - A_{11})\, r'_{31} \cdot r'_{32}.
\end{aligned}$$

Notice that in this case A'_{zz} *is not* invariant under *a* π rotation. If the goniometer axis does not coincide with a principal axis of the tensor, five of the six possible tensor elements can be determined from one orientation plot.

If the goniometer axis is however orthogonal to the magnetic field (i.e. $\vartheta = 90°$ or $r'_{33} = 0$) D and E vanish, which results in

$$A'_{zz} = A + B\cos 2\varphi + C\sin 2\varphi \text{ if } z_g \perp H_0 . \tag{2.49}$$

The special case discussed earlier in this section (Eq. (2.44)) can be recovered with $r'_{31} = 1$ and $r'_{32} = r'_{33} = 0$.

It may be particularly appropriate to express these rotations in terms of Wigner rotation matrices. Following Appendix B we may express the symmetric part of the secular spin interaction under the rotation $R'R$ according to Eq. (2.45) as

$$A'_{20} = [R'RA_{2q}R^{-1}R'^{-1}]_{q=0} \tag{2.50}$$

or

$$A'_{20} = \sum_{p,r} A_{2p} D^{(2)}_{pr}(\varphi, 0, 0) D^{(2)}_{r0}(\varphi', \vartheta, \psi) \tag{2.51}$$

$$A'_{20} = \sum_{p} A_{2p} e^{-ip\varphi'} d^{(2)}_{p0}(\vartheta) e^{-ip\varphi} \tag{2.52}$$

where the A_{2p} are spherical tensor elements as discussed in Appendix A.

Equation (2.52) can be immediately cast into the equivalent form of Eq. (2.48) with

$$
\begin{aligned}
A &= d_{00}^{(2)}(\vartheta)A_{20} && \text{(a)}\\
D &= d_{10}^{(2)}(\vartheta)[A_{21}e^{-i\varphi'} - A_{2-1}e^{i\varphi'}] && \text{(b)}\\
E &= (-i)d_{10}^{(2)}(\vartheta)[A_{21}e^{-i\varphi'} + A_{2-1}e^{i\varphi'}] && \text{(c)} \qquad (2.53)\\
B &= d_{20}^{(2)}(\vartheta)[A_{22}e^{-i2\varphi'} + A_{2-2}e^{i2\varphi'}] && \text{(d)}\\
C &= -id_{20}^{(2)}(\vartheta)[A_{22}e^{-i2\varphi'} - A_{2-2}e^{i2\varphi'}] && \text{(e)}
\end{aligned}
$$

where the corresponding tensor components A_{2q} may be obtained from Appendix A Eq. (A.12) or Eq. (A.17).

It should be noted that the goniometer axis is usually orthogonal to the magnetic field H_0 and Eq. (2.49) applies. However, Eq. (2.48) may be very useful for estimating the error if the orthogonality condition is not accurately fulfilled. On the other hand, it might be interesting to use a tilted goniometer axis, since in this case in principal already five of the six elements of the symmetric tensor can be determined with one orientation plot as mentioned before. If in addition the trace of the tensor is known, the full tensor can be determined in one orientation plot. This procedure has been applied by Haeberlen and co-workers in one case of a proton shielding tensor [6] (see Section 5.3). In the following we assume that the single crystal is rotated about an axis perpendicular to the magnetic field. Three of the six tensor elements can be determined from one orientation plot in this case. It would be sufficient to obtain one more orientation plot about an axis which gives non-degenerate results in order to determine the full tensor. However, in practice it is more convenient to rotate the crystal about three different orthogonal axes. In this case the general strategy for obtaining the full tensor can be outlined as follows

$$
\begin{array}{cccc}
\tilde{\mathbf{A}} & \mathbf{A}^* & \mathbf{A} & \mathbf{A}' \\
(1,2,3) \xrightarrow{R_p(\alpha',\beta',\gamma')} & (a,b,c) \xrightarrow{R_i(\alpha,\beta,\gamma)} & (x_g, y_g, z_g) \xrightarrow{R_g(\varphi, 90, 90)} & (x,y,z)
\end{array}
$$

principal axis system — crystal frame — goniometer frame — Lab. frame

where R_p is the transformation from the principal axis system to the crystal frame. $R_i (i = 1, 2, 3 \ldots)$ are the three or more different transformations from the crystal frame to the goniometer frame, depending on how the crystal is mounted on the goniometer axis. Finally R_g is the transformation from the fixed goniometer to the laboratory frame where φ is the angle between the goniometer x_g axis and H_0 (z axis) as indicated in Fig. 2.3.

From each orientation plot three different tensor elements of $\mathbf{A}^*$ can be determined once the corresponding R_i is known. If all the six tensor elements of the symmetric part of $\mathbf{A}^*$ are evaluated by two or more orientation plots, a matrix diagonalization finally yields **A** with the principal elements A_{11}, A_{22} and A_{33} and the direction cosine of the principal axes with respect to the crystal frame (*a, b, c*) [7].

Applying the transformation R_i and R_g leads to

$$A'_{zz} = [R_g R_i \mathbf{A}^* R_i^{-1} R_g^{-1}]_{33} \tag{2.54}$$

or

$$A'_{zz} = A + B\cos 2\varphi + C\sin 2\varphi$$

where

$$A = \frac{1}{2}\sum_{k,l} A^*_{kl}\,(r_{1k}r_{1l} + r_{2k}r_{2l}) \tag{2.55a}$$

$$B = \frac{1}{2}\sum_{k,l} A^*_{kl}(r_{1k}r_{1l} - r_{2k}r_{2l}) \tag{2.55b}$$

$$C = \frac{1}{2}\sum_{k,l} A^*_{kl}(r_{1k}r_{2l} + r_{2k}r_{1l}). \tag{2.55c}$$

The r_{ij} are the known matrix elements of the R_i rotation matrix, depending on how the crystal is mounted on the goniometer. Since the parameters *A, B* and *C* are determined experimentally, a set of linear equations results which can be brought into the form

$$\mathbf{Y} = \mathbf{a} \cdot \mathbf{X} \tag{2.56}$$

where

$$\mathbf{Y} = (A_1, B_1, C_1, A_2, B_2, C_2 \ldots) \tag{2.57}$$

is a vector, defined by the parameters *A, B, C* from *n* different orientation plots and

$$\mathbf{X} = (A^*_{11}, A^*_{12}, A^*_{13}, A^*_{22}, A^*_{23}, A^*_{33}) \tag{2.58}$$

is a vector spanned by the six independent tensor elements A^*_{ij} in the crystal frame and **a** is a $n \times 6$ matrix which can be obtained from Eq. (2.55).

The vector **X** is obtained by solving the set of linear equations or by a matrix inversion procedure

$$\mathbf{X} = \mathbf{a}^{-1}\mathbf{Y}. \tag{2.59}$$

These relations are used in high resolution NMR in solids, to determine the chemical shift tensor in single crystals. As we have noted before only the symmetric part of the chemical shift tensor is determined in this way. A representative example of an orientation plot of the line shift in a single crystal in the case of ^{19}F in CF_3COOAg obtained in a four-pulse experiment is shown in Fig. 2.4. A single line is observed due to rapid

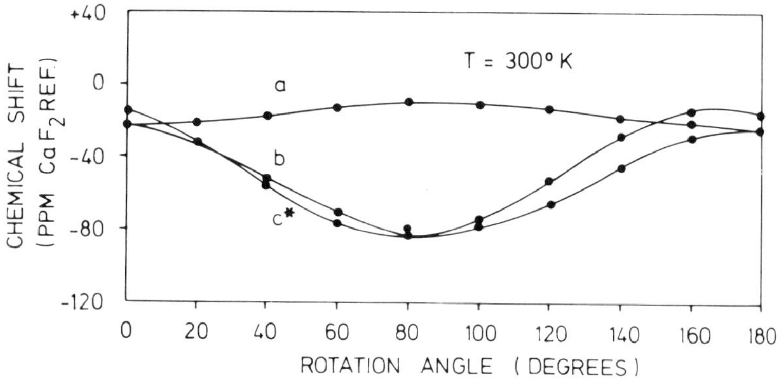

Fig. 2.4. Chemical shift of ^{19}F in solid CF_3COOAg at T = 300 °K versus the goniometer rotation angle, as obtained in a multiple-pulse experiment (WHH-4). The labels a, b, c* refer to the mutually orthogonal crystallografic axes (see Ref. [7])

motion of the CF_3-group at room temperature. Rotation of the single crystal was performed around three different axes closely related to the unit cell axes. The principal components and the principal axes of the shielding tensor were evaluated by Griffin *et al.* [7] following the procedure outlined above (see also Section 5.4).

2.4 Powder Spectrum Line Shape

Very often single crystals are not available or exceedingly difficult to grow (solid H_2, for example).

In this case a powder sample has to be investigated and the question arises what information about the shift tensor can be gained.

Following Eq. (2.40), the observed frequency ω_{zz} of a spectral line can be expressed by the Euler angles (α, β) which relate the principal axes of the tensor ($\omega_{11}, \omega_{22}, \omega_{33}$) to the laboratory frame as

$$\omega_{zz} = \omega_{11}\cos^2\alpha\sin^2\beta + \omega_{22}\sin^2\alpha\sin^2\beta + \omega_{33}\cos^2\beta. \tag{2.60}$$

For an axially symmetric tensor we write

$$\omega_{zz} = (\omega_{\parallel} - \omega_{\perp})\cos^2\beta + \omega_{\perp} \tag{2.61}$$

where β is the angle between the 3-axis and the direction of the magnetic field H_0 (z-axis) and where

$$\omega_{\parallel} = \omega_{33};\ \omega_{\perp} = \omega_{11} = \omega_{22}$$

leading to the total anisotropy

$$\Delta\omega = \omega_{\parallel} - \omega_{\perp}. \tag{2.62}$$

If $I(\omega)$ is the intensity of the NMR signal at the frequency ω and $p(\Omega)\,d\Omega$ is the probability of finding the tensor orientation in the range between the solid angle Ω and $\Omega + d\Omega$, we may write

$$p(\Omega)d\Omega = I(\omega)d\omega.$$

In a powder all angles Ω are equally probable i.e. $p(\Omega) = 1$, leading to

$$I(\omega) = \left|\frac{d\Omega}{d\omega}\right|. \tag{2.63}$$

In the case of axial symmetry $d\Omega = \sin\beta d\beta$ we arrive at the line shape function, using Eqs. (2.61) and (2.63) as

$$I(\omega) = \frac{1}{2}\left[(\omega_{\parallel} - \omega_{\perp})(\omega - \omega_{\perp})\right]^{-1/2}. \tag{2.64}$$

The general case with $\omega_{33} > \omega_{22} > \omega_{11}$ (convention) is slightly more complicated and has been calculated by Bloembergen and Rowland [8] as

$$I(\omega) = \pi^{-1}(\omega - \omega_{11})^{-1/2}(\omega_{33} - \omega_{22})^{-1/2} \cdot K(m). \tag{2.65a}$$

with

$$m = (\omega_{22} - \omega_{11})(\omega_{33} - \omega)/(\omega_{33} - \omega_{22})(\omega - \omega_{11})$$

for

$$\omega_{33} \geqslant \omega > \omega_{22}$$

and

$$I(\omega) = \pi^{-1}(\omega_{33} - \omega)^{-1/2}(\omega_{22} - \omega_{11})^{-1/2} K(m) \tag{2.65b}$$

with

$$m = (\omega - \omega_{11})(\omega_{33} - \omega_{22})/(\omega_{33} - \omega)(\omega_{22} - \omega_{11})$$

for

$$\omega_{22} > \omega \geqslant \omega_{11}$$

$I(\omega) = 0$ in case $\omega > \omega_{33}$ and $\omega < \omega_{11}$.

$K(m)$ is the complete elliptic integral of the first kind

$$K(m) = \int_0^{\pi/2} d\varphi[1 - m^2 \sin^2\varphi]^{-1/2}. \tag{2.66}$$

The different line shapes according to Eqs. (2.64, 2.65) are plotted in Fig. 2.5 for two different cases. Usually a residual line broadening is taken into account by convolution of $I(\omega)$ with a broadening function (Lorentzian or Gaussian).

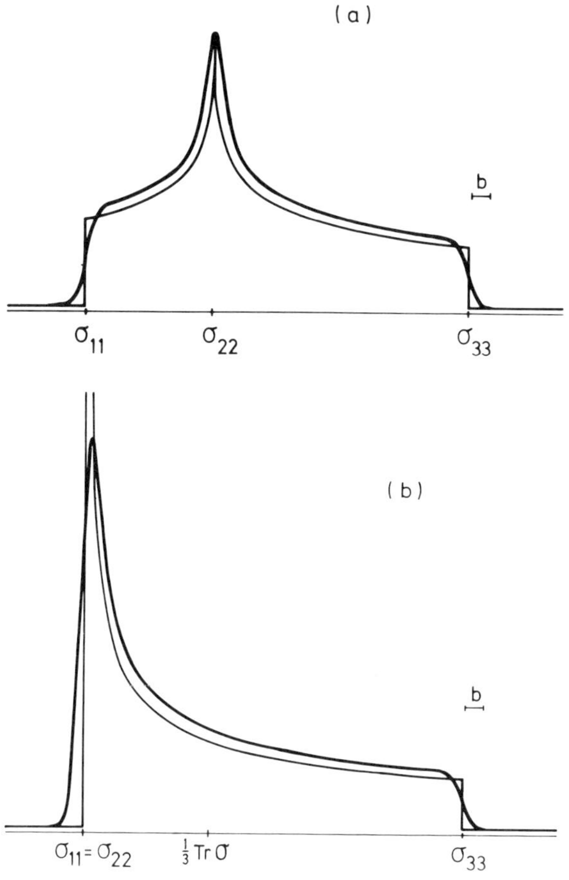

Fig. 2.5. Schematic representation of theoretical powder line shapes for second rank tensor interactions (i.e., chemical shift tensor, Knightshift tensor, *g*-tensor etc.) according to Eqs. (2.64, 2.65). The theoretical curves are convoluted with Lorentzian broadening functions, whose width *b* is indicated. (a) arbitrary second rank tensor, (b) axially symmetric second rank tensor

It is evident from Fig. 2.5 and the corresponding formulas for the powder spectrum how the principal shielding components (σ_{11}, σ_{22}, σ_{33}) can be obtained directly from the spectrum. However no information is obtained about the orientation of the principle axes of the tensor. One has to resort to theoretical arguments, symmetry considerations, consistency checks with related compounds or other molecular properties in order to assign the principal axes tentatively. Such other properties can be e.g. molecular rotations which are known to proceed about certain symmetry axes. One case where such knowledge has been utilized to elucidate the orientation of the principal axes has been reported in Ref. [*9*]. Fig. 2.6 shows the powder spec-

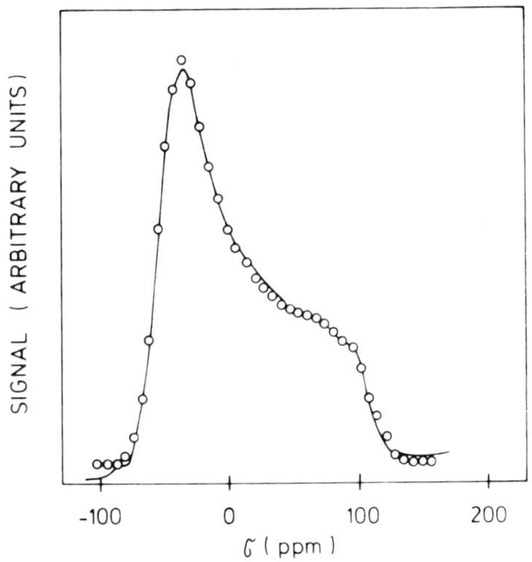

Fig. 2.6. ^{19}F spectrum of C_6F_6 obtained in a 4-pulse experiment (WHH-4) at 40 K. The solid line is the experimental trace while the open circles are calculated points, obtained by convolution of the theoretical powder line shape Eq. (2.64) with a Gaussian broadening function [*9*]

trum of ^{19}F in C_6F_6 which displays an axially symmetric shielding tensor. Above $T = 100\ ^\circ K$ the molecule is known to reorient rapidly about its 6-fold axis and the shielding tensor is expected to be axially symmetric about this axis. At a temperature of $T = 50\ ^\circ K$ however this motion is slowed down to a rate very much below the shielding anisotropy and the "rigid" tensor should grow out of the spectrum. The big surprise at the time, however, was the fact that the spectrum was about the same at this low temperature. This leaves as the only possible conclusion, that the shielding tensor is in fact axially symmetric with the unique axis being the 6-fold axis of the molecule [*9*].

The same approach has been taken in several other cases and is very helpful in assigning tentatively the principal axes to the molecular frame.

2.5 Specimen Rotation

In this technique the sample is spun by a turbine with high revolution about an axis which is tilted by an angle ϑ with respect to the magnetic field H_0 (see Fig. 2.7). All interactions become time dependent and if the rotation rate ω_r is larger than the linewidth, sidebands are produced in the spectrum (see Fig. 2.8). The separation of the sidebands from the centerline is ω_r and $2 \cdot \omega_r$. In the following we assume that the sidebands are well separated and we deal with the centerline only, thus observing the time average of the timedependent interaction Hamiltonian in first order.

We face the same situation as in Section 2.3, if we assume the goniometer axis being rapidly rotated and we can immediately apply Eqs. (2.45) and (2.48) by setting $\varphi = \omega_r \cdot t$.

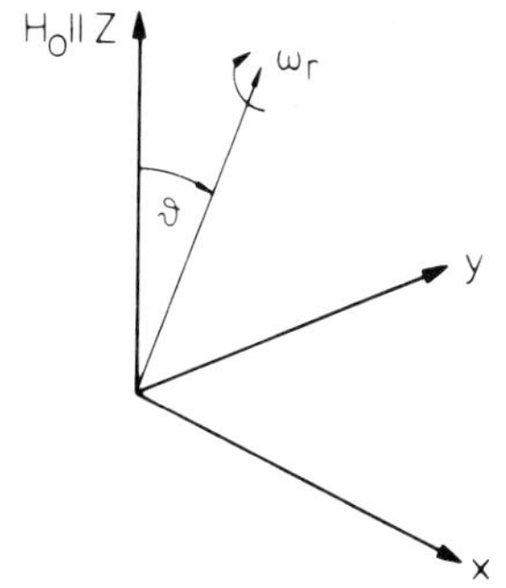

Fig. 2.7. Schematic drawing of the rotation axis (rotation rate ω_r) tilted with respect to the static magnetic field H_0 by an angle ϑ

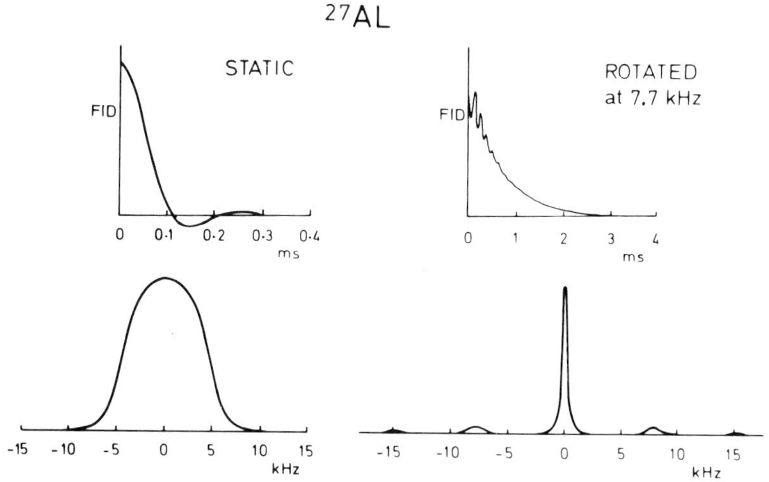

Fig. 2.8. Decay and spectrum of ^{27}Al in a powder sample of aluminium, with and without rotation at the "magic angle", according to Andrew *et al.* [16]. The generation of sidebands and the narrowing of the centerline is clearly demonstrated

The time average of Eq. (2.48) leaves only

$$A'_{zz} = A = \frac{1}{2}(A_{11} + A_{22})(r'^2_{31} + r'^2_{32}) + A_{33}r'^2_{33}$$

where the 3-axis is now the rotation axis.
With

$$r'_{31} = \cos\varphi' \sin\vartheta ; r'_{32} = \sin\varphi' \sin\vartheta ; r'_{33} = \cos\vartheta$$

we obtain

$$\overline{A'_{zz}} = \frac{1}{2}\,\mathrm{Tr}\{A_{ij}\}\sin^2\vartheta + \frac{1}{2}(3\cos^2\vartheta - 1)A_{33} \qquad (2.67a)$$

or

$$\overline{A'_{zz}} - \frac{1}{3}\mathrm{Tr}\{A_{ij}\} = \frac{1}{2}(3\cos^2\vartheta - 1)A^*_{33} \tag{2.67b}$$

where

$$A^*_{33} = A_{33} - \frac{1}{3}\mathrm{Tr}\{A_{ij}\}.$$

Different cases can be discussed:

(i) $\vartheta = 0 : \overline{A'_{zz}} = A_{33}$

(ii) $\vartheta = \pi/2 : \overline{A'_{zz}} = \frac{1}{2}(A_{11} + A_{22})$

(iii) $\vartheta = \vartheta m = 54°44' : A'_{zz} = \frac{1}{3}\mathrm{Tr}\{A_{ij}\}$

where the "magic angle" ϑ_m is defined by $\cos^2\vartheta_m = 1/3$.

The last case is obviously the most interesting one as was first realized by E. R. Andrew and co-workers [*10, 11*] and I. Lowe [*12*].

In this case only the trace of the second rank tensor interaction survives, i.e. all tensor interactions which have zero trace like dipole-dipole and quadrupole-interaction vanish. This is true also for the heteronuclear dipolar interaction, since we operate in real space and no extra decoupling technique has to be applied. The measurable quantities in the "magic angle" specimen rotation method are thus the isotropic chemical shift, Knight-shift and the scalar coupling. E. R. Andrew and co-workers have applied this method extensively to study these parameters in solids [*13, 14*]. One example of this technique has been presented in the introduction (Fig. 1.3) where the J coupling between the ^{19}F and the ^{75}As in $KAsF_6$ reveals a four line spectrum for the ^{19}F nuclei [*15*].

Figure 2.8 shows the application of this technique to a metal powder [*16*], namely ^{27}Al. The development of sidebands with increasing rate of revolution combined with a narrowing of the center line is clearly visible. It was possible in this experiment to determine the isotropic Knight shift of ^{27}Al with respect to $AlCl_3$ solution with more than an order of magnitude improvement in precision. The result of 1640 ± 1 ppm agrees within the combined errors with the value of 1636 ± 3 ppm as obtained by multiple pulse experiments [*17*], to be discussed later (Section 5.6). The "magic angle" specimen rotation method has been successfully applied to polymers and componds of biological interest also [*18, 19*]. Babka *et al.* have resolved proton NMR spectra in lecithin and related membrane model systems [*19b*].

Figure 2.9 shows an example for egg yolk lecithin. The line narrowing techniques may fail however in these compounds, since the slow motions create a rapid spin lattice relaxation rate which may limit the resolution appreciably as is discussed in more detail in Chapter 6. For further reading on the application of "magic angle" rotation methods we refer the interested reader to the review papers on this subject by E. R. Andrew [*13, 14*] and the references cited therein.

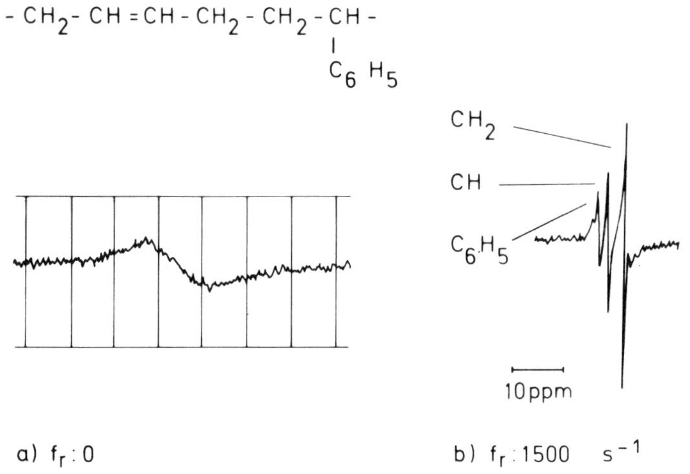

Fig. 2.9. Application of the "magic angle" rotation method to egg yolk lecithin [*19*]. Different proton resonance lines due to CH_2, CH and C_6H_5 groups can be distinguished

So far we have neglected the antisymmetric part of the tensor interaction. In the case of indirect dipolar coupling an antisymmetric part can occur in the secular Hamiltonian. For completeness we write down here the expression for the secular part of $\mathcal{H}_{IS}$ in a specimen rotation experiment. Using the spherical tensor representation we obtain readily:

(i) homonuclear coupling $\omega_{0I} = \omega_{0S}$

$$\mathcal{H}'_{IS} = J\,\mathbf{I}\cdot\mathbf{S} - \frac{1}{2}\cos\vartheta (D_{x'y'} - D_{y'x'})(I_xS_y - I_yS_x)$$
$$+ \frac{1}{2}(3\cos^2\vartheta - 1)\cdot\frac{1}{2}(D_{z'z'} - J)(3I_zS_z - \mathbf{I}\cdot\mathbf{S}) \tag{2.68a}$$

where the primed axes refer to the sample rotating frame.

(ii) heteronuclear coupling $\omega_{0I} \neq \omega_{0S}$

$$\mathcal{H}'_{IS} = [J(1 - \frac{3}{4}(3\cos^2\vartheta - 1)) + \frac{3}{4}(3\cos^2\vartheta - 1)\,D_{z'z'}]I_zS_z \tag{2.68b}$$

where

$$J = \frac{1}{3}\,\mathrm{Tr}\{D_{ij}\}.$$

Under the "magic angle" condition $\vartheta = \vartheta_m = 54°44'$ we obtain

(i) $\omega_{0I} = \omega_{0S}$:

$$\mathcal{H}'_{IS} = J\,\mathbf{I}\cdot\mathbf{S} - \frac{1}{2\sqrt{3}}(D_{x'y'} - D_{y'x'})(I_xS_y - I_yS_x) \tag{2.69a}$$

and

(ii) $\omega_{0I} \neq \omega_{0S}$: $\mathcal{H}'_{IS} = J\,\mathbf{I}\cdot\mathbf{S}$. (2.69b)

Notice, however, that the antisymmetric part does not contribute to magnetic transitions in first order, i.e. any contribution of this part of the Hamiltonian to the spectrum is due to a second order effect and therefore in general negligible.

2.6 Rapid Anisotropic Molecular Rotation

It has been realized for a long time that even in solids, especially "organic solids", rapid molecular reorientation can occur at high rates (e.g. up to 10^{13} sec^{-1}) at intermediate temperatures (e.g. 100 K to 500 K). The best known examples are benzene and methyl groups. The variation of relaxation rates and the change in the second moment of the resonance line have been extensively studied in the past. We concentrate here on the spectrum or line shape under condition of rapid anisotropic molecular reorientation.

As in the preceeding sections, the interaction Hamiltonian $\mathcal{H}_{\text{int}}$ is devided into a time averaged Hamiltonian $\mathcal{H}'_{\text{int}}$ (secular) and a time dependent Hamiltonian (non-secular) $\mathcal{H}''_{\text{int}}(t)$

$$\mathcal{H}_{\text{int}} = \mathcal{H}'_{\text{int}} + \mathcal{H}''_{\text{int}}(t). \tag{2.70}$$

In this section we assume the random process governing the motion with the correlation time τ_c, to be rapid, i.e. $1/\tau_c \gg \|\mathcal{H}_{\text{int}}\|$. In this sense only secular contributions have to be taken into account and it is sufficient to calculate the average Hamiltonian only. We can readily use the equations of the preceeding section. If we consider the rotation of two homonuclear spins I about an axis orthogonal to their internuclear vector ($\vartheta_{ij} = 90°$) as an example, we find from Eq. (2.68a) for their dipolar interaction

$$\mathcal{H}'_{II} = \frac{1}{4}(3\cos^2\vartheta - 1) \cdot \frac{\gamma_I^2 \hbar}{r^3} (3I_{z1}I_{z2} - \mathbf{I}_1 \cdot \mathbf{I}_2) \tag{2.71}$$

where ϑ is the angle between the rotation axis and the magnetic field H_0. Equation (2.71) again reflects the well known fact that the line splitting vanishes at the "magic angle" ($\vartheta_m = 54°44'$).

We turn now to the impact rapid anisotropic molecular reorientation has on the chemical shift anisotropy, since this is one of the quantities usually observed in high resolution NMR in solids. It is obvious that the component of the shielding tensor in the direction of the rotation axis is unchanged, whereas the components perpendicular to the rotation axis are averaged, leading to an axially symmetric tensor with the rotation axis as the symmetry axis [*9*]. The trace of the tensor is, of course, unchanged. The relevant equation describing this fact is Eq. (2.67) of the preceeding section which we repeat here for the convenience of the reader in a slightly different notation

$$\overline{S_{zz}^*} = \frac{1}{2}(3\cos^2\vartheta - 1)\, S_{ZZ}^* \tag{2.72}$$

where

$$S_{ii}^* = S_{ii} - \frac{1}{3}\mathrm{Tr}\{\mathbf{S}\}$$

and the Z-axis corresponds to the rotation axis. In this expression all values are referred to the trace of the tensor. The axial symmetry is quite evident, leaving as the motionally narrowed anisotropy

$$\overline{\Delta S} = S_{\parallel} - S_{\perp} = \frac{3}{2} S_{ZZ}^* \qquad (2.73)$$

where

$$S_{ZZ}^* = S_{11}^* \cos^2\alpha \sin^2\beta + S_{22}^* \sin^2\alpha \sin^2\beta + S_{33}^* \cos^2\beta$$

with the tensor elements S_{11}^*, S_{22}^*, and S_{33}^* and the Euler angles (α, β) defining the transformation from the principal axis system of the tensor (1, 2, 3) to the molecular rotating frame (Z-axis).

Equation (2.73) may be rewritten in the following way [9]:

$$\overline{\Delta S} = \frac{1}{2}(3\cos^2\beta - 1)[S_{33} - \frac{1}{2}(S_{11} + S_{22}) + \frac{3}{4}(S_{11} - S_{22}) \sin^2\beta \cos 2\alpha. \qquad (2.74)$$

This equation is valid for the "starred" and "unstarred" values of S. If the shift tensor is axially symmetric in the "rigid" case, as in the case of ^{19}F in C_6F_6, Eq. (2.74) reduces to

$$\overline{\Delta S} = \frac{1}{2}(3\cos^2\beta - 1)\Delta S$$

leaving the shift tensor unchanged if $\beta = 0$ (e.g. C_6F_6). If the rotation occurs about an axis perpendicular ($\beta = 90°$) to the symmetry axis of an axially symmetric shielding tensor, we obtain for example $\overline{\Delta S} = -\frac{1}{2}\Delta S$, with the old $S_\perp$ component being the new unique component of the averaged tensor.

Those motionally averaged powder spectra have been observed in multiple pulse experiments [9] and it was possible from the powder spectra of ^{19}F in CF_3COOAg at different temperatures to assign the direction of the principle axes to the molecular frame tentatively. This assignment was confirmed in a later single crystal investigation [7].

Let us adopt a simple model for the shielding tensor of the ^{19}F spins in a fluoromethyl group as $S_{11} = -1$, $S_{22} = 0$, $S_{33} = +1$ with the 3-axis parallel to the CF bond and the 1-axis perpendicular to the CF bond in the CCF plane. This is realistic besides a factor of about 70 ppm in the S_{ii} values [7]. If we apply Eq. (2.73) with the angles $\beta = 70.5°$ and $\alpha = 0°$ we obtain $\overline{\Delta S} = -1.166$, with the trace unchanged. Figure 2.10 shows the corresponding powder spectrum. Similar spectra were obtained experimentally in the case of ^{19}F in CF_3COOAg [9].

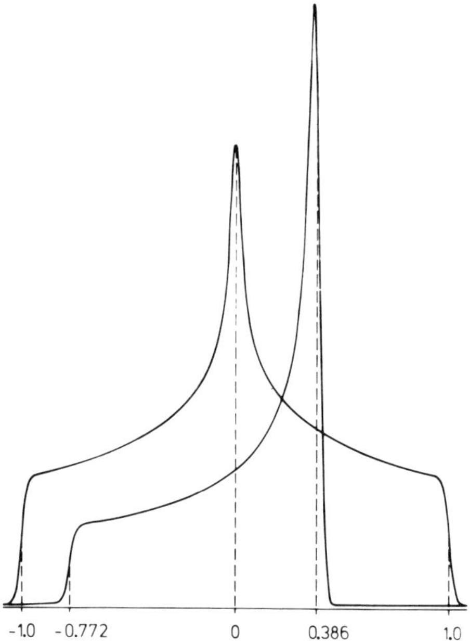

Fig. 2.10. Powder line shape for a model shift tensor with $S_{11} = -1$, $S_{22} = 0$, $S_{33} = +1$, corresponding to a fluoromethyl group, with the 3-axis parallel to the CF-bond and the 1-axis perpendicular to the CF-bond in the CCF plane. Rapid rotation about the threefold axis of the methyl group leads to an axially symmetric tensor according to Eq. (2.73) as indicated by the corresponding powder pattern

2.7 Line Shapes in the Presence of Molecular Reorientation

In the preceeding section we have treated the most simple case of how the line shape changes under rapid molecular rotations. Now we are going to deal with the more general case, i.e. molecular reorientation at an arbitrary rate. This problem has been dealt with in a number of recent publications [*20–28*]. Especially interesting is the region where the time scale of the motion is comparable with the size of the interaction to be studied. One can hope to differentiate between different types of molecular reorientation in this case.

We shall concentrate here on NMR spectra governed by shielding tensors, although it is evident that the same arguments can be applied to other types of interaction. There is particularly a close relationship to esr line shapes in liquids governed by g-tensors. The usefulness of the lineshape analysis in this case applied to spectra displaying molecular motion has been demonstrated recently by Sillescu and co-workers [*24, 25*]. On the other hand Spiess *et al.* [*27, 28*] have applied this analysis, to NMR powder spectra for the first time (see Fig. 2.11). The powder spectrum $I(\omega)$

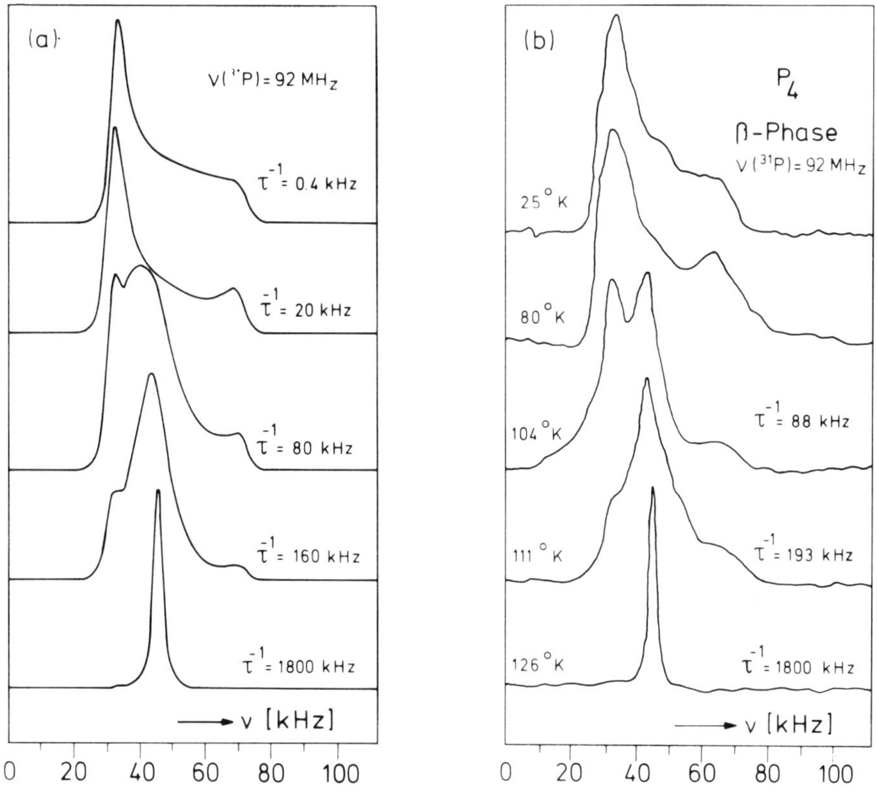

Fig. 2.11 a and b. Calculated (a) and measured (b) powder line shapes for a random jumping tetrahedron with jump rates τ^{-1} according to Spiess *et al.* [27, 28]. The ^{31}P spectra are obtained from solid white phosphorus in the β-phase at various temperatures. The values of the jump rates τ^{-1} for the experimental spectra were obtained from T_1 data

due to a tensor interaction can be obtained by integration of the single crystal spectrum $I'(\omega, \Omega)$ over the solid angle Ω

$$I(\omega) = \frac{1}{\mathcal{N}} \int_\Omega d\Omega I'(\omega, \Omega) \tag{2.75}$$

with the integration constant $\mathcal{N}$
where

$$\mathcal{N} = 4\pi \text{ if } d\Omega = \sin\vartheta d\vartheta d\varphi$$

or

$$\mathcal{N} = 8\pi^2 \text{ if } d\Omega = \sin\vartheta d\vartheta d\varphi d\psi .$$

The motion of the molecule is introduced by considering a stationary random process with a rate constant κ where the molecular reorientation is performed by a rotational jump $\Omega \to \Omega'$. In this rotational jump it is assumed that the molecule spends an average time τ in a fixed orientation Ω, whereupon it jumps to a new orientation in a time $\tau' \ll \tau$.

It is further assumed that there is no correlation of the orientation Ω and Ω' before and after the rotational jump. Thus the average residence time τ of the molecule in one orientation Ω equals the rotational correlation time κ^{-1}. The total number of different sites which can be occupied may be n. For the time evolution of $\Omega(t)$ we assume a Markoff process.

If $P(\Omega_0/\Omega,t)$ is the probability of finding the molecule at the orientation Ω, given it had the orientation Ω_0 a time t earlier, we can express $P(\Omega_0/\Omega,t)$ by the master equation [22]

$$\frac{d}{dt} P(\Omega_0/\Omega,t) = \int \pi(\Omega', \Omega) P(\Omega_0/\Omega',t) d\Omega' \tag{2.76}$$

with the initial condition

$$P(\Omega_0/\Omega,0) = \delta(\Omega - \Omega_0). \tag{2.77}$$

$\pi(\Omega', \Omega)$ plays the role of a memory function which "keeps in mind" the history of Ω over which $P(\Omega_0/\Omega',t)$ has to be integrated. In the limit of $t \to 0$ in Eq. (2.76), i.e.

$$P(\Omega_0/\Omega', t \to 0) = \delta(\Omega' - \Omega_0)$$

we arrive at a definition for the transition probability per unit time $\pi(\Omega', \Omega) d\Omega$ as

$$\pi(\Omega', \Omega) = \left[\frac{d}{dt} P(\Omega'/\Omega, t)\right]_{t=0}. \tag{2.78}$$

If $\kappa \cdot t$ is the probability that Ω' will have changed into Ω during time t and correspondingly $1 - \kappa \cdot t$ is the probability that Ω' did not change, we may write

$$P(\Omega'/\Omega, t) = (1 - \kappa \cdot t)\delta(\Omega' - \Omega) + W(\Omega)\,\kappa \cdot t \tag{2.79}$$

where $W(\Omega)$ is the a priori probability of finding the molecule in the orientation Ω. Using the definition for $\pi(\Omega', \Omega)$ according to Eq. (2.78) we obtain readily from Eq. (2.79)

$$\pi(\Omega', \Omega) = \quad [W(\Omega) - \delta(\Omega' - \Omega)]. \tag{2.80}$$

In the case of finite approximation this results in

$$\pi(\Omega_m, \Omega_n) = \quad [W(\Omega_n) - \delta_{mn}] \tag{2.81}$$

where jumping with the rate κ between k different sites labelled by Ω_i ($i = 1, 2, \ldots k$) is considered.

Let us calculate now the NMR spectrum $I(\omega)$ [see Eq. (2.75)] due to this rotational jump process. We may define the spectrum $I(\omega)$ as being real part of the Fourier transform of the free induction decay $G(t)$ as

$$I(\omega) = \mathrm{Re} \int_0^\infty dt G(t) e^{-i\omega t} \tag{2.82}$$

where

$$G(t) = \frac{1}{\mathcal{N}} \int_\Omega d\Omega G(t, \Omega). \tag{2.83}$$

Eq. (2.82) applies also to $I'(\omega, \Omega)$ if $G(t)$ is exchanged by $G(t, \Omega)$. $G(t, \Omega)$ itself is obtained by summing up all components of a k-dimensional vector $Q(Q_1, Q_2, ..., Q_k)$ as

$$G(t, \Omega) = \mathbf{Q} \cdot \mathbf{1} = \sum_i Q_i \tag{2.84}$$

where $\mathbf{1}$ is a k dimensional column vector containing only the number 1 for each element i.e. $\mathbf{1} = (1, 1, ... 1)$ and

$$Q_i \underset{\text{def}}{=} \mathbf{Q}(t, \Omega_i). \tag{2.85}$$

The equation of motion for the Q_i in this k-site exchange problem is readily obtained under the condition that each spectral δ-line at the frequency $\omega_i = \omega(\Omega_i)$ contributes to the spectrum. The equation of motion may be written as

$$\frac{d}{dt} Q_i = i\omega(\Omega_i) Q_i + \sum_j \pi(\Omega_j, \Omega_i) Q_j \tag{2.86}$$

or expressed by the k-dimensional $\mathbf{Q}$ vector as

$$\frac{d}{dt} \mathbf{Q} = (i\overset{\rightrightarrows}{\omega} + \overset{\rightrightarrows}{\pi})\mathbf{Q}. \tag{2.87}$$

Here $\overset{\rightrightarrows}{\omega}$ is a diagonal matrix with $\omega_{ij} = \omega_i \delta_{ij}$ and where the jump matrix $\overset{\rightrightarrows}{\pi}$ with the matrix elements $\pi_{ij} = \pi(\Omega_i, \Omega_j)$ has been defined in Eq. (2.81) describing the jumping rate from orientation Ω_i to Ω_j. We obtain as a formal solution of Eq. (2.87)

$$\mathbf{Q}(t) = \mathbf{Q}(0) \exp[(i\overset{\rightrightarrows}{\omega} + \overset{\rightrightarrows}{\pi})t]. \tag{2.88}$$

The initial state vector $\mathbf{Q}(0)$ is identical with the row vector of the a priori probabilities

$$\mathbf{Q}(0) \equiv \mathbf{W} = (W(\Omega_1), W(\Omega_2), ... W(\Omega_n)) \tag{2.89}$$

leading according to Eqs. (2.84, 2.85–2.88) to

$$G(t, \Omega) = \mathbf{W} \cdot \exp[(i\overset{\rightrightarrows}{\omega} + \overset{\rightrightarrows}{\pi})t] \cdot \mathbf{1}. \tag{2.90}$$

Performing the Fourier transform of Eq. (2.90) according to Eq. (2.82) leads to

$$I'(\omega, \Omega) = \mathrm{Re}\{\mathbf{W} \cdot \mathbf{A}^{-1} \cdot \mathbf{1}\} \tag{2.91}$$

where

$$\mathbf{A} = i(\overset{\rightrightarrows}{\omega} - \omega\mathbf{E}) + \overset{\rightrightarrows}{\pi} \tag{2.92}$$

and $\mathbf{E}$ is a $k\times k$ unit matrix. The powder spectrum $I(\omega)$ can be readily obtained by averaging over the full solid angle Ω. However, the singularities in Eq. (2.91) may cause severe problems, when calculating powder spectra numerically. Fake spikes and peaks may occur in the powder spectrum which have no physical significance. On the other hand, the calculation of spectra according to Eq. (2.91) may become very time consuming even on a fast digital computer. Gordon and McGinnis [*29*] proposed therefore a procedure, which uses the so-called QR transformation [*30*] to diagonalize rapidly and stably the non-Hermitian matrix $i\overset{\rightrightarrows}{\omega} + \overset{\rightrightarrows}{\pi}$ as

$$\mathbf{S}^{-1}(i\overset{\rightrightarrows}{\omega} + \overset{\rightrightarrows}{\pi})\,\mathbf{S} = \overset{\rightrightarrows}{\lambda} \tag{2.93}$$

where $\lambda_{ij} = \lambda_i \delta_{ij}$ is the eigenvalue matrix.

If we start from Eq. (2.90) with

$$G(t, \Omega) = \mathbf{W} \cdot \mathbf{S} \cdot \mathbf{S}^{-1} \exp[(i\overset{\rightrightarrows}{\omega} + \overset{\rightrightarrows}{\pi})t]\mathbf{S} \cdot \mathbf{S}^{-1} \cdot \mathbf{1} \tag{2.94}$$

we readily obtain

$$I(\omega, \Omega) = \mathrm{Re}\{\mathbf{W} \cdot \mathbf{S} \cdot \int_0^\infty \exp[(\overset{\rightrightarrows}{\lambda} - i\omega\mathbf{E})t']dt' \cdot \mathbf{S}^{-1} \cdot \mathbf{1}\} \tag{2.95}$$

which can be rewritten as

$$I(\omega, \Omega) = \mathrm{Re} \sum_j \frac{(\mathbf{W} \cdot \mathbf{S})_j (\mathbf{S}^{-1} \cdot \mathbf{1})_j}{i\omega - \lambda_j}. \tag{2.96}$$

Thus the whole line shape problem of k lines is reduced to a single sum over the k lines once the diagonalization has been carried out. However, not only the diagonalization procedure has to be performed, but also the transformation matrix $\mathbf{S}$ has to be determined. Averaging over the full solid angle Ω leads to the powder spectrum.

This integration again might introduce numerical instabilities, since the proper integration steps have to be selected. We therefore propose a different procedure which first calculates the free induction decay according to Eq. (2.94) and then averages $G(t, \Omega)$ as

$$G(t) = \frac{1}{\mathcal{N}} \int_\Omega d\Omega G(t, \Omega).$$

The final Fourier transform of $G(t)$ to yield $I(\omega)$ is always numerically stable and can be performed, using a fast algorithm.

As an illustrative example let us consider a two site problem with the frequencies $\omega_1 = \omega(\Omega_1)$ and $\omega_2 = \omega(\Omega_2)$.

By using Eqs. (2.81) and (2.92) we write

$$\overrightarrow{\overrightarrow{\pi}} = \frac{\kappa}{2}\begin{pmatrix} -1 & 1 \\ 1 & -1 \end{pmatrix}$$

and

$$\mathbf{A} = \begin{pmatrix} i(\omega_1 - \omega) - \kappa/2 & \kappa/2 \\ \kappa/2 & i(\omega_2 - \omega) - \kappa/2 \end{pmatrix}.$$

With $\mathbf{W} = (1/2, 1/2)$ and $\mathbf{1} = (1, 1)$ the following expression for the spectrum according to Eq. (2.91) results

$$I(\omega) = \frac{\kappa(\omega_1 - \omega_2)}{4(\omega_1 - \omega)^2(\omega_2 - \omega)^2 + \kappa^2(\omega_1 + \omega_2 - 2\omega)^2} \tag{2.97}$$

which has been discussed in detail by Abragam [*1*].

The same result is, of course, obtained when Eq. (2.96) is used. As mentioned above it might be numerically more stable to sum up free induction decays when additional integration has to be performed over different orientations. We therefore calculate $G(t, \Omega)$ according to Eqs. (2.93, 2.94) as [*31*]

$$G(t, \Omega) = e^{\frac{i}{2}(\omega_1 + \omega_2)t} \frac{1}{2}[K_+ e^{-\frac{1}{2}(1-c)\kappa t} + K_- e^{-\frac{1}{2}(1+c)\kappa t}] \tag{2.98}$$

where

$$K_\pm = 1 \pm 1/c$$

and

$$c = [\kappa^2 - (\omega_1 - \omega_2)^2]^{1/2}/\kappa$$

and where the following transformation matrix has been used

$$\mathbf{S} = \begin{pmatrix} s & 1 \\ 1 & -s \end{pmatrix} \tag{2.99}$$

with

$$s = c + i(\omega_1 - \omega_2)/\kappa. \tag{2.100}$$

The complete free induction decay and the corresponding spectrum $I(\omega)$ are readily obtained when the Ω dependence of ω_1 and ω_2 is known.

Suppose there is a flipping of the molecule or a molecular group about an angle δ i.e. the interaction tensor connected with this molecule is flipped by an angle δ about an arbitrary axis. To determine the secular part of the interaction connected with this flipping, we write

$$\mathcal{H}'_{\text{int}} = A_{00}T_{00} + A_{20}T_{20}$$

which for interactions linear in the spin variable may be rewritten as

$$\mathcal{H}'_{\text{int}} = [\tfrac{1}{3}\operatorname{Tr}\{A_{ij}\} + \frac{2}{\sqrt{6}} A_{20}(\delta)] H_0 I_z. \tag{2.101}$$

The two characteristic resonance frequencies ω_1 and ω_2 can now be expressed as

$$\begin{aligned} \omega_1 &= \omega_0 + \omega(\delta = 0) \\ \omega_2 &= \omega_0 + \omega(\delta) \end{aligned} \tag{2.102}$$

where

$$\omega_0 = \frac{1}{3}\operatorname{Tr}\{A_{ij}\} \text{ and } \omega(\delta) = \frac{2}{\sqrt{6}} A_{20}(\delta).$$

Suppose $R''(\alpha, \beta, \gamma)$ is the transformation from the principal axis system (1, 2, 3) to the molecular flipping frame (x_M, y_M, z_M) with the z_M-axis being the flipping axis and where $R'_z(\delta)$ represents the flippling about the z_M-axis by an angle δ. A further transformation $R(\varphi, \vartheta, \psi)$ has to be applied to transform from the molecular frame (x_M, y_M, z_M) to the laboratory frame (x, y, z). The complete transformation strategy is outlined as follows

$$\underset{(1,2,3)}{\tilde{\mathbf{A}}} \xrightarrow{R''(\alpha,\beta,\gamma)} \mathbf{A}_{M'} \xrightarrow{R'_z(\delta)} \underset{(x_M, y_M, z_M)}{\mathbf{A}_M} \xrightarrow{R(\varphi,\vartheta,\psi)} \underset{(x,y,z)}{\mathbf{A}'}$$

and the total transformation may be written as

$$R_{\text{total}} = R(\varphi, \vartheta, \psi) R'_z(\delta) R''(\alpha, \beta, \gamma). \tag{2.103}$$

The powder average has to be performed over the angles $(\varphi, \vartheta, \psi)$ whereas (α, β, γ) represent the relation of the principle axis system to the flipping axis. Using the transformation properties of irreducible spherical tensors as outlined in Appendix A we obtain

$$A_{20}(\delta) = \sum_{q,p} \tilde{A}_{2q} D^{(2)}_{qp}(\alpha, \beta, \gamma) e^{-ip\delta} D^{(2)}_{po}(\varphi, \vartheta, \psi) \tag{2.104}$$

or

$$A_{20}(\delta) = \sum_{q,\,p} \tilde{A}_{2q} d^{(2)}_{qp}(\beta) d^{(2)}_{po}(\vartheta) e^{-ip(\gamma+\varphi+\delta)}. \tag{2.105}$$

Using (see Appendix A)

$$\tilde{A}_{20} = \frac{3}{\sqrt{6}}(S_{33} - \frac{1}{3}\operatorname{Tr}\{S_{ij}\}) = \frac{3}{\sqrt{6}} S^*_{33}$$

$$\tilde{A}_{2\pm1} = 0 \text{ and } \tilde{A}_{2\pm2} = \frac{1}{2}(S_{11} - S_{22})$$

we arrive at

$$\omega(\delta) = \frac{2}{\sqrt{6}} A_{20}(\delta) = S_{33}^{*} \sum_{p=-2}^{+2} d_{op}^{(2)}(\beta) d_{po}^{(2)}(\vartheta) e^{-ip(\gamma+\varphi+\delta)}$$

$$+ \frac{1}{2}(S_{11} - S_{22}) \sum_{p=-2}^{+2} (d_{2p}^{(2)}(\beta) e^{-i2\alpha} + d_{-2p}^{(2)}(\beta) e^{i2\alpha})$$

$$\cdot d_{po}^{(2)}(\vartheta) e^{-i(\gamma+\varphi+\delta)}. \tag{2.106}$$

Further reduction of Eq. (2.106) is straightforwardly possible using the Wigner rotation matrices as listed in Appendix B and results in

$$\begin{aligned}\omega(\delta) = {} & S_{33}^{*}[P_2(\cos\beta)\, P_2(\cos\vartheta) - \frac{3}{4}(\sin 2\beta \sin 2\vartheta \cos(\gamma + \varphi + \delta) \\ & - \sin^2\beta \sin^2\vartheta \cos 2(\gamma + \varphi + \delta))] \\ & + \sqrt{\tfrac{3}{8}}\,(S_{11} - S_{22})[P_2(\cos\vartheta) \sin^2\beta \cos 2\alpha + \sin\beta \sin 2\vartheta \\ & \cdot (\cos\beta \cos 2\alpha \cos(\gamma + \varphi + \delta) - \sin 2\alpha \sin(\gamma + \varphi + \delta) \\ & + \sin^2\vartheta(\frac{1}{2}(1 + \cos^2\beta) \cos 2\alpha \cos 2(\gamma + \varphi + \delta) \\ & - \cos\beta \sin 2\alpha \sin 2(\gamma + \varphi + \delta))]\end{aligned} \tag{2.107}$$

where P_2 is the Legendre Polynomial. (α, β, γ) are the Euler angles of the flipping axis with respect to the principal axis system of the tensor, with δ being the flipping angle. (ϑ, φ) are the Euler angles of the magnetic field H_0 with respect to the molecular frame containing the flipping axis as the z-axis.

Equation (2.107) becomes particularly simple in the case of axial symmetry $(S_{11} = S_{22})$. As a simple example let us consider a water molecule (H_2O or DHO) in a solid, like in gypsum. The proton shielding tensor is supposed to be axially symmetric about the bond direction and the water molecule is assumed to perform 180° jumps about an axis, bisecting the HOH-angle (β=54°44′).

Under these conditions Eq. (2.107) reduces to

$$\omega(\delta) = S_{33}^{*}[\frac{1}{2}\sin^2\vartheta \cos 2(\varphi + \delta) - \frac{1}{\sqrt{2}}\sin 2\vartheta \cos(\varphi + \delta)]. \tag{2.108a}$$

With the two conditions $\delta = 0, 180°$ we obtain

$$\omega_1 - \omega_2 = S_{33}^{*}\sqrt{2}\sin 2\vartheta \cos\varphi \tag{2.108b}$$

$$\frac{1}{2}(\omega_1 + \omega_2) = S_{33}^{*} \sin^2\vartheta \cos 2\varphi. \tag{2.108c}$$

If these relations are inserted into Eqs. (2.98) and (2.97) the free induction decay and the corresponding spectrum can be calculated for different orientations of the magnetic field and for different flipping rates κ of the molecule. In the case of a powder sample in addition a powder average over (ϑ, φ) has to be performed. It may be convenient to define a normalized flipping rate

$$\eta = \frac{\kappa}{\Delta S} = \frac{2\kappa}{3S_{33}^*} . \tag{2.109}$$

Figure 2.12 represents a computer calculation of powder line shapes for different values of η for a water molecule, flipping by 180° about its two-fold axis [*31*].

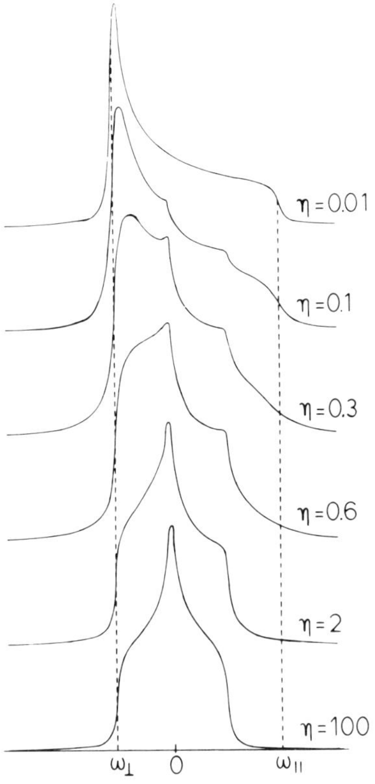

Fig. 2.12. Powder lineshapes of an axially symmetric shielding tensor, performing 180° jumps about an axis, which makes an angle of 54°44″ with respect to the symmetry axis of the tensor (Becker [*31*]). Here η is the relative jump rate with respect to the anisotropy $\omega_{\|} - \omega_{\perp}$. This is considered to be a realistic model for "crystal water", like in $CaSO_4 \cdot 2H_2O$

Notice, that a completely non-axially symmetric pattern evolves at a very high flipping rate η. This can be rationalized easily by considering the fact that the tensor component orthogonal to the molecular plane (which is $\omega_{\perp}$) and the one along the flipping axis ($\omega = 0$) are not affected by the flipping motion. Since the trace of the tensor must be unchanged the third component should appear at $\omega_{\|}/2$.

Alexander *et al.* [*32*] and Pines *et al.* [*33*] have investigated powder lineshapes due to planar rotational jumps, both theoretically and experimentally. Figure 2.13 shows those spectra for hexamethylbenzene at various jumping rates about the sixfold axis [33]. There is a characteristic difference between sixfold jumps and rotational diffusion which is borne out by the spectra. Clearly, the experimental spectra seem to favour the jump model.

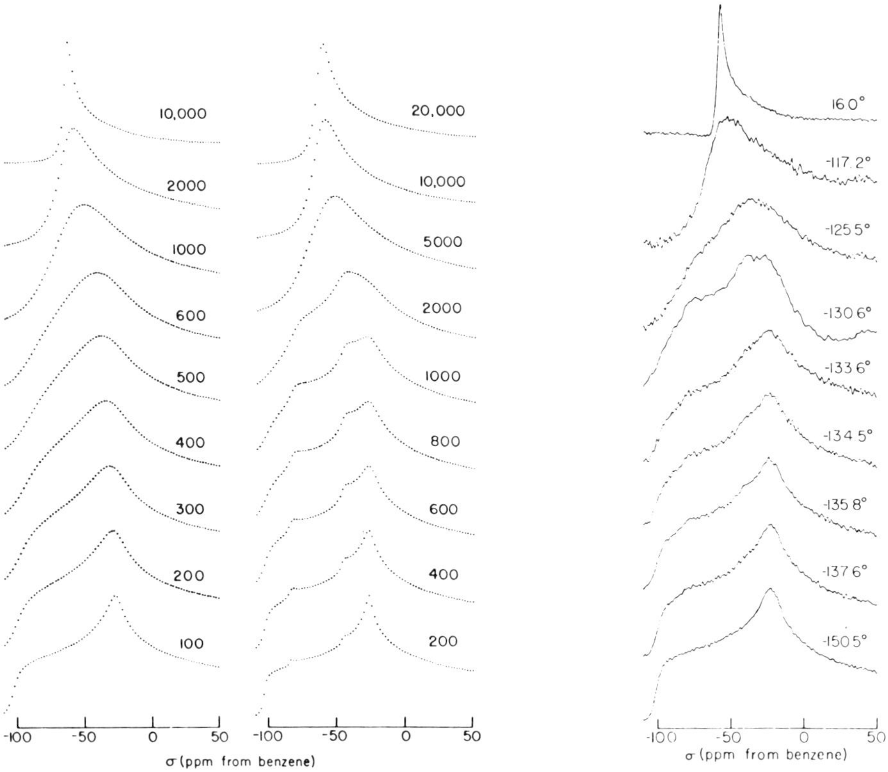

Fig. 2.13. Left part of the powder pattern of the ^{13}C spectrum of hexamethylbenzene at different jumping (or rotation) rates, according to Pines *et al.* [33]. Left: Calculated spectra assuming rotational diffusion with different rates. Middle: Calculated spectra assuming six fold jumps about the C_6 axis with different jumping rates in Hertz. $\nu(^{13}C) = 45$ MHz. Right: Experimental spectra at various temperatures in degrees Celsius. (Courtesy of A. Pines)

3. Multiple-Pulse NMR Experiments

Multiple-pulse experiments have been applied to nuclear spins since the early days of NMR, beginning with the two pulse Hahn-echo [*1*], followed by three pulse echoes and the Carr-Purcell sequence [*2*], being the first cyclic multiple-pulse experiment at all. Later on modifications like the Gill-Meiboom [*3*] modification have been designed in order to reduce the influence of pulse errors which often have an accumulative effect in multiple-pulse sequences. Although these pulse sequences were first applied to liquids, it was soon realized, that similar pulse sequences could be applied to solids as well [*4–8*]. Even though only partial refocusing of the initial state was achieved, substantial insight into the spin dynamics of solids was gained in these experiments. Modified Carr-Purcell sequences applied to solids by Ostroff and Waugh [*4*] and by Mansfield and Ware [*5*] were unexpectedly successful in achieving a considerably lengthened decay of transverse magnetization. This effect contradicts the spin temperature hypothesis and was unexpected at first sight. Much effort was applied in designing more and more efficient multiple-pulse sequences. Since these sequences were promising in achieving a long transverse decay, but leaving resonance offset and chemical shift interactions effective, high resolution spectra in solids were expected. This development culminated in the four-pulse sequence by Waugh, Huber and Haeberlen [*9*], which is the most efficient basic multiple-pulse sequence up to date. Many modifications of this sequence have been proposed recently for correcting pulse errors.

Before entering the more detailed description of these sequences we shall summarize some facts about the transverse response of nuclear magnetization, following a pulsed excitation [*10*].

The spin system is assumed to be in thermal equilibrium initially, described by the Boltzman spin density matrix

$$\rho_B = \exp(-\beta\mathcal{H})/\mathrm{Tr}\{\exp(-\beta\mathcal{H})\} \tag{3.1}$$

where $\beta = \hbar/kT$ and $\mathcal{H}$ is the stationary Hamiltonian of the system. At time $t = 0$ an rf δ pulse at the Larmor frequency may be applied in the y direction of the rotating frame preparing an initial superradiant state as

$$\rho(0) = \mathbf{P}_y(\vartheta)\rho_B\mathbf{P}_y^{-1}(\vartheta) \tag{3.2}$$

where

$$\mathbf{P}_y(\vartheta) = \exp(-i\vartheta \cdot I_y).$$

Here ϑ is the rotation angle of the δ pulse. In the case of the Bloch decay ϑ equals usually $\pi/2$.

With the Zeeman interaction $\mathcal{H}_z$ being the dominant contribution in high fields we obtain

$$\rho(0) = \exp(-\beta\mathcal{H}_x)/\mathrm{Tr}\{\exp(-\beta\mathcal{H}_z)\} \tag{3.3}$$

where

$$\mathcal{H}_x = \mathbf{P}_y(\pi/2)\,\mathcal{H}_z\mathbf{P}_y^{-1}(\pi/2) = -\omega_0 I_x.$$

The initial state $\rho(0)$ is often expanded and truncated, according to the high temperature approximation as [*11*]

$$\rho(0) = (1 + \beta\omega_0 I_x)/\mathrm{Tr}\{\mathbf{1}\} \tag{3.4}$$

where

$$\mathrm{Tr}\{\mathbf{1}\} = (2I + 1)^N$$

is the total number of states if only a single spin species I is present.

This approximation is by no means straightforward and has to be justified by the corresponding expansion of observables. In NMR, where macroscopic observables like energy, magnetization etc. which average over a large number of states are treated only, this approximation is always legitimate down to $T = 1$ K.

The superradiant state, prepared by the initiating pulse, now decays due to the spin interaction $\mathcal{H}_i(t)$, according to the Liouville-v. Neumann Equation

$$\frac{d}{dt}\rho(t) = -i[\mathcal{H}_i(t), \rho(t)] \tag{3.5}$$

which has the formal solution

$$\rho(t) = \mathbf{L}(t)\,\rho(0)\,\mathbf{L}^+(t) \tag{3.6}$$

with

$$\mathbf{L}(t) = T\exp[-i\int_0^t dt'\,\mathcal{H}_i(t')]. \tag{3.7}$$

T: Dyson time ordering operator and where $\mathbf{L}^+(t)$ is the adjoint of $\mathbf{L}(t)$ with $\mathbf{L}^+(t) = \mathbf{L}^{-1}(t)$ because $\mathcal{H}_i(t)$ is an Hermitian operator.

The propagator $\mathbf{L}(t)$ which describes the time evolution of the system will be subject to detailed discussion in later sections. Actually this propagator $\mathbf{L}(t)$ is the important parameter, which will be determined in multiple-pulse experiments in the following sections, to describe the time evolution of the transverse magnetization [*12*].

The transverse decay (Bloch decay) in the x direction of the rotating frame is now obtained as (see Appendix E)

$$G_x(t) = \mathrm{Tr}\{I_x\rho(t)\}/\mathrm{Tr}\{I_x\rho(0)\} \tag{3.8}$$

or by using Eq. (3.6)

$$G_x(t) = \mathrm{Tr}\{I_x \mathbf{L}(t)\rho(0)\mathbf{L}^{-1}(t)\}/\mathrm{Tr}\{I_x\rho(0)\} \tag{3.9}$$

which leads in the high temperature approximation to

$$G_x(t) = \mathrm{Tr}\{I_x \mathbf{L}(t) I_x \mathbf{L}^{-1}(t)\}/\mathrm{Tr}\{I_x{}^2\}. \tag{3.10}$$

Such a decay function is schematically drawn in Fig. 3.1. If dipolar interaction among many spins is involved, $G_x(t)$ cannot be calculated exactly and approximations as developed in many body theories have to be applied to Eq. (3.10) in order to describe $G_x(t)$ approximately.

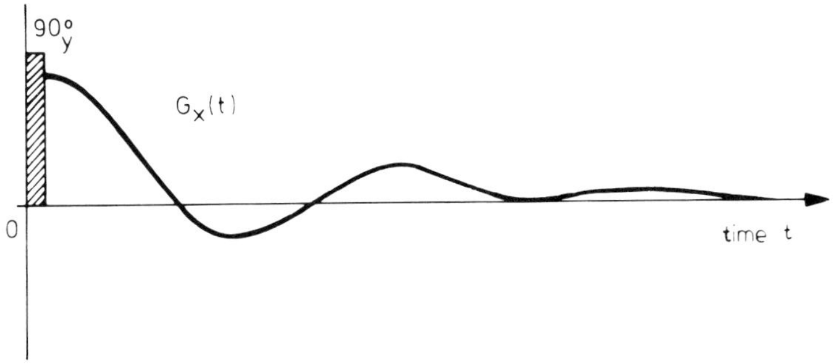

Fig. 3.1. Schematic representation of a Bloch-decay or free induction decay (FID) $G_x(t)$ of a spin system, excited by a 90° pulse in the y-direction of the rotating frame

As shown by van Vleck, a moment expansion of $G_x(t)$ can be performed exactly with [*13*]

$$G_x(t) = \sum_{n=0}^{\infty} \frac{(-it)^n}{n!} M_n. \tag{3.11}$$

Before proceeding with the discussion of approximations to $G_x(t)$ we introduce a shorthand notation for convenience, to be used later.

Introducing the Liouville operator [*14*] (see Appendix E)

$$\hat{\mathcal{H}} \underset{\text{def.}}{=} [\mathcal{H}, \ldots] \tag{3.12}$$

which is a superoperator acting on state vectors in Liouville space, we can rewrite the Liouville-v. Neumann Equation Eq. (3.5) as

$$\frac{d}{dt} | \rho(t)) = -i\hat{\mathcal{H}}_i(t) | \rho(t)) \tag{3.13}$$

with the formal solution

$$|\rho(t)) = \hat{\mathbf{L}}(t)\,|\rho(0)) \tag{3.14}$$

where

$$\hat{\mathbf{L}}(t) = T\exp\left[-i\int_0^t dt'\hat{\mathcal{H}}_i(t')\right].$$

Defining the scalar product of two vectors in Liouville space by [*14*]

$$(\mathbf{A}\,|\,\mathbf{B}) \underset{\text{def.}}{=} \mathrm{Tr}\{\mathbf{A}^+\mathbf{B}\} \tag{3.15}$$

where $\mathbf{A}^+$ is the adjoint of $\mathbf{A}$, we can write the Bloch decay $G_x(t)$ as (see Appendix E)

$$G_x(t) = \frac{(I_x\,|\rho(t))}{(I_x\,|\rho(0))} \tag{3.16}$$

or

$$G_x(t) = \frac{(I_x\,|\hat{\mathbf{L}}(t)|\rho(0))}{(I_x\,|\rho(0))}\,. \tag{3.17}$$

If $\mathcal{H}_i(t)$ is not explicitly time dependent we obtain with

$$\hat{\mathbf{L}}(t) = \exp(-i\hat{\mathcal{H}}_i t) = \sum_{n=0}^{\infty}\frac{(-it)^n}{n!}\hat{\mathcal{H}}_i^n \tag{3.18}$$

$$G_x(t) = \sum_{n=0}^{\infty}\frac{(-it)^n}{n!}\frac{(I_x\,|\hat{\mathcal{H}}_i^n\,|\rho(0))}{(I_x\,|\rho(0))} \tag{3.19}$$

which defines the moments as

$$M_n = \frac{(I_x\,|\hat{\mathcal{H}}_i^n\,|\rho(0))}{(I_x\,|\rho(0))} \tag{3.20}$$

or in the high temperature approximation as

$$M_n = \frac{(I_x\,|\hat{\mathcal{H}}_i^n\,|I_x)}{(I_x\,|I_x)}\,. \tag{3.21}$$

Equation (3.21) together with Eqs. (3.12) and (3.15) leads to

$$M_2 = -\mathrm{Tr}\{[\mathcal{H}_i, I_x]^2\}/\mathrm{Tr}\{I_x{}^2\} \tag{3.22a}$$

and

$$M_4 = \mathrm{Tr}\{[\mathcal{H}_i, [\mathcal{H}_i, I_x]]^2\}/\mathrm{Tr}\{I_x{}^2\} \tag{3.22b}$$

which are the familiar second and fourth moment [*10*].

If the corresponding line shape is symmetrical about the frequency $\omega_0 = 0$ (central moments vanish) all odd moments vanish and $G_x(t)$ may be expressed as

$$G_x(t) = 1 - \frac{M_2}{2!}t^2 + \frac{M_4}{4!}t^4 - + \dots . \tag{3.23}$$

Higher order moments are usually difficult to evaluate, although in the case of a simple cubic lattice all moments up to the eighth moment have been calculated. But in most cases just the second and fourth moment are known.

Since the moment expansion does not converge rapidly, the second and fourth moment describe only the very beginning of the decay function $G_x(t)$. That is why much effort has been put in approximative methods recently, to utilize the knowledge of low order moments most efficiently for predicting the full line shape of the spectrum (see Appendix E).

The simplest approximation would be to guess the line shape, or to determine it experimentally and to fit it by known moments. Two commonly used line shapes are the:

(i) Gaussian line shape

$$G(t) = \exp\left(-\frac{M_2}{2}t^2\right) \tag{3.24a}$$

with the cosine transform of $G(t)$ as

$$I(\omega) = \frac{1}{(2\pi M_2)^{1/2}} \exp\left(-\omega^2/2M_2\right) \tag{3.24b}$$

where the corresponding half width at half height equals

$$\delta_0 = [2\ln 2 \cdot M_2]^{1/2} \cong 1.18\, M_2^{\ 1/2} . \tag{3.25}$$

(ii) Truncated Lorentzian line shape [10, 11]

$$G(t) = \exp(-t/T_2) \text{ for } t > t^* \tag{3.26a}$$

and

$$I(\omega) = \frac{1}{\pi}\frac{\delta_0}{\omega^2 + \delta_0^2} \text{ for } \omega \leqslant \omega^* \tag{3.26b}$$

where

$$\delta_0 = 1/T_2 \tag{3.27}$$

is the half width at half height of the spectrum.

Since the second and higher order moments diverge for a Lorentzian line shape, any experimentally observed line shape can be Lorentzian only in a limited frequency range. This justifies the limits in Eq. (3.26).

If $I(\omega)$ vanishes outside $\pm\omega^*$ we arrive at the following moments for the truncated Lorentzian line shape [*10, 11*]:

$$M_2 = \frac{2\omega^*\delta}{\pi} \quad \text{and} \quad M_4 = \frac{2\omega^{*3}\delta}{3\pi}$$

which leads to the half width at half height as

$$\delta_0 = \frac{\pi}{2\sqrt{3}} \cdot \left[\frac{M_2}{\mu}\right]^{1/2} \tag{3.28}$$

where

$$\mu = M_4/M_2^2 .$$

Since the direct approximation of $G(t)$ by assuming a functional shape which suits $G(t)$ cannot be made in general and on the other hand is very unsatisfactory from a theoretical point of view, more subtle methods have been designed which use an approximation at a less sensitive level. This is the "memory function approach" [*15–20*] which is extensively discussed in Appendix E.

In the case of a symmetric line shape function (vanishing central moments) the following exact integro-differential equation for the decay function $G(t)$ is obtained from the Liouville-v. Neumann Equation (see Appendix E)

$$\frac{d}{dt} G(t) = -\int_0^t dt' K(t-t')G(t') = -\int_0^t d\tau K(\tau)G(t-\tau) \tag{3.29}$$

where $K(t-t')$ is the so-called "memory function" which describes the memory $G(t')$ has of its early history in time development, weighted by $K(t-t')$. A schematic drawing if this functional behavior is presented in Fig. 3.2. Since $G(t)$ is less sensitive to approximations on $K(\tau)$, rather than to approximations on $G(t)$ itself, we have reached a higher level of approximation. The next step further would be to express

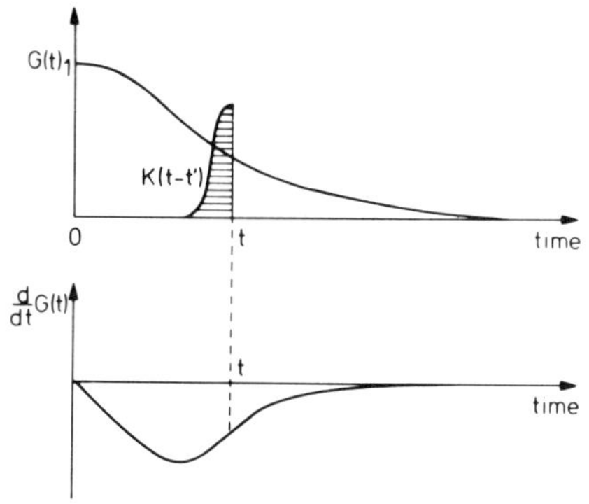

Fig. 3.2. Schematic drawing of the "memory function" approach to the calculation of FID's

$K(\tau)$ again by an integro-differential equation like Eq. (3.29) whose memory function $K'(t-t')$ is approximated instead of $K(\tau)$ itself.

The simplest approximation to $K(t-t')$ would be a δ function approximation i.e. no memory (correlation time $\tau_c = 0$)

$$K(t-t') = K_0 \delta(t-t') \tag{3.30}$$

leading to

$$G(t) = \exp(-t/T_2)$$

where

$$1/T_2 = K_0 = M_2^{1/2} \tag{3.31}$$

and M_2 is the second moment of $G(t)$. If $K(t-t')$ is much more rapidly decaying than $G(t)$ i.e. the correlation time τ_c of $K(\tau)$ is much shorter than the decay time T_2, we may use a less stringent approximation than Eq. (3.30), namely

$$\frac{d}{dt} G(t) = -\int_0^\infty d\tau K(\tau) \cdot G(t) = -\frac{1}{T_2} G(t) \tag{3.32a}$$

where

$$\frac{1}{T_2} = \int_0^\infty d\tau K(\tau). \qquad \text{for } \tau_c \ll T_2 \quad \text{"short correlation limit"}$$

If we make a less severe approximation than Eq. (3.32a) by taking the upper limit of the integral to be t instead of infinity we arrive at the Andersson-Weiss [*18c*] Model

$$\frac{d}{dt} G(t) = -\int_0^t d\tau K(\tau) G(t). \tag{3.32b}$$

The exact line shape $I(\omega)$ can be obtained from Eq. (3.29) with

$$I(\omega) = \int_0^\infty dt G(t) \cos\omega t$$

by formal integration as [*16*]

$$I(\omega) = \frac{G(0) K'(\omega)}{[\omega - K''(\omega)]^2 + [K'(\omega)]^2} \tag{3.33}$$

where

$$K'(\omega) = \int_0^\infty dt K(t) \cos\omega t$$

and

$$K''(\omega) = \int_0^\infty dt K(t) \sin\omega t.$$

The half width δ_0 at half height of the lineshape $I(\omega)$ according to Eq. (3.33) can be obtained by iteration from

$$\delta = K''(\delta) + [2K'(0)K'(\delta) - K'^2(\delta)]^{1/2}. \tag{3.34}$$

If $G(t)$ and $K(t)$ are given by the following moment expansion [*21*]

$$G(t) = G(0) \sum_{n=0}^{\infty} (-1)^n \frac{t^{2n}}{(2n)!} M_{2n} \tag{3.35a}$$

and

$$K(t) = K(0) \sum_{n=0}^{\infty} (-1)^n \frac{t^{2n}}{(2n)!} N_{2n} \tag{3.35b}$$

it is straightforward to obtain the following relations by comparing both sides of Eq. (3.29): [*16, 21, 22*]

$$K(0) = M_2; N_2 = M_2(M_4/M_2^2 - 1)$$

$$N_4 = M_2^2(M_6/M_2^3 - 2M_4/M_2^2 + 1) \text{ etc.} \tag{3.36}$$

See also Appendix E for a more rigorous discussion.

So far no approximation has been used in Eqs. (3.33–3.36). A convenient approximation which is still quite flexible in describing completely different line shapes including Gaussian and Lorentzian shapes, would be to use a Gaussian shape for $K(t)$ [*21–23*].

$$K(t) = K(0) \exp [-\frac{N_2}{2} t^2]. \tag{3.37}$$

Only the second moment of $K(t)$ i.e. the second and fourth moment of $G(t)$ is needed. The corresponding line-shape is discussed in more detail in Appendix E. As demonstrated in Appendix E a simple expression for the linewidth δ of the corresponding spectral line is obtained by

$$\delta = \sqrt{\frac{\pi}{2}} [\frac{M_2}{\mu - 1}]^{1/2} \tag{3.38}$$

which is valid for large values of $\mu = M_4/M_2^2$. In general Eq. (3.38) will be multiplied by a function of the order of one, which will depend on the functional form used for the memory function. In order to cover a wide range of different line shapes it is shown in Appendix E that the following expression for the universal line width δ is more appropriate

$$\delta = \sqrt{\frac{\pi}{2}} [\frac{M_2}{\mu - 1.87}]^{1/2}. \tag{3.39}$$

With the help of these approximative methods it is thus possible to calculate line-shapes and the corresponding linewidth of "many body systems", i.e. where the time evolution of the system is governed by e.g. many body dipolar interactions. The above formulas may be used to calculate the residual linewidth in high resolution NMR in solids, as will be exploited in the next chapters.

3.1 Idealized Multiple-Pulse Sequences

In this and some of the following sections we make the assumptions: (i) spin lattice relaxation is neglected (ii) the applied rf pulses are assumed to be δ pulses.

Let us first discuss the well-known

a) Carr-Purcell Sequence [2]

As shown in Fig. 3.3 a preparation pulse P_0 (usually a $\pi/2$ pulse) is applied to initiate the first coherent state, which decays during time τ to some intermediate state due to the static secular spin interactions. At time τ a π pulse P_1 refocusses all the spins (only if the interactions are linear in the spin variable) to produce an echo at 2τ. The state at 2τ can now be considered as a new initial coherent state whose decay can be inverted by another π pulse a time τ later and so fourth (see Fig. 3.3).

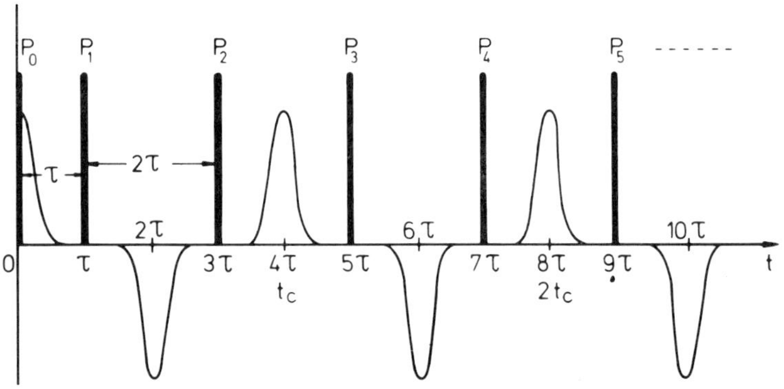

Fig. 3.3. Schematic drawing of a Carr-Purcell sequence, generated by P_0 being a $\frac{\pi}{2}$ pulse and the π pulse P_i, $i \geqslant 1$. Notice, that the phase of the spin echoes is alternating

With

$$\mathbf{P}_0 = e^{-i\frac{\pi}{2}I_y};\ \mathbf{P}_k = e^{-i\pi I_y}\ k = 1, 2 \ldots n$$

and

$$D(\tau) = e^{-i\mathcal{H}_i\tau}$$

we obtain for the spin density matrix at $t = 2\tau$

$$|\rho(2\tau)) = \hat{\mathbf{D}}(\tau)\hat{\mathbf{P}}_1\hat{\mathbf{D}}(\tau)|\rho(0)) \tag{3.40a}$$

or

$$|\rho(2\tau)) = \hat{\mathbf{L}}(2\tau)|\rho(o))$$

where

$$\mathbf{L}(2\tau) = \mathbf{D}(\tau)\mathbf{P}_1\mathbf{D}(\tau) \tag{3.40b}$$

and

$$|\rho(0)) = \hat{\mathbf{P}}_0|\rho_B).$$

The time evolution operator $\mathbf{L}$ (2τ) in Eq. (3.40) may be rewritten as

$$\mathbf{L}(2\tau) = \mathbf{D}(\tau)\widetilde{\mathbf{D}}(\tau)\mathbf{P}_1 \tag{3.41}$$

where

$$\widetilde{\mathbf{D}}(\tau) = \mathbf{P}_1\mathbf{D}(\tau)\mathbf{P}_1^{-1} = e^{-i\widetilde{\mathcal{H}}_i\tau}$$

with

$$\widetilde{\mathcal{H}}_i = \mathbf{P}_1\mathcal{H}_i\mathbf{P}_1^{-1}.$$

Let us assume $\widetilde{\mathcal{H}}_i = -\mathcal{H}_i$ i.e. the pulses P_k invert the interaction Hamiltonian in the case of δ pulses which is consistent with interactions, which depend linearly on the spin variable (e.g. shielding Hamiltonian); we obtain:

$$\mathbf{D}(\tau)\widetilde{\mathbf{D}}(\tau) = \mathbf{1} \text{ and } \mathbf{L}(2\tau) = \mathbf{P}_1.$$

In general

$$|\rho(n2\tau) = \hat{\mathbf{L}}(n2\tau)|\rho(0)) \tag{3.42}$$

where

$$\mathbf{L}(n2\tau) = \prod_{k=1}^{n}\mathbf{P}_k = \mathbf{P}_1^n$$

which results in

$$\mathbf{L}(n2\tau) = [\exp(-i\pi I_y)]^n = \exp(-in\pi I_y) \tag{3.43}$$

or with the initial state $|\rho(o)) = |I_x)$ in

$$|\rho(n2\tau)) = \cos n\pi |I_x) = (-1)^n |I_x) \tag{3.44}$$

displaying the alternating sign of the echo observed in Carr-Purcell experiments, when phase sensitive detection is used. Notice that identical states of the density matrix occur after each "cycle" containing two pulses. This leads us to defining a "cycle" by

$$\prod_{k=1}^{n} \mathbf{P}_k = 1 \tag{3.45}$$

if the lowest possible n is chosen.

We realize that $n = 2$ for the Carr-Purcell sequence with a "cycle time" $t_c = 4\tau$.

An accumulation of pulse length errors may spoil the beauty of the classical Carr-Purcell experiment. Assuming

$$\mathbf{P}_k = e^{-i(\pi+\epsilon)I_y}$$

we see, that because of

$$\prod_{k=1}^{n} \mathbf{P}_k = e^{-in\epsilon I_y} e^{-in\pi I_y} \tag{3.46}$$

the pulse length error ϵ accumulates, leading to a destruction ot the echo envelope.

Gill and Meiboom [*3*] suggested a modification which cures this disorder by a 90° phase shift of the initializing pulse with respect to the cyclic pulses as follows

$$\mathbf{P}_0 = e^{-i\frac{\pi}{2}I_y}; \mathbf{P}_k = e^{-i\pi I_x}. \tag{3.47}$$

Since $\mathbf{P}_k$ commutes with the initial state, we now have

$$\prod_{k=1}^{n} \hat{\mathbf{P}}_k |\rho(0)) = |\rho(0))$$

and only the error due to the violation of the condition $\tilde{\mathcal{H}}_i = -\mathcal{H}_i$ is still effective. If the rotation angle is close to π, this is far less dramatic than without the 90° phase shift. Notice that the echo phase is no longer alternating.

We turn next to the

*b) Phase Alternated Sequence (*PAS*)* [*4–7*]

which is merely a series of equally spaced δ pulses alternating in phase (see Fig. 3.4). An initializing pulse $\mathbf{P}_0$, as in the Carr-Purcell sequence, may be applied but is not a necessity.

With

$$\mathbf{P}_1 = e^{-i\beta I_y}; \mathbf{P}_2 = \mathbf{P}_1^{-1}$$

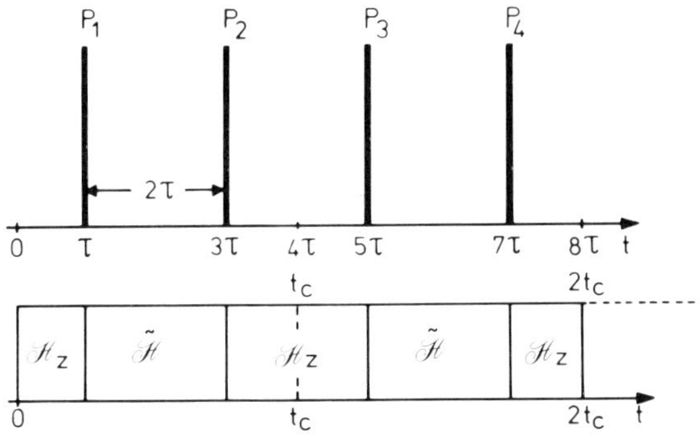

Fig. 3.4. Phase alternating sequence with two pulses per cycle alternating in phase. The interaction Hamiltonian in the "toggling rotating frame" at different times during the cycle is schematically drawn

and $\mathbf{P}_k\mathbf{P}_{k+1} = \mathbf{1}$

we obtain for the propagator $\mathbf{L}(t_c)$ at the cycle time $t_c = 4\tau$

$$\mathbf{L}(t_c) = \mathbf{D}(2\tau)\mathbf{P}_2\mathbf{D}(2\tau)\mathbf{P}_1 \tag{3.48}$$

or

$$\mathbf{L}(t_c) = e^{-i\mathcal{H}_i 2\tau}\, e^{-i\widetilde{\mathcal{H}}_i 2\tau} \tag{3.49}$$

where

$$\widetilde{\mathcal{H}}_i = \mathbf{P}_1^{-1}\mathcal{H}_i\mathbf{P}_1.$$

$\mathbf{L}(t_c)$ may be formally rewritten as

$$\mathbf{L}(t_c) = e^{-i\mathcal{H}_i^* t_c} \tag{3.50}$$

where $\mathcal{H}_i^*$ is defined by equating Eqs. (3.49) and (3.50).

The spin density at any integer multiple of the cycle time can be expressed as

$$|\rho(Nt_c)) = \hat{\mathbf{L}}(Nt_c)|\rho(0)) \tag{3.51}$$

where

$$\mathbf{L}(Nt_c) = \mathbf{L}^N(t_c) = \exp(-i\mathcal{H}_i^* Nt_c). \tag{3.52}$$

As stated before, we are mainly concerned with the time evolution of the magnetization, the issue is to find expressions for $\mathbf{L}(t_c)$. The mathematical effort is considerably reduced by exploiting the cyclic property of the multiple-pulse sequences, i.e. the time evolution operator of the system has to be evaluated over one cycle only, in order to describe the total response of the system at any integer multiple of the cycle time.

Using Eq. (3.49), we obtain in the case of a shift Hamiltonian

$$\mathcal{H}_i = \sum_k \omega_k I_{zk}$$

$$\mathbf{L}(t_c) = \exp(-i\sum_k 2\omega_k \tau I_{zk}) \cdot \exp[-i\sum_k 2\omega_k \tau(I_{zk}\cos\beta + I_{xk}\sin\beta)] \tag{3.53}$$

where β is the rotation angle of the pulses.

The general system response can be calculated exactly, using this propagator [Eq. (3.53)], but it is more elucidating to discuss the response in the limit of small pulse spacing 2τ.

In this limit, i.e.

$$\|\mathcal{H}_1\|\tau; \|\mathcal{H}_2\|\tau \ll 1$$

using Eq. (3.53)
and

$$\exp(-i\mathcal{H}_i\tau)\exp(-i\mathcal{H}_2\tau) \cong \exp(-i(\mathcal{H}_1 + \mathcal{H}_2)\tau)$$

with

$$\mathbf{L}(t_c) = \exp(-i\mathcal{H}^* t_c) \qquad t_c = 4\tau$$

we find for the "average" Hamiltonian $\mathcal{H}^*$

$$\mathcal{H}^* = \sum_k \frac{\omega_k}{2}[I_{zk}(1 + \cos\beta) + I_{xk}\sin\beta]. \tag{3.54}$$

An effective quantization axis in the x,z plane is defined by Eq. (3.54), which makes an angle ϑ with the z axis and where

$$\tan\vartheta = \sin\beta(1 + \cos\beta).$$

The effective frequency ω_{ke} which describes the oscillations of the decay can be straightforwardly derived from Eq. (3.54) as

$$\omega_{ke} = S\omega_k$$

where

$$S = [(1 + \cos\beta)/2]^{1/2} = |\cos(\beta/2)| \tag{3.55}$$

is the "scaling factor" by which all the spectral line shifts are scaled. This effect has been named "chemical shift concertina" and was discussed in detail by Ellett and Waugh [*24*].

By varying the rotation angle β of the rf pulses different scaling factors can be obtained, showing up as a simulation of different smaller static magnetic fields although the spectrometer is working at a high field. These features are demonstrated in Fig. 3.5 and 3.6.

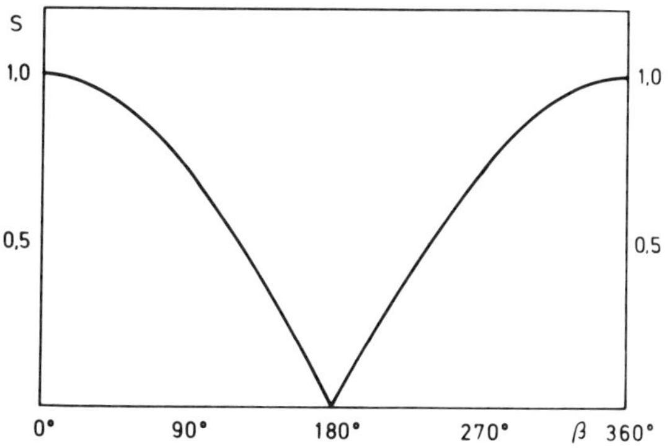

Fig. 3.5. Theoretical scaling factor S of interactions, which are linear in the spin variable (e.g. chemical shift, Knight-Shift etc.) versus the rotation angle of the rf pulse in the phase alternating sequence according to Eq. (3.55)

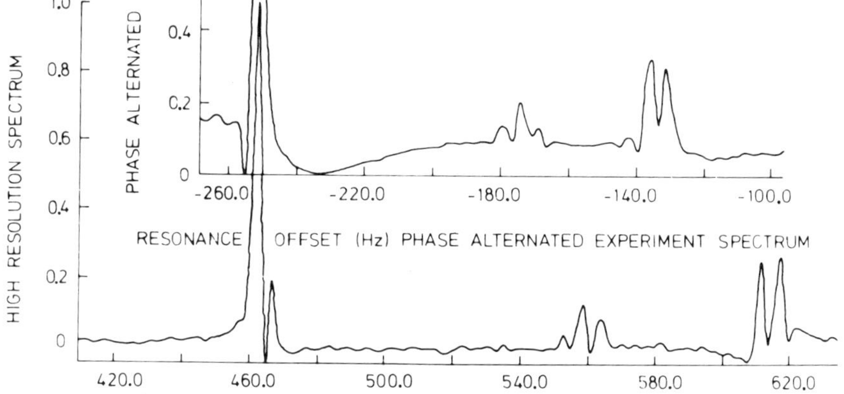

Fig. 3.6. Application of the phase-alternated sequence to a high resolution NMR spectrum by J. D. Ellett and J. S. Waugh [*24*] to demonstrate the scaling effect

We have only treated the simple shift Hamiltonian so far, since the theoretical results are especially easy to visualize. If we turn to secular dipolar interactions with

$$\mathcal{H}_{Dz} = \sum_{i<j} A_{ij}(3I_{zi}I_{zj} - \mathbf{I}_i \cdot \mathbf{I}_j) \quad \text{with } A_{ij} = \frac{1}{2}\gamma_I^2 \hbar(1 - 3\cos^2\vartheta_{ij})/r_{ij}^3$$

we immediately realize that $\mathcal{H}_{Dz}$ is invariant under a π rotation, i.e. it cannot be inverted by a π pulse. As a consequence we expect a completely different behavior as compared with the shift Hamiltonian.

For a phase alternated 90° pulse sequence, we obtain according to Eq. (3.49)

$$\mathbf{L}(t_c) = \exp(-i\mathcal{H}_{Dz}2\tau) \cdot \exp(-i\mathcal{H}_{Dx}2\tau)$$

where

$$\mathcal{H}_{Dx} = \exp(-i\frac{\pi}{2}I_y)\,\mathcal{H}_{Dz}\exp(i\frac{\pi}{2}I_y)$$

or

$$\mathcal{H}_{Dx} = \sum_{i<j} A_{ij}(3I_{xi}I_{xj} - \mathbf{I}_i \cdot \mathbf{I}_j).$$

In the limit $\|\mathcal{H}_D\|\tau \ll 1$, the one cycle propagator $\hat{\mathbf{L}}(t_c)$ may be written as

$$\mathbf{L}(t_c) = \exp[-it_c(\mathcal{H}_{Dz} + \mathcal{H}_{Dx})/2]$$

or

$$\mathbf{L}(t_c) = \exp[it_c\frac{1}{2}\mathcal{H}_{Dy}] \tag{3.56}$$

where

$$\mathcal{H}_{Dx} + \mathcal{H}_{Dy} + \mathcal{H}_{Dz} = 0$$

has been used.

The dipolar interaction is consequently reduced by a factor of two, leading to twice the free induction decay time. If the initial conditon is prepared in the y direction by a $\pi/2$ x pulse no decay results due to $\mathbf{L}(t_c)$, since $|I_y)$ is invariant under the operation of $\mathbf{L}(t_c)$ as given by Eq. (3.56). This is the case of spin-locking. Ostroff and Waugh [*4*] and also Mansfield and Ware [*5*] have analyzed this sequence in more detail. If shift interaction and dipolar interaction are present at the same time, shift interactions are resolved only up to the point where the largest shift is just resolved [*4*].

Since both groups Mansfield and co-workers and Waugh and co-workers have extensively investigated the behavior of dipolar interaction in this two pulse sequence, we shall refer to it in later sections as MW-2 sequence. The corresponding 90° pulse sequence without phase alternation which is actually a four pulse sequence will be referred to as MW-4.

3.2 The Four-Pulse Sequence [9] (WHH-4)

The basic multiple-pulse experiment with vanishing average dipolar and quadrupolar Hamiltonian, retaining shift interactions is the four-pulse experiment, proposed by J. S. Waugh, L. M. Huber and U. Haeberlen [9] (see Fig. 3.7). The rf pulses P_i are

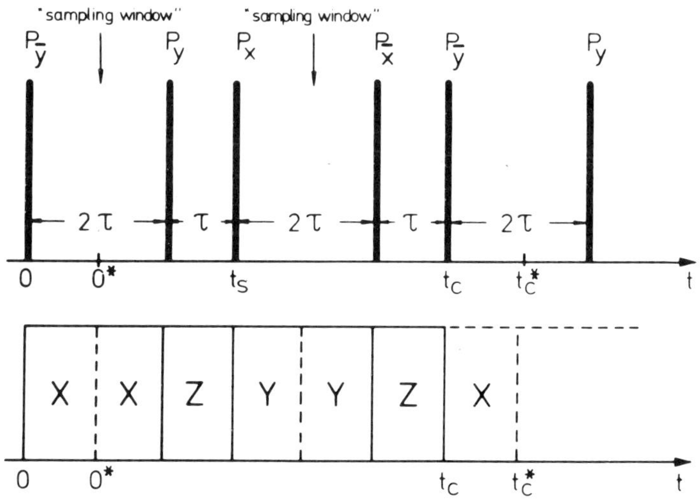

Fig. 3.7. Schematical drawing of the four-pulse sequence (WHH-4). The pulses are assumed to be 90° pulses applied to the spins in different directions of the rotating frame as indicated. Here $\bar{x}$ means $-x$ etc. The interaction Hamiltonian in the "toggling rotating frame" is labelled in short form by capital letters as explained in the text. The magnetization is usually sampled during one of the large windows

assumed to be δ pulses with a rotation angle of 90°. If P_x represents the rotation operator about the x direction of the rotating frame, this sequence may be symbolized in short form by $(P_{\bar{y}}-2\tau-P_y-\tau-P_x-2\tau-P_{\bar{x}}-\tau-)_n$, where the sequence with the cycle time t_c is repeated n times. But it is also legitimate to visualize the cycle from $0^* \to t_c^*$ in short form $(-\tau-P_y-\tau-P_x-2\tau-P_{\bar{x}}-\tau-P_{\bar{y}}-\tau)_n$ as a four-pulse cycle which is repeated n times. In both cases $t_c = t_c^* = 6\tau$, where τ is the closest pulse spacing. The magnetization is usually sampled once every cycle during one of the large "windows" following the $\bar{y}$ or x pulse (see Fig. 3.7).

The one cycle propagator $\mathbf{L}(t_c)$ of the four-pulse experiment can be written down immediately, following Fig. 3.7 as:

$$\mathbf{L}(t_c) = \mathbf{D}_z(\tau)\mathbf{P}_{\bar{x}}\mathbf{D}_z(2\tau)\mathbf{P}_x\mathbf{D}_z(\tau)\mathbf{P}_y\mathbf{D}_z(2\tau)\mathbf{P}_{\bar{y}} \tag{3.57}$$

where

$$\mathbf{P}_\alpha = \exp\left(-i\frac{\pi}{2}I_\alpha\right)$$

and

$$\mathbf{D}_\alpha(\tau) = \exp(-i\,\mathcal{H}_\alpha \tau);\ \alpha = \pm x;\ \pm y;\ \pm z.$$

After applying the transformation operation produced by the rf pulses P_α which is usually referred to as going into the "toggling rotating frame" [*25*], we obtain

$$\mathbf{L}(t_c) = \mathbf{D}_z(\tau)\mathbf{D}_y(2\tau)\mathbf{D}_z(\tau)\mathbf{D}_x(2\tau). \tag{3.58}$$

By applying a short form notation for the "switching Hamiltonian" in the "toggling rotating frame" as

$$\begin{aligned} &\mathbf{X} \underset{\text{def}}{=} \mathcal{H}_x = \mathbf{P}_y \mathcal{H}_z \mathbf{P}_y^{-1} \\ &\mathbf{Y} \underset{\text{def}}{=} \mathcal{H}_y = \mathbf{P}_x^{-1} \mathcal{H}_z \mathbf{P}_x \\ &\mathbf{Z} \underset{\text{def}}{=} \mathcal{H}_z \end{aligned} \tag{3.59}$$

we can characterize the Hamiltonian, switching through the different states **X**, **Y**, **Z** as shown in Fig. 3.7. The one cycle propagator $\mathbf{L}(t_c)$ may be conveniently expressed in the Mansfield notation [*26*] as

$$\mathbf{L}(t_c) = [2\mathbf{X}, \mathbf{Z}, 2\mathbf{Y}, \mathbf{Z}]. \tag{3.60}$$

Figure 3.7 depicts also another cycle with the cycle time t_c^*. This cycle is recognized immediately as being symmetrical by writing down its one cycle propagator in the Mansfield notation [*26*] as

$$\mathbf{L}(t_c^*) = [\mathbf{X}, \mathbf{Z}, \mathbf{Y}; \mathbf{Y}, \mathbf{Z}, \mathbf{X}] = [\mathbf{X}, \mathbf{Z}, \mathbf{Y}][\mathbf{Y}, \mathbf{Z}, \mathbf{X}] \tag{3.61a}$$

or

$$\mathbf{L}(t_c^*) = [\![\mathbf{X}, \mathbf{Z}, \mathbf{Y}]\!] \text{ reflection symmetry [26]}. \tag{3.61b}$$

Aspects of symmetry [*26*] in the one cycle propagator will be discussed later in connection with higher order correction terms in the Magnus expansion (see Section 3.4). In this section we are mainly interested in the average Hamiltonian [*12*] as discussed before. With

$$\overline{\mathcal{H}} = \frac{1}{t_c} \int_0^{t_c} dt'\, \widetilde{\mathcal{H}}(t')$$

we obtain from Eqs. (3.60) and (3.61)

$$\overline{\mathcal{H}} = \frac{1}{3}(\mathcal{H}_x + \mathcal{H}_y + \mathcal{H}_z). \tag{3.62}$$

The following cases are discussed:
(i) dipolar interaction (and quadrupolar interaction)

$$\bar{\mathcal{H}}_D = 0 \text{ since } \mathcal{H}_{Dx} + \mathcal{H}_{Dy} + \mathcal{H}_{Dz} = 0$$

(ii) shift interaction

$$\bar{\mathcal{H}}_S = \frac{1}{3} \sum_k (\delta_k + \Delta\omega)(I_{xk} + I_{yk} + I_{zk}) \tag{3.63}$$

where $\Delta\omega$ is the resonance offset.

Eq. (3.63) may be written in short form as

$$\bar{\mathcal{H}}_S = \frac{1}{\sqrt{3}} \sum_k (\delta_k + \Delta\omega) I_{k111} \tag{3.64}$$

where

$$I_{111} = (I_x + I_y + I_z)/\sqrt{3}$$

and the "scaling factor" equals $1/\sqrt{3}$ by which all shift interactions, including the resonance offset are scaled.

Some representative examples of the application of the four-pulse experiment (WHH-4) are given in Figs. 3.8 to 3.15. A spectrum is obtained in a WHH-4 experiment in the following way:

(i) The magnetization is sampled once every cycle, conveniently in one of the large "windows". The resulting decay is stored in some sort of memory.

(ii) The decay is Fourier-transformed by means of a computer and phase shifted to obtain the proper "absorption signal".

The first single crystal spectrum of ^{19}F in a non-cubic solid using the WHH-4 sequence was obtained by L. M. Stacey, R. W. Vaughan, and D. D. Elleman [*27*]. Figure 3.8 represents a ^{19}F multi-line spectrum, obtained from a single crystal of CF_3COOAg at low temperature (50 °K) [*28*]. The rotation of the fluoro-methyl group is essentially frozen at this temperature (jumping rate is small compared to the line width), resulting in six different lines from the two non-equivalent CF_3-groups in the CF_3COOAg dimer [*28*]. Upon crystal rotation all these lines shift according to the different ^{19}F shielding tensors which have been analyzed in Ref. [*28*]. The rotation plot for a rotation about the c^* crystalline axis is shown in Fig. 3.9. By rotation about the two other crystalline axes a and b the ^{19}F shielding tensors for the three different fluorine of the methyl group have been determined [*28*]. See Section 5.4 for a detailed discussion. The first chemical shielding tensor powder spectrum [*29*] as obtained by the four-pulse sequence is shown in Fig. 3.10. The shift tensor of ^{19}F in fluoranil has three distinct tensor elements which are readily obtained from the powder pattern. The principal axes of the tensor are, however, undetermined from such an experiment. These can be assigned unambiguously only by a single crystal study. However, a tentative assignment may be possible by checking the consistency with related compounds, such as substituted fluorobenzenes (see Section 5.4).

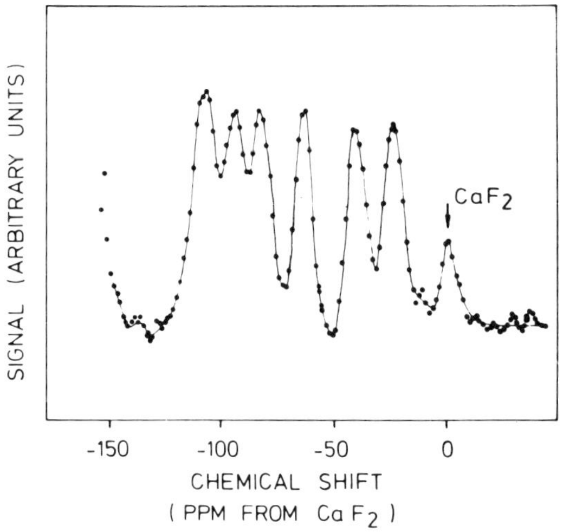

Fig. 3.8. Four-pulse spectrum of ^{19}F in a single crystal of CF_3COOAg at 40 K (methyl group "rotation rate" much less than line width) obtained by Griffin *et al.* [*28*]. The six lines from the molecular dimer are clearly distinguished. This spectrum corresponds to a rotation angle of 100° in Fig. 3.9

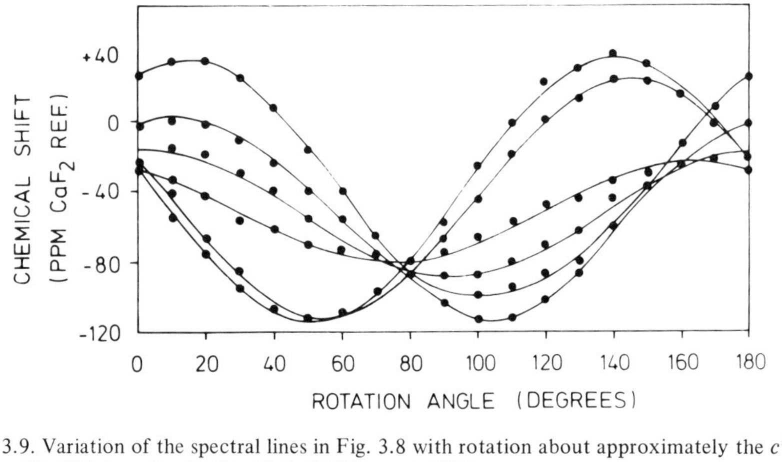

Fig. 3.9. Variation of the spectral lines in Fig. 3.8 with rotation about approximately the c^* axis of the CF_3COOAg unit cell. $T = 40$ K (Griffin *et al.* [*28*])

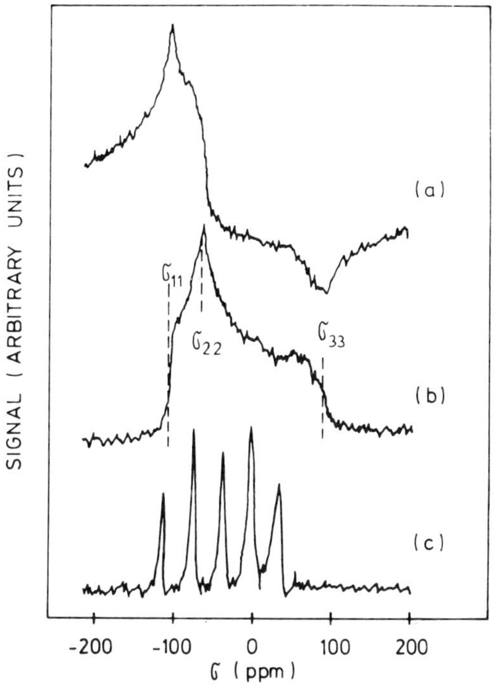

Fig. 3.10 a–c. ^{19}F powder spectra of fluoranil, $C_6F_4O_2$ at 300 K. (a) and (b) are the dispersion and absorption signal respectively, with the tensor elements σ_{11}, σ_{22}, and σ_{33} as indicated, while (c) gives the calibration markers obtained with liquid benzene [*30*]

^{19}F shielding tensors were investigated in the pioneer era of multiple-pulse experiments because of their large values [*30–34*]. However, recently also other nuclei have been studied, including protons, which have a very small shift anisotropy [*35–44*]. Besides a few exceptions, most of these detailed investigations of proton shielding tensors were done by Haeberlen and co-workers [*35–38*].

A representative example of this work is demonstrated in Figs. 3.11 and 3.12. It turned out that protons in carboxyl groups displayed the largest values of shielding anisotropies among the protons. The shielding tensor is mostly axially symmetric with the unique axis being the "hydrogen bond" direction [*25*]. Its size is about 20 ppm and extends over the entire range of isotropic shift values, as demonstrated in the case of "squaric acid" [*39*] in Fig. 3.13. For a further discussion of proton shielding tensors see Section 5.3. Also metals have been investigated using the WHH-4 experiments [*45*] (see Figs. 3.14 and 3.15). A special obstacle in these cases is the skin effect which attenuates and phase shifts the rf field, when penetrating into the metal. This can be circumvented by using powder samples with a grain size less than the skin depth. Since aluminium is a cubic metal, second rank tensor shift interactions vanish and the four-pulse experiment may be effectively used to reduce the

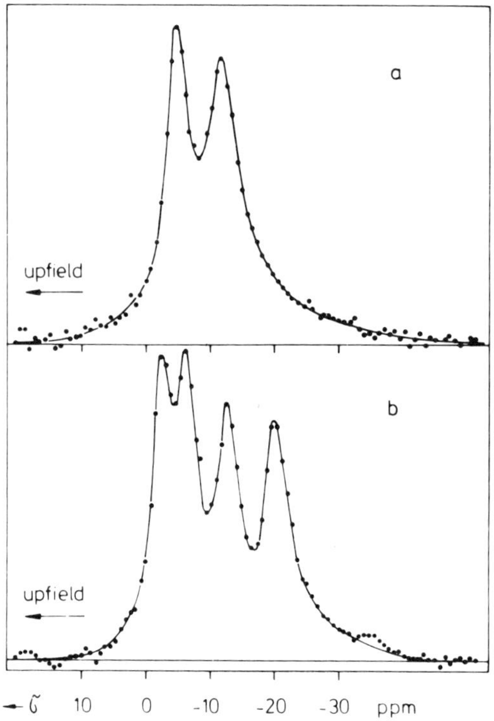

Fig. 3.11 a and b. Multiple pulse spectra of protons in a single crystal of malonic acid according to Haeberlen *et al.* [*36*]. (a) Degenerate spectrum, where the two lines of the CH_2 group coalesce into a single line (left) and the same holds for the two carboxyl (OH) protons (right). (b) Arbitrary orientation, where all four lines are visible

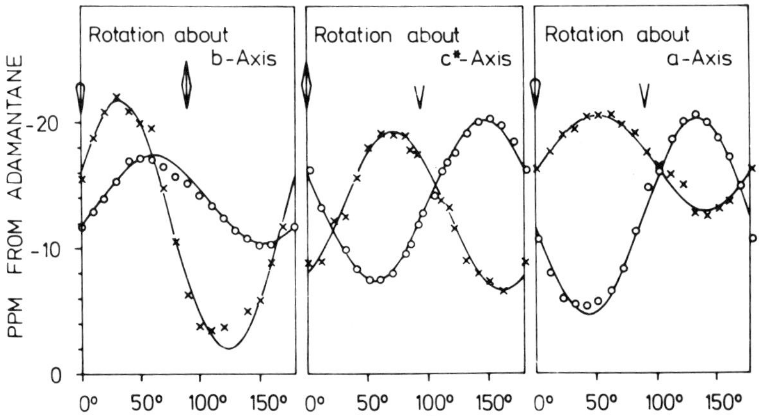

Fig. 3.12. Rotation patterns for the carboxyl proton lines in malonic acid (see Fig. 3.11) for three different crystal orientations. The markers indicate equivalent orientations of H_0 (Haeberlen *et al.* [*36*])

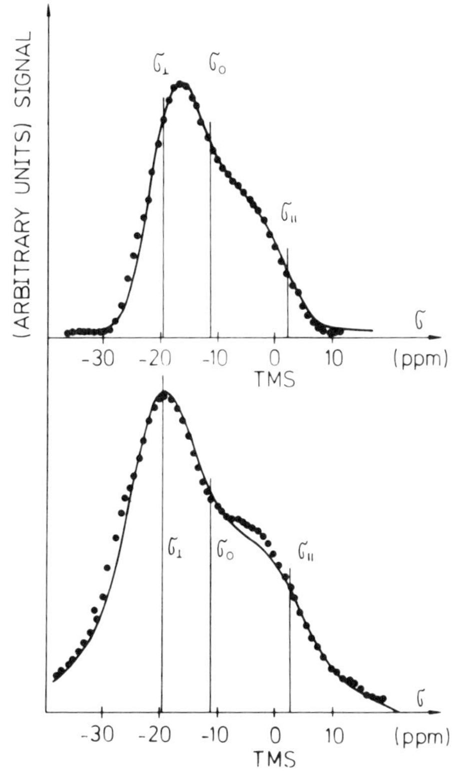

Fig. 3.13. Powder spectra of protons in squaric acid obtained in a four-pulse experiment by Raber *et al.* [*39*]. Notice, that the spectrum covers more than the isotropic shift range of protons. Absorption spectrum (top) and power spectrum (bottom) are plotted for comparison

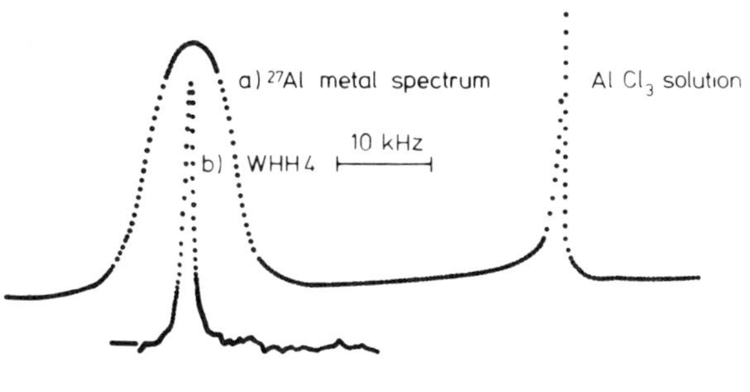

Fig. 3.14 a and b. Ordinary (a) and four-pulse spectrum (b) of ^{27}Al in aluminium powder, containing an aqueous solution of $AlCl_3$. Dipolar interaction between ^{27}Al nuclei and quadrupole interactions due to lattice defects are effectively reduced in the four-pulse experiment (see Ref. [*45*])

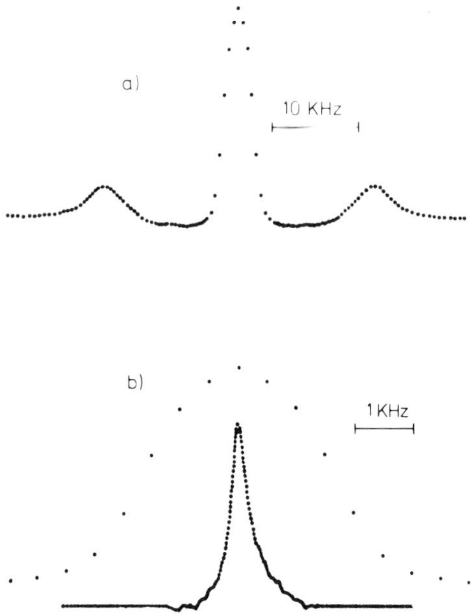

Fig. 3.15 a and b. Spectrum of ^{9}Be in beryllium powder (a) displaying the characteristic quadrupole powder spectrum. (b) Blow up of the central line with the corresponding four-pulse spectrum inserted (Ref. [*45*])

dipolar as well as the strain produced quadrupolar line width [*46*] as shown in Fig. 3.14. Similarly in Fig. 3.15, the large quadrupole splitting in ^{9}Be has been reduced considerably [*45*]. Although Knight shift anisotropy could arise in principal in beryllium, none has been observed. The isotropic Knight shifts, however, have been measured with respect to aqueous solutions of $AlCl_3$ and $BeCl_2$ respectively as [*45*]: 1636 ± 3 ppm for ^{27}Al and -10 ± 3 ppm for ^{9}Be. The Knight shift of aluminium is in accord with the high precision values of 1640 ± 1 ppm as obtained by the sample rotation method (see Section 2.5).

Since dipolar interactions and quadrupolar interactions vanish in the multiple-pulse experiment, broadenings of the line due to shift interactions as produced by e.g. paramagnetic impurities can be effectively studied using this method.

One difficulty which one encounters in determining shifts by multiple-pulse experiments in metals is due to the skin effect, which decreases the finite rotation angle of the pulses for those nuclei which are further away from the surface [*47*]. A distribution of rotation angles results over the sample which leads to a different scaling factor (see Section 3.5) for the metal and e.g. a liquid sample, even if an "internal standard" is used. It is a necessity therefore to determine the line shift versus resonance offset for both the metal and the reference at the same time. The correct resonance frequency is then determined from the intersection of the two lines (which may have a different slope!) with the off-resonance shift axis. The following puzzling situation can occur, which has been observed in ^{9}Be, that the distance between the liquid and the metal resonance line changes with resonance offset and even changes sign i.e. there are cases where the reference line is observed at the same side of the metal line, no matter what the spectrometer frequency is, above or below resonance. For a further discussion of the influence of the skin effect on NMR pulse experiments, see Ref. [*47*]. So far only homonuclear spin systems with only I spins present have been discussed.

This is, however, not the general case and the line narrowing efficiency may deteriorate substantially if a second spin species S is present. In a high magnetic field only the secular part of the I and S spin interaction survives, as was shown in Section 2.2

$$\mathcal{H}'_{IS} = \sum_{i,j} B_{ij} I_{zi} S_{zj}. \qquad \text{with } B_{ij} = \gamma_I \gamma_S \hbar (1 - 3\cos^2 \vartheta_{ij})/r_{ij}^3$$

This I-S interaction is of course altered when a multiple-pulse experiment is applied to the I spins. It is e.g. in a WHH-4 experiment scaled by a factor $1/\sqrt{3}$ which is not a bargain in terms of resolution. Thinking in terms of an average Hamiltonian, we would have to quench the average of $S_z(t)$ by some means of irradiation at the Larmor frequency of the S spins. A continous irradiation of the S spins at their Larmor frequency with an rf field H_1 in say the x direction would result in

$$S_z(t) = S_z \cos \omega_1 t - S_y \sin \omega_1 t$$

with

$$\omega_1 = \gamma H_1$$

which gives $\overline{S_z(t)} = 0$. This method is extensively applied in double resonance experiments, as discussed in Section 4.2 and 4.4. However, the reader may convince himself that his does not in general lead to a vanishing average I-S Hamiltonian in a WHH-4 experiment i.e. the condition for vanishing I-S interaction has to be stated more correctly as

$$\overline{\mathcal{H}'_{IS}(t)} = \sum_{i,j} B_{ij} \overline{I_{zi}(t) S_{zj}(t)} = 0.$$

One way of achieving this is to apply a π pulse every WHH-4 cycle to the S spins as outlined in Fig. 3.16 [*48*]. Other means have been discussed by Haeberlen [*25*]. The first experiment of this type was performed on $^{23}Na^{19}F$, where a WHH-4 experiment was applied to the ^{19}F spins while the ^{23}Na spins were irradiated by π pulses once every cycle [*48*] (see Fig. 3.17). Without the decoupling pulses the ^{19}F resonance was still too broad to be observed on the scale of Fig. 3.17. This technique has been extensively used in investigations of other heteronuclear spin systems by multiple pulse NMR [*44, 49*]. Especially the determination of ^{19}F shielding tensors in partially fluorinated benzenes as discussed in Section 5.4 could not have been done without the decoupling of the protons [*49*]. Figure 3.18 represents such a case, where the different stages of line narrowing are clearly visible.

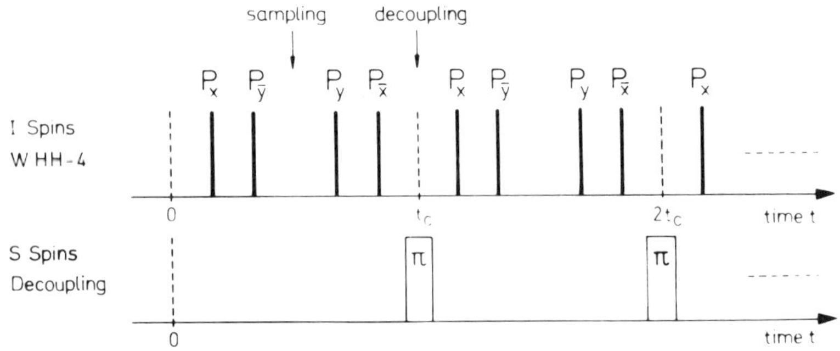

Fig. 3.16. Pulse timing of a WHH-4 sequence applied to the *I* spins, while applying decoupling pulses to the *S* spins once every four-pulse cycle

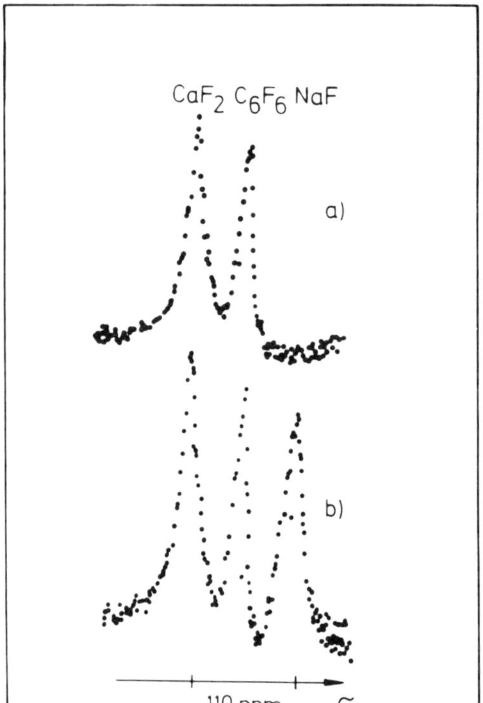

Fig. 3.17 a and b. Four-pulse spectra of ^{19}F in a sample containing a mixture of C_6F_6, CaF_2 and NaF. (a) Without decoupling pulses applied to the ^{23}Na spins. (b) Decoupling pulses applied to the ^{23}Na spins as outlined in Fig. 3.16 (Ref. [*48*])

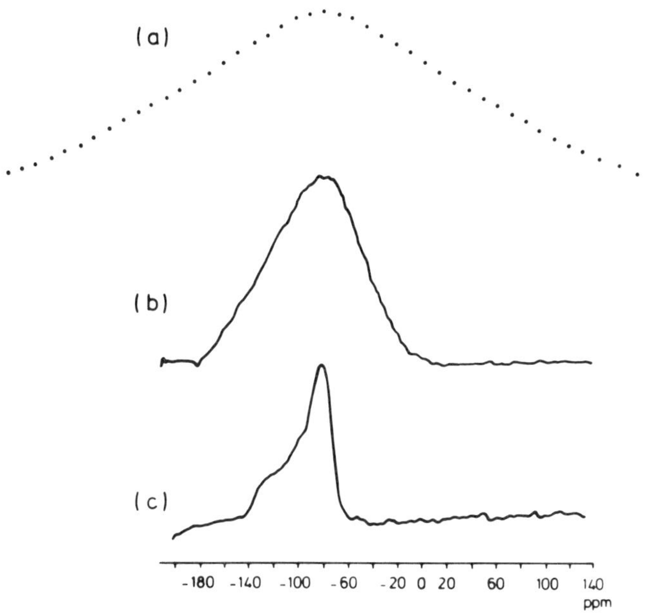

Fig. 3.18 a–c. Different stages of resolution enhancement, demonstrated in the case of ^{19}F resonance of a CF_3-group in $CF_3C_6H_4COOH$ at room temperature. (a) Fourier transform of FID. (b) Four-pulse spectrum. (c) Same as (b) but decoupling pulses are applied to the proton spins. Curve (c) shows the characteristic powder spectrum of a rapidly rotating fluoromethyl group. (Courtesy of D. Suwelack)

A shift of the resonance frequency occurs in all these decoupling experiments due to the non-secular contribution of the decoupling field. This effect is also termed "Bloch Siegert shift" [*50*], which originally accounted for the non-secular contribution of the counter-rotating component of a linear polarized rf field. This effect will be discussed in more detail in appendix D. Because of the second order character this shift is usually very small and amounts to about 10 ppm in the case of ^{19}F–^{1}H with ordinary rf fields (see Appendix D). This may seem to be appreciable. However, in NMR never absolute frequencies, but rather differences with respect to some reference line are determined, where the reference line is shifted by about the same amount, if it is recorded in the course of the same experiment. This of course is a must. The remaining Bloch Siegert shift between different lines in the spectrum is far below detectability, even if highest resolution is achieved [*59*].

The efficiency of spin decoupling and some more spin dynamical aspects are discussed in greater detail in Section 4.4.

3.3 Coherent Averaging Theory [*12, 25*]

In Sections 3.1 and 3.2 we have dealt with multiple pulse experiments simply by taking the average Hamiltonian over the cycle time. This certainly represents the correct Hamiltonian of the propagator in the limit of $\|\mathcal{H}_{int}\| t_c \to 0$. The cycle time, however, cannot be made arbitrarily small because of technical reasons. On the contrary, usually

$$\|\mathcal{H}_{\text{int}}\| t_c \cong 1$$

and any further reduction of t_c demands considerable effort. Coherent averaging theory [*12, 25*] now describes the coherent evolution of nuclear states, taking into account higher order terms i.e. interference terms of non-commuting operators in the cycle. Here we follow closely the paper by U. Haeberlen and J. S. Waugh [*12*], who have introduced this concept to multiple-pulse NMR. The general Hamiltonian in the rotating frame is split into two parts

$$\mathcal{H} = \mathcal{H}_{\text{ext}}(t) + \mathcal{H}_{\text{int}}$$

leading to the propagator

$$\mathbf{L}(t) = T\exp[-i\int_0^t dt' \, (\mathcal{H}_{\text{ext}}(t') + \mathcal{H}_{\text{int}})]. \tag{3.65}$$

T: Dyson time ordering operator.
This propagator may be split into a product

$$\mathbf{L}(t) = \mathbf{L}_1(t)\mathbf{L}_0(t) \tag{3.66}$$

where

$$\mathbf{L}_1(t) = T\exp[-i\int_0^t dt' \, \mathcal{H}_{\text{ext}}(t')] \tag{3.66a}$$

and

$$\mathbf{L}_0(t) = T\exp[-i\int_0^t dt' \widetilde{\mathcal{H}}(t')] \tag{3.66b}$$

and with

$$\widetilde{\mathcal{H}}(t) = \mathbf{L}_1^{-1}(t)\mathcal{H}_{\text{int}}\mathbf{L}_1(t). \tag{3.66c}$$

This product corresponds to the Heisenberg picture i.e. to an interaction representation or speaking colloquially to going into the "toggling frame". $\mathcal{H}_{\text{ext}}$ describes the spin interactions with external fields like

$$\mathcal{H}_{\text{ext}} = \Delta\omega I_z + \mathcal{H}_1(t)$$

where $\mathcal{H}_1(t)$ may result from the pulsed rf field and

$$\mathcal{H}_{\text{int}} = \mathcal{H}_D + \mathcal{H}_Q + \mathcal{H}_S + \ldots$$

represents the spin interactions with "internal fields". In general $\|\mathcal{H}_{\text{ext}}\| \gg \|\mathcal{H}_{\text{int}}\|$, although this is not necessary for Eq. (3.66) to be valid. It is, however, assuring rapid convergence, when approximating $\mathbf{L}_0(t)$ by a truncated cumulant expansion.

We now further assume, that the external field may be periodic [*12, 25, 51*] i.e.

$$\mathcal{H}_{\text{ext}}(t + nt_c) = \mathcal{H}_{\text{ext}}(t); n = 0, 1, 2, \ldots \tag{3.67}$$

from which follows

$$\mathbf{L}_1(nt_c) = \mathbf{L}_1^n(t_c) \tag{3.68}$$

and which also leads to a periodicity of $\widetilde{\mathcal{H}}(t)$ with

$$\widetilde{\mathcal{H}}(t) = \widetilde{\mathcal{H}}(t + nt_c) \tag{3.69}$$

and

$$\mathbf{L}_0(nt_c) = \mathbf{L}_0^n(t_c). \tag{3.70}$$

The general propagator for periodical external fields may now be expressed as:

$$\mathbf{L}(nt_c) = \mathbf{L}_1^n(t_c) \cdot \mathbf{L}_0^n(t_c).$$

If in addition the external field is cyclic, in the sense

$$\mathbf{L}_1(t_c) = 1 \tag{3.71}$$

the complete time evolution is described by the one cycle propagator $\mathbf{L}_0(t_c)$.
Eq. (3.71) is fulfilled in all phase-alternated multiple-pulse experiments, i.e. in all cases where $\mathcal{H}_{\text{ext}}(t) = -\mathcal{H}_{\text{ext}}(t_c - t)$. Eq. (3.71), however, is a too stringent condition, i.e. it is a sufficient, but not a necessary condition for the external field to be cyclic. A less stringent condition would be

$$\mathbf{L}_1^{-1}(t_c) I_i \mathbf{L}_1(t_c) = I_i \text{ with } i = x, y \text{ or } z \tag{3.72}$$

i.e. $\mathbf{L}_1(t_c)$ leaves the spin operator I_i invariant. This is for example the case in a series of not phase alternated pulses, where the rotation angle β of the rf pulses is an integer multiple of 2π. Condition Eq. (3.72), however, is not fulfilled if a resonance offset $\Delta\omega$ is included in $\mathcal{H}_{\text{ext}}$. Even in a phase alternated experiment the resonance offset part $\Delta\omega I_z$ has to be included into $\mathcal{H}_{\text{int}}$ for the external field $\mathcal{H}_{\text{ext}}(t)$ to be cyclic.

The one-cycle propagator $\mathbf{L}_0(t_c)$ was expressed by a *product* of unitary operators in Eqs. (3.49) and (3.58). Our goal, however, is to represent $\mathbf{L}_0(t_c)$ as

$$\mathbf{L}_0(t_c) = \exp[-it_c\mathcal{H}_e(t_c)] \tag{3.73}$$

where $\mathcal{H}_e(t_c)$ is the effective Hamiltonian which may be expanded as follows [*51*]

$$\mathcal{H}_e(t_c) = \sum_{k=0}^{\infty} \overline{\mathcal{H}}^{(k)}(t_c) = \overline{\mathcal{H}}^{(0)} + \sum_{k=1}^{\infty} \overline{\mathcal{H}}^{(k)}(t_c). \tag{3.74}$$

Here $\bar{\mathcal{H}}^{(0)}$ is the average Hamiltonian which has been used in Sections 3.1 and 3.2 exclusively. Although $\bar{\mathcal{H}}^{(0)}$ is independent of the cycle time t_c, all higher order terms depend on the cycle time explicitly and may be ordered with increasing exponent of the cycle time as

$$\bar{\mathcal{H}}^{(k)} = t_c^k \, \mathbf{F}_k \qquad k = 0, 1, 2, \ldots \tag{3.75}$$

where $\mathbf{F}_k$ is a Hermitian operator and is independent of the cycle time. The effective Hamiltonian $\mathcal{H}_e(t_c)$ may be obtained by equating Eqs. (3.73) and (3.66b) and comparing terms with increasing exponent of t_c. This results in the so-called Magnus-expansion [*12, 51–55*]

$$\mathbf{L}_0(t_c) = \exp\left[-it_c(\bar{\mathcal{H}}^{(0)} + \bar{\mathcal{H}}^{(1)} + \bar{\mathcal{H}}^{(2)} + \ldots)\right] \tag{3.76}$$

where

$$\bar{\mathcal{H}}^{(0)} = \frac{1}{t_c}\int_0^{t_c} dt\,\tilde{\mathcal{H}}(t) \tag{3.76a}$$

$$\bar{\mathcal{H}}^{(1)} = \frac{-i}{2t_c}\int_0^{t_c} dt_2 \int_0^{t_2} dt_1 \left[\tilde{\mathcal{H}}(t_2), \tilde{\mathcal{H}}(t_1)\right] \tag{3.76b}$$

$$\bar{\mathcal{H}}^{(2)} = -\frac{1}{6t_c}\int_0^{t_c} dt_3 \int_0^{t_3} dt_2 \int_0^{t_2} dt_1 \; [\tilde{\mathcal{H}}(t_3), [\tilde{\mathcal{H}}(t_2), \tilde{\mathcal{H}}(t_1)]]$$

$$+ [\tilde{\mathcal{H}}(t_1), [\tilde{\mathcal{H}}(t_2), \tilde{\mathcal{H}}(t_3)]] \tag{3.76c}$$

etc.

Notice, that the effective Hamiltonian in the Magnus expansion stays Hermitian at every stage of approximation, thus leaving $\mathbf{L}_0(t_c)$ always unitary, independent of the degree of approximation used.

For demonstrative purposes it is usually sufficient to consider the average Hamiltonian Eq. (3.76a) only, as we have done in the preceeding sections. In practice it turns out that no higher correction terms than the second term $\bar{\mathcal{H}}^{(2)}$ have to be taken into account. Otherwise one has to question the interaction representation chosen. Exploiting the symmetry of the "switching Hamiltonian" $\tilde{\mathcal{H}}(t)$ during one cycle it is possible to reduce the computational effort considerably, as has been first shown by Mansfield [*26*] and later by several other authors [*56, 57, 51, 58*]. We follow now the arguments of Wang and Ramshaw [*51*]. The one cycle propagator $\mathbf{L}_0(t_c)$ according to Eq. (3.73) and (3.74) is a unitary operator with

$$\mathbf{L}^+(t_c) = \mathbf{L}_0^{-1}(t_c) = \exp\left[it_c\mathcal{H}_e(t_c)\right]. \tag{3.77}$$

For a

(i) symmetric cycle, i.e. $\tilde{\mathcal{H}}(t) = \tilde{\mathcal{H}}(t_c - t)$ it follows

$$\mathbf{L}_0^+(t_c) = \mathbf{L}_0(-t_c) \tag{3.78}$$

which leads to

$$\exp[it_c \mathcal{H}_e(t_c)] = \exp[it_c\mathcal{H}_e(-t_c)]$$

or

$$\mathcal{H}_e(t_c) = \mathcal{H}_e(-t_c). \tag{3.79}$$

Expanding $\mathcal{H}_e(t_c)$ according to Eq. (3.74) leads to

$$\sum_{k=0}^{\infty} t_c^k \mathbf{F}_k = \sum_{k=0}^{\infty} (-t_c)^k \mathbf{F}_k$$

which results in $F_k = 0$ if k is odd, i.e. all correction terms $\overline{\mathcal{H}}^{(k)}$ vanish with $k = 1, 3, 5 \ldots$ odd.

Because of this fact, multiple pulse cycles are conveniently designed symmetrically, leaving only the average Hamiltonian and even order correction terms in the Magnus expansion to be looked at more closely. We shall take advantage of this fact in the subsequent sections. Furthermore it can be shown for a symmetrical cycle that [*56*]

$$\overline{\mathcal{H}}^{(2)}(t_c) = \overline{\mathcal{H}}^{(2)}(\tfrac{1}{2}t_c) \text{ if } \overline{\mathcal{H}}^{(0)} = 0 \tag{3.80}$$

which states that $\overline{\mathcal{H}}^{(2)}$ has to be computed over half the cycle only in this case, which may save some labour.

For an

(ii) antisymmetric cycle, i.e. $\widetilde{\mathcal{H}}(t) = -\widetilde{\mathcal{H}}(t_c-t)$ it follows

$$\mathbf{L}_0^+(t_c) = \mathbf{L}_0(t_c) \tag{3.81}$$

which leads to

$$\mathcal{H}_e(t_c) = -\mathcal{H}_e(t_c)$$

or

$$\mathcal{H}_e(t_c) = 0 \text{ and } \mathbf{L}_0(t_c) = 1. \tag{3.82}$$

There is no change of the initial state whatsoever in this case. This merely amounts to stating, that the initial state can always be recovered if the full Hamiltonian of the system can be inverted. This phenomenon is called the Loschmid demon and is the essential part of every spin echo experiment [*1, 57*].

3.4 Application of Coherent Averaging Theory to Multiple-Pulse Sequences

In this section we want to discuss the correction terms $\bar{\mathcal{H}}^{(1)}$ and $\bar{\mathcal{H}}^{(2)}$ to the average Hamiltonian $\bar{\mathcal{H}}^{(0)}$ for a) the phase-alternated sequence (MW-2), b) the spin-locking sequence (MW-4) and c) the four-pulse experiment (WHH-4).

*a) The Phase-Alternated Sequence (*MW-2*)* [*4–7*]

The phase-alternated sequence may be expressed as $(\tau - P_{\bar{y}} - 2\tau - P_y - \tau)_n$ or by the "switched Hamiltonian" $\tilde{\mathcal{H}}(t)$ as represented in Fig. 3.19.

The one cycle propagator $\mathbf{L}_0(t_c)$ can be expressed by

$$\mathbf{L}_0(t_c) = [\mathbf{Z}, 2\mathbf{X}, \mathbf{Z}] = [\![\mathbf{Z}, \mathbf{X}]\!]. \tag{3.83}$$

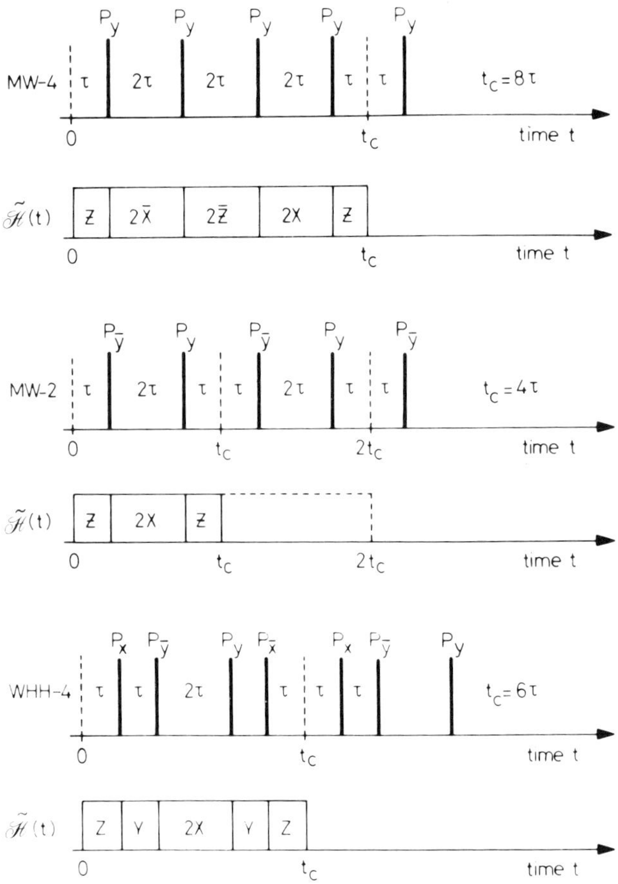

Fig. 3.19. Pulse timing and interaction Hamiltonian in the "toggling rotating frame" for the MW-4, MW-2, and the WHH-4 pulse sequences

We obtain immediately

$$\bar{\mathcal{H}}^{(0)} = \frac{1}{2}(\mathcal{H}_x + \mathcal{H}_z)$$

and in the special case with $\mathcal{H}$ being the dipolar Hamiltonian

$$\bar{\mathcal{H}}_D^{(0)} = -\frac{1}{2}\mathcal{H}_{Dy}.$$

In case the interaction is represented by a shift Hamiltonian

$$\bar{\mathcal{H}}_S^{(0)} = \frac{1}{2}\sum_k (\delta_k + \Delta\omega)(I_{xk} + I_{zk}) \tag{3.84}$$

which may be rewritten as

$$\bar{\mathcal{H}}_S^{(0)} = \sum_k \frac{1}{\sqrt{2}} (\delta_k + \Delta\omega) I_{101}^{(k)}$$

to emphasize the scaling factor $S = 1/\sqrt{2}$, and where

$$I_{101}^{(k)} = (I_{xk} + I_{zk})/\sqrt{2}.$$

Because of the symmetry of the cycle $(\tilde{\mathcal{H}}(t) = \tilde{\mathcal{H}}(t_c - t))$ all odd order correction terms vanish and the leading correction term equals $\bar{\mathcal{H}}^{(2)}$, which may be obtained by applying Eq. (3.76c) as

$$\bar{\mathcal{H}}^{(2)} = -\frac{\tau^2}{12}[(2\mathbf{X} + \mathbf{Z}), [\mathbf{X}, \mathbf{Z}]];\ t_c = 4\tau \tag{3.85}$$

which results in the case of shift interactions in

$$\bar{\mathcal{H}}_S^{(2)} = -\sum_k K(\delta_k + \Delta\omega)(2I_{zk} - I_{xk}) \tag{3.86}$$

where $K = \tau^2(\delta_k + \Delta\omega)^2/12$.

The total average shift Hamiltonian $\bar{\mathcal{H}}_S = \bar{\mathcal{H}}_S^{(0)} + \bar{\mathcal{H}}_S^{(2)}$ is obtained by combining Eqs. (3.84) and (3.86) to

$$\bar{\mathcal{H}}_S = \sum_k S(\delta_k + \Delta\omega)\, I_{\mu k} \tag{3.87}$$

with the scaling factor

$$S = \frac{1}{\sqrt{2}}(1 - 2K + 10K^2)^{1/2} \tag{3.88}$$

and

$$I_\mu = e^{-i\beta I_y} I_z e^{i\beta I_y}$$

where

$$\tan\beta = \frac{1 + 2K}{1 - 4K}. \tag{3.89}$$

In Fig. 3.20 the exact scaling factor obtained from an exact solution of the equation of motion using Eq. (3.53), is compared with this approximative method, including correction terms up to second order. A similar calculation has been made by Ellet and Waugh [*24*], there however the symmetry of the cycle was not realized.

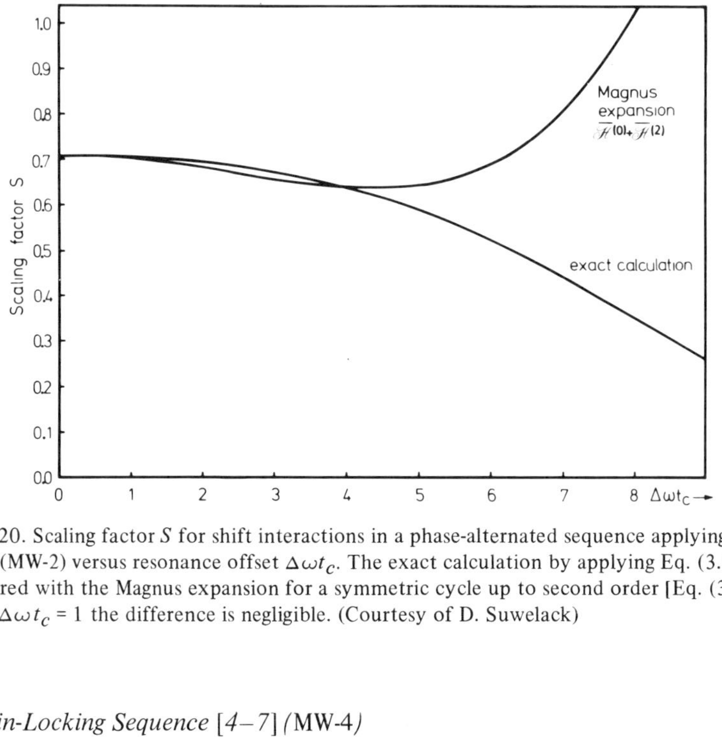

Fig. 3.20. Scaling factor S for shift interactions in a phase-alternated sequence applying 90° pulses (MW-2) versus resonance offset $\Delta\omega t_c$. The exact calculation by applying Eq. (3.53) is compared with the Magnus expansion for a symmetric cycle up to second order [Eq. (3.88)]. Up to $\Delta\omega t_c = 1$ the difference is negligible. (Courtesy of D. Suwelack)

b) Spin-Locking Sequence [*4–7*] *(*MW-4*)*

The spin-locking sequence as shown in Fig. 3.19 and represented by $(\tau - P_y - 2\tau - P_y - 2\tau - P_y - 2\tau - P_y - \tau)_n$ is similar to the phase-alternated sequence, besides the phase alternation. This leads to a cycle time of $t_c = 8\tau$ and a slightly changed "switched Hamiltonian" with the propagator

$$\mathbf{L}(t_c) = [\mathbf{Z}, 2\overline{\mathbf{X}}, 2\overline{\mathbf{Z}}, 2\mathbf{X}, \mathbf{Z}] \tag{3.90}$$

where $\overline{\mathbf{X}}$ for example, means a minus sign in front of each I_x occuring in the Hamiltonian. Since the dipolar Hamiltonian is invariant under such a change of sign, we have exactly the same expression for $\overline{\mathcal{H}}_D^{(k)}$ as in the phase alternated experiment where especially all odd order correction terms vanish, because of the symmetry of the cycle. The spin-locking cycle, however, is not a symmetrical cycle in the above defined sense for shift interactions, since $\overline{\mathbf{X}}$ and $\overline{\mathbf{Z}}$ are different from $\mathbf{X}$ and $\mathbf{Z}$ respectively. It can be shown in a straightforward manner, that

$$\mathcal{H}^{(1)} = -\frac{i\tau}{4}[\overline{\mathbf{Z}}, (\overline{\mathbf{X}} - \mathbf{X})] \tag{3.91}$$

and we leave it as an exercise for the interested reader, to derive the correction terms of the average shift and dipolar Hamiltonian and the corresponding cross correction terms, using Eq. (3.91)

*c) The Four-Pulse Sequence (*WHH-4*) [9, 12]*

Here we use a slightly different cycle than in Section 3.2, which is shown in Fig. 3.19. This cycle possesses reflection symmetry as expressed by the one cycle propagator

$$\mathbf{L}(t_c) = [\mathbf{Z}, \mathbf{Y}, \mathbf{X}; \mathbf{X}, \mathbf{Y}, \mathbf{Z}] = [\![\mathbf{Z}, \mathbf{Y}, \mathbf{X}]\!] \tag{3.92}$$

and it follows, that

$$\overline{\mathcal{H}}^{(0)} = \frac{1}{3}(\mathbf{X} + \mathbf{Y} + \mathbf{Z})$$

with

$$\overline{\mathcal{H}}_D^{(0)} = 0$$

and

$$\overline{\mathcal{H}}_S^{(0)} = \sum_k \frac{1}{\sqrt{3}} (\delta_k + \Delta\omega)\, I_{111}$$

as obtained in Section 3.2. Because of the reflection symmetry $\overline{\mathcal{H}}^{(1)}$ and all other odd order correction terms vanish, leaving as the leading correction term

$$\overline{\mathcal{H}}^{(2)} = -\frac{\tau^2}{18}\{[(\mathbf{Z} + 2\mathbf{X} + 2\mathbf{Y}), [(\mathbf{X} + \mathbf{Y}), \mathbf{Z}]] - [(\mathbf{Y} + 2\mathbf{X}), [\mathbf{Y}, \mathbf{X}]]\} \tag{3.93}$$

which in the case of the dipolar interaction reduces to [*12*]

$$\overline{\mathcal{H}}_D^{(2)} = \frac{\tau^2}{18}[(\mathcal{H}_{Dx} - \mathcal{H}_{Dz}), [\mathcal{H}_{Dy}, \mathcal{H}_{Dx}]] \tag{3.94}$$

and in the case of shift interactions to

$$\overline{\mathcal{H}}_S^{(2)} = -\frac{\tau^2}{18}\sum_k (\delta_k + \Delta\omega)^3 (I_{yk} - 2I_{xk} + 4I_{zk}). \tag{3.95}$$

The interested reader may want to calculate cross terms between shift and dipolar interactions, which can be done straightforwardly by using Eq. (3.93). It should be noted, that $\overline{\mathcal{H}}_D^{(2)}$ is responsible for the ultimate resolution of the WHH-4 experiment *on-resonance*, if no pulse imperfections are present, whereas, cross terms between the shift Hamiltonian and the dipolar Hamiltonian may become important *off-resonance* [*25*].

*d) Eight-Pulse Sequences (*HW-8 [*12*] *and* MREV-8 [*26, 33, 60, 70*]*)*

Haeberlen and Waugh [*12*] realized very soon, that using their four-pulse sequence (WHH-4) as a mother cycle, a wealth of daughter cycles could be designed to suit different purposes. Some of them are "a beauty" others are not. The basic WHH-4 cycle is used as a subcycle and repeated with different pulse settings (phase, width, timing etc.) to form a 2^n pulse cycle, where n went up to as high a value as $n = 6$ (theoretically).[*61, 59*] The idea behind this procedure is to design compensation schemes for the different deseases real pulse sequences suffer from, such as rf inhomogeneity, pulse width errors, phase errors etc. It was Mansfield [*26*] who realized that symmetrization of pulse sequences and cascading cycles with different symmetries could lead to cancellation of errors.

We are going to discuss here merely eight-pulse cycles, namely the Haeberlen-Waugh [*12*] (HW-8) and another cycle, which was cryptically proposed by Mansfield [*26*] and later independently discovered by Rhim, Elleman and Vaughan [*33, 60*] to which we will therefore refer as MREV-8 sequence. Figure 3.21 represents the pulse timing and the "switching Hamiltonian" for these two pulse sequences. Of course, other sequences are possible by changing the initial condition, however, we do not regard those to be *different*.

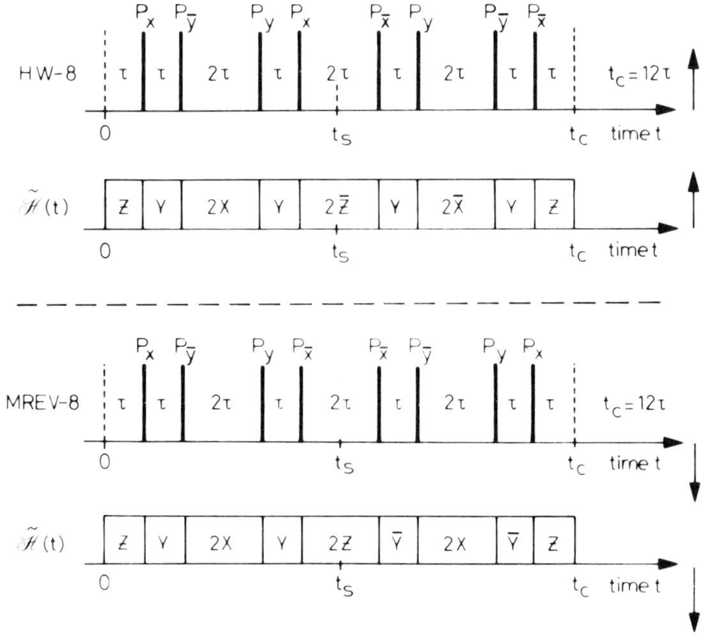

Fig. 3.21. Pulse timing and interaction Hamiltonian in the "toggling rotating frame" for the HW-8 and the MREV-8 pulse sequences

The one cycle propagator $L(t_c)$ is immediately written down according to Fig. 3.21 as

$$\text{HW-8: } \mathbf{L}(t_c) = [\mathbf{Z}, \mathbf{Y}, 2\mathbf{X}, \mathbf{Y}, \overline{\mathbf{Z}}][\overline{\mathbf{Z}}, \mathbf{Y}, 2\overline{\mathbf{X}}, \mathbf{Y}, \mathbf{Z}] \tag{3.96}$$

$$\text{MREV-8: } L(t_c) = [\mathbf{Z}, \mathbf{Y}, 2\mathbf{X}, \mathbf{Y}, \mathbf{Z}][\mathbf{Z}, \overline{\mathbf{Y}}, 2\mathbf{X}, \overline{\mathbf{Y}}, \mathbf{Z}]$$

$$\text{or} \quad L(t_c) = [\![\mathbf{Z}, \mathbf{Y}, \mathbf{X}]\!]\,[\![\mathbf{Z}, \overline{\mathbf{Y}}, \mathbf{X}]\!] \tag{3.97}$$

where $t_c = 12\tau$.

It is straightforward to show that $\overline{\mathcal{H}}_D^{(0)} = 0$, in both cases and

$$\text{HW-8: } \overline{\mathcal{H}}_S^{(0)} = \frac{1}{3}\sum_k (\delta_k + \Delta\omega) I_{010} \tag{3.98}$$

where $I_{010} = I_{yk}$, scaling factor $S = 1/3$

and

$$\text{MREV-8: } \overline{\mathcal{H}}_S^{(0)} = \frac{\sqrt{2}}{3}\sum_k (\delta_k + \Delta\omega) I_{101} \tag{3.99}$$

where

$$I_{101} = (I_{xk} + I_{zk})/\sqrt{2}, \text{ scaling factor } S = \sqrt{2}/3.$$

For the dipolar interaction the bars in Eqs. (3.96, 3.97) are of no significance since $\mathcal{H}_{Dz} = \mathcal{H}_{D-z}$ and both cycles are symmetric leading to $\overline{\mathcal{H}}_D^{(n)} = 0$ for all n which are odd, especially $\overline{\mathcal{H}}_D^{(1)} = 0$. This is not so for the shift Hamiltonian, since the cycle symmetry is broken, according to Eqs. (3.96, 3.97) for Hamiltonians which are not invariant under π rotation:

This first correction term is readily obtained as

$$\text{HW-8: } \overline{\mathcal{H}}^{(1)} = \frac{-i\tau}{12}[(\mathbf{X} - \overline{\mathbf{X}}), (\mathbf{Z} - \overline{\mathbf{Z}}) - 2\mathbf{Y}];\ t_c = 12\tau \tag{3.100}$$

$$\text{MREV–8: } \overline{\mathcal{H}}^{(1)} = \frac{-i\tau}{6}[(\mathbf{X} + \mathbf{Z}), (\mathbf{Y} - \overline{\mathbf{Y}})]. \tag{3.101}$$

The corresponding corrections to the shift interaction

$$\text{HW-8: } \overline{\mathcal{H}}_S^{(1)} = \frac{-\tau}{6}\sum_k (\delta_k + \Delta\omega)^2 (I_{yk} + I_{zk}) \tag{3.102}$$

$$\text{MREV-8: } \overline{\mathcal{H}}_S^{(1)} = \frac{\tau}{3}\sum_k (\delta_k + \Delta\omega)^2 (I_{zk} - I_{xk}) \tag{3.103}$$

and crossterms between shift interactions and dipolar interactions are readily obtained. The second correction term $\overline{\mathcal{H}}_D^{(2)}$ for dipolar interactions need to be calculated over the subcycle [**Z**, **Y**, **X**] only and is thus identical for the WHH-4, HW-8 and the MREV-8 sequence, namely

$$\bar{\mathcal{H}}_D^{(2)} = -\frac{\tau^2}{18}\,[(\mathcal{H}_{Dx} - \mathcal{H}_{Dz}), [\mathcal{H}_{Dx}, \mathcal{H}_{Dy}]]. \tag{3.104}$$

In summarizing we note, that nothing seems to be gained by using the eight-pulse cycles compared with the WHH-4 cycle, on the contrary, shift Hamiltonians are even scaled further than in the four-pulse experiment. It will turn out, however, later, when we treat rf inhomogeneity (see Section 3.7) that the eight-pulse experiments are in some respects superior to the WHH-4 sequence.

3.5 Arbitrary Rotations in Multiple-Pulse Experiments

So far we always have considered idealized pulse sequences by assuming the rf pulses to be δ pulses, usually with a rotation angle $\beta = \pi/2$ or π.

Although we have treated the phase-alternated sequence in Section 3.1 for arbitrary β, we still assumed the pulses to be δ pulses there. Moreover, we have not taken into account their effect on the dipolar interactions. In this section we are going to lift all these constraints. Arbitrary rotations are conveniently treated with the interaction Hamiltonians expressed by irreducible tensor operators (see Appendix A) as [*62–64*]

$$\mathcal{H}_{\text{int}} = \sum_k \sum_{q=-k}^{k} (-1)^q A_{k,-q} T_{kq} \tag{3.105}$$

which in the special case of
(i) shift interactions (see Appendix A) is written as

$$\mathcal{H}_S = \sum_{q=-1}^{1} (-1)^q A_{1,-q} T_{1q} \tag{3.106}$$

where

$$A_{10} = H_{Sz};\, A_{1\pm 1} = \mp \frac{1}{\sqrt{2}}(H_{Sx} \pm iH_{Sy}) \tag{3.107}$$

and

$$T_{10} = I_z;\, T_{1\pm 1} = \mp \frac{1}{\sqrt{2}}(I_x \pm iI_y). \tag{3.108}$$

With $\mathbf{H}_0 = (0, 0, H_0)$ we can write

$$H_{Sx} = S_{xz}H_0;\, H_{Sy} = S_{yz}H_0;\, H_{Sz} = S_{zz}H_0.$$

In the case of
(ii) dipolar interaction (see Appendix A) we obtain

$$\mathcal{H}_D = \sum_{i<j} \sum_{q=-2}^{2} (-1)^q A_{2,-q}^{(i,j)}\, T_{2q}^{(i,j)} \tag{3.109}$$

where

$$A_{20}^{ij} = -dC_{20}^{(i,j)};\ A_{2\pm1}^{(i,j)} = \mp dC_{2\pm1}^{(i,j)};\ A_{2\pm2}^{(i,j)} = -dC_{2\pm2}^{(i,j)} \tag{3.110}$$

with $d = \sqrt{6}\gamma_I^2\hbar/r_{ij}^3$ and where the modified spherical harmonics C_{kq} are defined by the spherical harmonics Y_{kq} as

$$C_{kq} = \left[\frac{4\pi}{2k+1}\right]^{1/2} Y_{kq} \tag{3.111}$$

and

$$\begin{aligned} T_{20}^{(i,j)} &= \frac{1}{\sqrt{6}}(3I_{zi}I_{zj} - \mathbf{I}_i \cdot \mathbf{I}_j) \\ T_{2\pm1}^{(i,j)} &= \mp\frac{1}{2}(I_{zi}I_{\pm j} + I_{\pm i}I_{zj}) \\ T_{2\pm2}^{(i,j)} &= \frac{1}{2}I_{\pm i}I_{\pm j}. \end{aligned} \tag{3.112}$$

In the following we drop the indices i, j on the tensor operators for convenience. The external Hamiltonian performing the rotational operation in the rotating frame may be expressed as:

$$\mathcal{H}_{\text{ext}} = -\vec{\omega}(t) \cdot \mathbf{I} = -\omega_1(t)\, \mathbf{n} \cdot \mathbf{I} \tag{3.113}$$

where $\mathbf{n}$ is a unit vector in the direction of the applied field. The rotation is now described by the unitary operator

$$\mathbf{L}_1(t) = \exp[-i\beta(t)\, \mathbf{n} \cdot \mathbf{I}] \tag{3.114}$$

where

$$\beta(t) = \int_0^t dt' \omega_1(t').$$

In multiple-pulse experiments the time evolution operator $\mathbf{L}_0(t)$ which governs the nuclear response is expressed by the "modulated" or "switched" Hamiltonian

$$\widetilde{\mathcal{H}}(t) = \mathbf{L}_1^{-1}(t)\, \mathcal{H}_{\text{int}} \mathbf{L}_1(t)$$

as shown in Section 3.3. Since we operate in spin space, the rotation operation is applied to the T_{qk} only, and if only secular contributions are taken into account we may write

$$\widetilde{\mathcal{H}}(t) \sim \widetilde{T}_{ko}(t)$$

where

$$\widetilde{T}_{kq}(t) = \mathbf{L}_1^{-1}(t) T_{kq} \mathbf{L}_1(t). \tag{3.115}$$

In the following we discuss the secular part $\widetilde{T}_{ko}(t)$ when rf irradiation is applied in the x, y plane of the rotating frame. In this case (see Appendix B) [*62–64*]

$$\widetilde{T}_{k0}(t) = \sum_{q=-k}^{k} T_{kq} D_{q0}^{(k)}(\alpha, \beta(t), -\alpha) \tag{3.116}$$

with the Wigner matrices [*62–64*]

$$D_{q0}^{(k)}(\alpha, \beta(t), -\alpha) = e^{-iq\alpha} d_{q0}^{(k)}(\beta) \tag{3.117}$$

where α is the angle between the rf field direction and the y axis and $\beta(t)$ is the rotating angle during rf irradiation
(i) Irradiation in y direction ($\alpha = 0$) [*64*]

$$\widetilde{T}_{k0}(t) = \sum_{q=-k}^{k} T_{kq} d_{q0}^{(k)}(\beta)$$

with

$$\widetilde{T}_{10}(t) = T_{10}\cos\beta(t) + \frac{1}{\sqrt{2}}(T_{1-1} - T_{1+1})\sin\beta(t) \tag{3.118}$$

and

$$\widetilde{T}_{20}(t) = T_{20}\frac{1}{2}(3\cos^2\beta(t) - 1) + \sqrt{\tfrac{3}{2}}(T_{2-1} - T_{21})\sin\beta(t)\cos\beta(t)$$

$$+ \sqrt{\tfrac{3}{8}}(T_{22} + T_{2-2})\sin^2\beta(t) \tag{3.119}$$

(ii) irradiation in x direction ($\alpha = -\pi/2$)

$$\widetilde{T}_{k0}(t) = \sum_{q=-k}^{k} T_{kq} e^{iq\pi/2} d_{q0}^{(k)}(\beta(t)) \tag{3.120}$$

from where it follows, that the terms with $q = \pm 1$ in (i) have to be multiplied by $\pm i$, whereas the terms with $q = \pm 2$ have to be multiplied by -1.

The average Hamiltonian during rf irradiation is defined as [*64*]

$$\overline{T}_{k0}(\beta_1) = \frac{1}{\beta_1}\int_0^{\beta_1} d\beta\, \widetilde{T}_{k0}(\beta) \tag{3.121}$$

where

$$\beta_1 = \int_0^{t_w} dt\, \omega_1(t) \tag{3.122}$$

with the pulse width t_w.

We are now well equipped to analyse the phase-alternated sequence (MW-2) and the four-pulse (WHH-4) cycle with arbitrary pulse width. According to Fig. 3.22 we may use the following parameters:

$$t_c = \tau_1 + \tau_2 + 2t_w \qquad \text{cycle time} \tag{3.123a}$$

$$\delta = 2t_w/t_c \qquad \text{duty factor} \tag{3.123b}$$

$$\kappa = \frac{\tau_1 + t_w}{\tau_2 + t_w}\,. \qquad \text{pulse timing} \tag{3.123c}$$

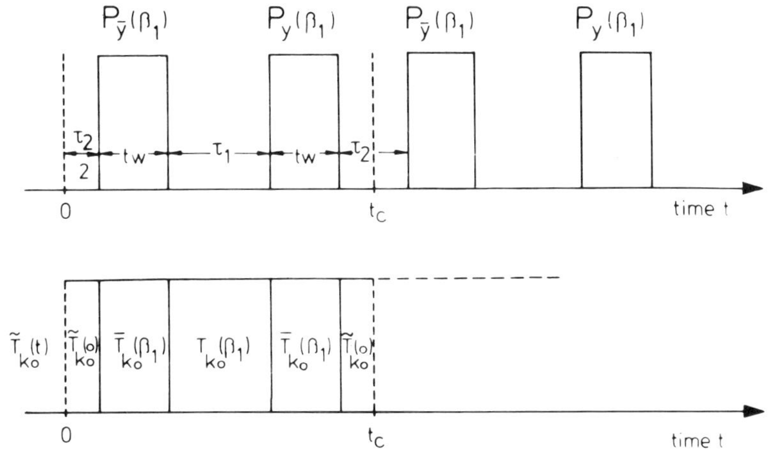

Fig. 3.22. Pulse timing and tensor representation of the interaction Hamiltonian in a phase alternated sequence with arbitrary pulse width

The cycle has reflection symmetry which leads to $\bar{\mathcal{H}}^{(1)} = 0$. In the following we discuss the average Hamiltonian only, which is straightforwardly obtained for an arbitrary direction of the irradiation, as [*64*])

$$\overline{T}_{ko} = \frac{1}{t_c}\{\tau_2\,\widetilde{T}_{ko}(0) + \tau_1\,\widetilde{T}_{ko}(\beta_1) + 2t_w\,\overline{T}_{ko}(\beta_1)\}\,. \tag{3.124}$$

Using the above parameters this may be expressed as (there is a slight change in parameters compared with Ref. [*64*])

$$\overline{T}_{ko} = \frac{2-(1+\kappa)\delta}{2(1+\kappa)}\,\widetilde{T}_{ko}(0) + \frac{2\kappa-(1+\kappa)\delta}{2(1+\kappa)}\,\widetilde{T}_{ko}(\beta_1) + \delta\,\overline{T}_{ko}(\beta_1) \tag{3.125}$$

or by writing this in compact form as

$$\overline{T}_{ko} = \sum_{q=-k}^{k} T_{kq}\,\overline{D_{q0}^{(k)}(\alpha,\beta_1,-\alpha)}. \tag{3.126}$$

The following abbreviations are used in the above equation for $\alpha = 0$

$$D_0 = \overline{D_{00}^{(2)}(\beta_1)};\ D_1 = \mp\overline{D_{\pm 10}^{(2)}(\beta_1)};\ D_2 = \overline{D_{\pm 20}^{(2)}(\beta_1)} \tag{3.127}$$

$$C_0 = D_{00}^{(1)}(\beta_1);\ C_1 = \pm D_{\pm 10}^{(1)}(\beta_1) \tag{3.128}$$

with

$$D_0 = \frac{3}{4}\cos\beta_1\left[\delta\frac{\sin\beta_1}{\beta_1} + \left(\frac{2\kappa}{1+\kappa} - \delta\right)\cos\beta_1\right] + \frac{2-\kappa}{2(1+\kappa)} \tag{3.129a}$$

$$D_1 = \sqrt{\tfrac{3}{8}}\,\sin\beta_1\left[\delta\frac{\sin\beta_1}{\beta_1} + \left(\frac{2\kappa}{1+\kappa} - \delta\right)\cos\beta_1\right] \tag{3.129b}$$

$$D_2 = \frac{1}{2}\sqrt{\tfrac{3}{8}}\left[\sin\beta_1\left(\frac{2\kappa}{1+\kappa} - \delta\right) - \delta\frac{\cos\beta_1}{\beta_1}\right]\cdot\sin\beta_1 + \frac{1}{2}\sqrt{\tfrac{3}{8}}\,\delta \tag{3.129c}$$

$$C_0 = \delta\frac{\sin\beta_1}{\beta_1} + \frac{1}{2}\left(\frac{2\kappa}{1+\kappa} - \delta\right)\cos\beta_1 + \frac{1}{2}\left(\frac{2}{1+\kappa} - \delta\right) \tag{3.130a}$$

$$C_1 = \frac{\delta}{\sqrt{2}}(1 - \cos\beta_1)/\beta_1 = \frac{1}{2\sqrt{2}}\left(\frac{2\kappa}{1+\kappa} - \delta\right)\sin\beta_1. \tag{3.130b}$$

Equations (3.129, 3.130), were misprinted in Ref. [*64*] and are expressed here for convenience with a slightly different parameter κ.

The average dipolar Hamiltonian in the WHH-4 experiment ($\kappa = 2$) can now be expressed as

$$\overline{T}_{10} = C_0 T_{10} + C_1\left[\frac{1}{2}(1-i)\,T_{1-1} - \frac{1}{2}(1+i)\,T_{11}\right] \tag{3.131a}$$

$$\overline{T}_{20} = D_0 T_{20} + D_1\left[\frac{1}{2}(1-i)\,T_{2-1} - \frac{1}{2}(1+i)\,T_{21}\right]. \tag{3.131b}$$

In order to digest these formulas, we take as a simple example a case which we have treated already in Section 3.2, i.e. $\delta = 0$, $\kappa = 1$, $\beta = \pi/2$.

We obtain from Eqs. (3.129) and (3.130) in this case

$$D_0 = \frac{1}{4};\ D_1 = 0;\ D_2 = \frac{1}{2}\sqrt{\tfrac{3}{8}}$$

and

$$C_0 = \frac{1}{2};\ C_1 = \frac{1}{2\sqrt{2}}.$$

$\overline{T}_{k0}$ may now be calculated by using Eqs. (3.126–3.128).

In the case of dipolar interaction this results in

$$\overline{T}_{20} = -\frac{1}{2} \cdot \frac{1}{\sqrt{6}}(3I_{yi}I_{yj} - \mathbf{I}_i \cdot \mathbf{I}_j)$$

or

$$\overline{\mathcal{H}}_D^{(0)} = -\frac{1}{2}\mathcal{H}_{Dy}$$

and in the case of shift interactions in

$$\overline{T}_{10} = \frac{1}{2}(I_z + I_x)$$

or

$$\overline{\mathcal{H}}_S^{(0)} = \sum_k \frac{1}{2}(\delta_k + \Delta\omega)(I_{zk} + I_{xk}).$$

This is identical with the results obtained in Section 3.1 under the δ pulse assumption. Equations (3.129) and (3.130), however, are much more powerful, since they describe the general case and can be readily applied to different multiple-pulse sequences.

The considered cycle in Fig. 3.22, is a subcycle of the WHH-4 sequence, if the subsequent cycle is performed under irradiation in the x direction. As noted before, Eqs. (3.124), (3.129), and (3.130) are still valid in this case, the terms with $q = \pm 1$ have to be multiplied by $\pm i$, however, whereas the terms with $q = \pm 2$ have to be multiplied by -1. This cancels the D_2 term in the average over the full cycle and the D_1 and D_0 term have to be considered only. For high resolution purposes the dipolar terms D_0 and D_1 must vanish. This may be achieved by making the square bracket in Eqs. (3.129a,b) zero. This is possible under the condition $\kappa = 2$, which appears to be the characteristic timing of the four-pulse cycle which we shall call the "ideal timing". A "dipolar scaling factor" $S_D = \|\overline{\mathcal{H}}_D^{(0)}\| / \|\mathcal{H}'_D\| = (D_0^2 + D_1^2)^{1/2}$ may be defined, which is plotted versus the total rotation angle β_1 in Fig. 3.23 for different duty factors δ [*64*]. As follows from Eq. 3.129 D_0 and D_1 always vanish for the same value of β_1. The corresponding condition for β_1 can be easily derived from Eqs. (3.129a,b) for $\kappa = 2$ to be [*64*]

$$\delta(1 - \tan\beta_0/\beta_0) = \frac{4}{3}; \beta_0 \geqslant 90^\circ. \tag{3.131c}$$

This condition is plotted in Fig. 3.24, demonstrating that for every duty factor δ up to $\delta = 2/3$ a rotation angle β_0 can be found which leads to a vanishing average dipolar Hamiltonian. Expanding $\tan\beta_0$ for $\beta_0 = \pi/2 + \epsilon$ leads to

$$\delta = \frac{2}{3}\pi\epsilon$$

which has been used before to account for the influence of finite pulse width [*12*].

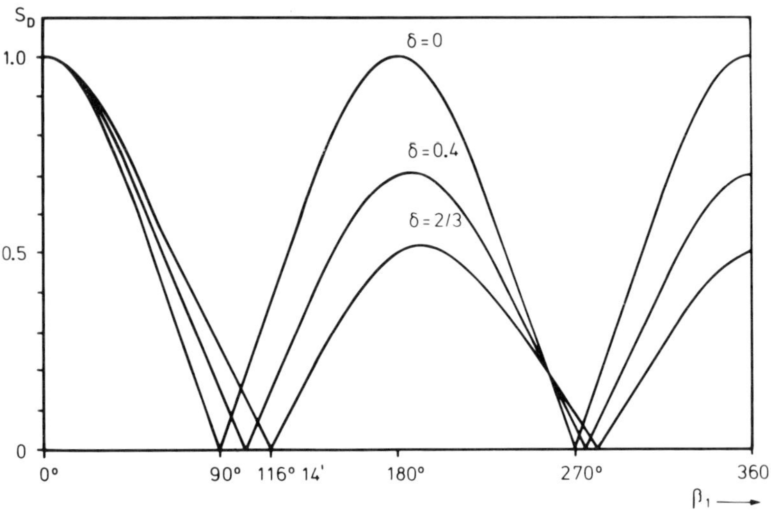

Fig. 3.23. Dipolar scaling factor $S_D = \|\bar{\mathcal{H}}_D^{(0)}\|/\|\mathcal{H}'_D\|$ in a four-pulse experiment versus the rotation angle β_1 of the rf pulses for different values of the duty factor δ. For each value $0 \leqslant \delta \leqslant 2/3$ an angle β_1 can be found, where $S_D = 0$. Notice, that rf inhomogeneity (variation of β_1 over the sample) becomes less important with increasing δ. We remark further, that S_D is invariant to a π or 2π pulse only if $\delta = 0$

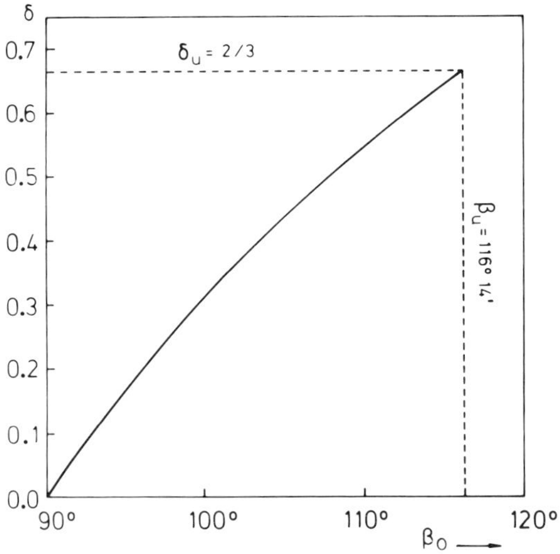

Fig. 3.24. Relation between the duty factor of a four-pulse cycle and the rotation angle β_0 for vanishing S_D, according to Eq. (3.131c) (see Ref. [*64*])

The upper pulse width to be used in the four-pulse experiment is according to Eq. (3.131c) given by ($\delta = 2/3$)

$$\tan\beta_u = -\beta_u$$

which leads to [*64*]

$$\beta_u = 116^\circ\ 14'\ 21''.$$

The scaling factor of the shift interaction is changed correspondingly, and may be obtained from Eq. (3.130a,b) by

$$S = [C_0^2 + C_1^2]^{1/2} = [\mathrm{Tr}\{\overline{T}_{10}^2\}/\mathrm{Tr}\{T_{10}^2\}]^{1/2}. \tag{3.132}$$

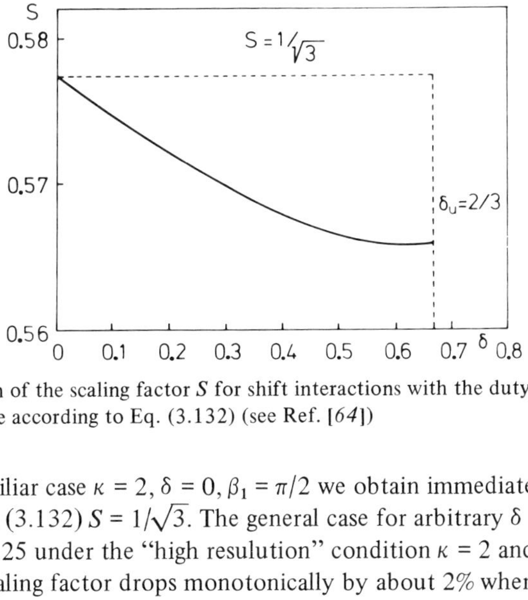

Fig. 3.25. Variation of the scaling factor S for shift interactions with the duty factor δ of a four-pulse sequence according to Eq. (3.132) (see Ref. [*64*])

Treating the familiar case $\kappa = 2$, $\delta = 0$, $\beta_1 = \pi/2$ we obtain immediately by using Eqs. (3.131) and (3.132) $S = 1/\sqrt{3}$. The general case for arbitrary δ up to $\delta = 2/3$ is plotted in Fig. 3.25 under the "high resulution" condition $\kappa = 2$ and obeying Eq. (3.131c). The scaling factor drops monotonically by about 2% when δ approaches $\delta_u = 2/3$, i.e. its effect on the scaling factor is negligible. We have demonstrated that coherent averaging takes place also during the rf pulse in multiple-pulse experiments, and conditions can be found for the average dipolar Hamiltonian to vanish. Thus finite pulse width is not to be considered as a pulse imperfection, as long as all pulses have the same width and the same rotation angle β_0.

A similar treatment has been given by Haeberlen [*25*] and H. Ernst *et al.* [*65*] using a slightly different approach.

3.6 Second Averaging

In the preceeding sections we have shown that the time evolution operator governing the nuclear response during a multiple-pulse sequence can be expressed by a single average Hamiltonian $\overline{\mathcal{H}}$, which may include all orders of correction. This was achieved by imposing a cyclic property on the rf pulses, where the cycle time t_c has to fulfil the condition

$$\text{(i)} \quad \frac{1}{t_c} \gg \|\mathcal{H}_{\text{int}}\|$$

in order to ensure a rapid convergence of the Magnus expansion. So far we always have included the offset Hamiltonian $\Delta\omega I_z$ in the interaction Hamiltonian.

We are now going to separate the offset Hamiltonian (modified by the multiple-pulse sequence) from the other terms of the interaction Hamiltonian

$$\overline{\mathcal{H}} = \overline{\Delta\omega} I_{\bar{\mu}} + \overline{\mathcal{H}}_{\text{int}} \tag{3.133}$$

since it may constitute a new "Zeeman term" with the new quantization axis $\bar{\mu}$, under the condition

$$\text{(ii)} \quad \overline{\Delta\omega} \gg \overline{\mathcal{H}}_{\text{int}}.$$

In this case now the modified offset Hamiltonian acts as an external field and the spins are precessing around this "average offset field" i.e., around the μ axis with the cycle time $t_\Delta = 2\pi/\overline{\Delta\omega}$.

The offset term may be factored out of $\overline{\mathcal{H}}$ [Eq. (3.124)] and coherent averaging theory (see Section 3.3) can be readily applied to

$$\tilde{\overline{\mathcal{H}}}_{\text{int}} = e^{-i\overline{\Delta\omega} I_{\bar{\mu}}\cdot t}\, \overline{\mathcal{H}}_{\text{int}} e^{i\overline{\Delta\omega} I_{\bar{\mu}}\cdot t}. \tag{3.134}$$

Again averaging over the cycle time t_Δ results in the "second averaged" Hamiltonian $\overline{\overline{\mathcal{H}}}{}_{\text{int}}^{(0)}$ as characterized by the two bars. However, this procedure is legitimate only under the constraints of the conditions (i) and (ii), summarized into the following condition for second averaging:

$$\frac{1}{t_c} \gg \overline{\Delta\omega} \gg \|\overline{\mathcal{H}}_{\text{int}}\|. \tag{3.135}$$

The main consequence of "second averaging" is that it provides another means of *manipulation* in spin space which is capable of averaging all interactions, which are orthogonal to the direction of the "effective off-resonance field". With the corresponding cycle time t_Δ we write

$$\overline{\overline{\mathcal{H}}}{}_{\text{int}}^{(0)} = \frac{1}{t_\Delta} \int_0^{t_\Delta} dt'\, \tilde{\overline{\mathcal{H}}}_{\text{int}}(t'). \tag{3.136}$$

Many of the residual interaction Hamiltonian in multiple-pulse experiments, such as $\overline{\mathcal{H}}{}_D^{(2)}$ and some Hamiltonian due to pulse imperfections etc., are averaged to zero according to Eq. (3.136). Thus enhanced resolution is observed in multiple pulse experiments, when the spectrometer frequency is shifted off-resonance.

Second averaging has been dealt with by several authors [*66, 67*]. We find, however, the approach of Pines and Waugh [*67*] especially appealing and shall follow along those lines.

We define a unit spin vector $\vec{\mu}$ along the z axis and apply to it the time evolution operator $L_1(t)$, see Eq. (3.66), to obtain the motion of the vector $\vec{\mu}$ as

$$\vec{\mu}(t) = \mathbf{L}_1^{-1}(t)\,\vec{\mu}\,\mathbf{L}_1(t).$$

The average unit spin vector $\vec{\bar{\mu}}$ may now be expressed as

$$\vec{\bar{\mu}} = \int_0^{t_c} dt\,\vec{\mu}(t) / \left|\int_0^{t_c} dt\,\vec{\mu}(t)\right|. \tag{3.137}$$

Since the offset Hamiltonian transforms like $\vec{\mu}$ it is evident, that

$$\overline{\Delta\omega} = S \cdot \Delta\omega \tag{3.138}$$

where the scaling factor is

$$S = \frac{1}{t_c} \int_0^{t_c} dt\,\vec{\mu}(t). \tag{3.139}$$

On the other hand $\vec{\bar{\mu}}$ may be obtained by a rotation

$$\vec{\bar{\mu}} = \mathbf{R}(\varphi, \vartheta, \psi)\vec{\mu}\,\mathbf{R}^{-1}(\varphi, \vartheta, \psi) \tag{3.140}$$

where R may be derived from Eq. (3.137).

Using the rotation R we can define any Hamiltonian with respect to the $\vec{\bar{\mu}}$ axis by

$$\mathcal{H}_{\bar{\mu}} = \mathbf{R}\,\mathbf{H}_z\,\mathbf{R}^{-1}. \tag{3.141}$$

The average interaction Hamiltonian $\overline{\mathcal{H}}_{\text{int}}$ usually is not "parallel" to $\mathcal{H}_{\bar{\mu}}$ as is demonstrated in Fig. 3.26. However, $\overline{\mathcal{H}}_{\text{int}}$ may be separated into a "parallel" and an "orthogonal" component (see Fig. 3.27)

$$\overline{\mathcal{H}}_{\text{int}} = \mathcal{H}_{\parallel} + \mathcal{H}_{\perp} \tag{3.142}$$

or by using the projector $\mathbf{P}_{\bar{\mu}}$

$$\overline{\mathcal{H}}_{\text{int}} = P_{\bar{\mu}}\overline{\mathcal{H}}_{\text{int}} + (1 - P_{\bar{\mu}})\,\overline{\mathcal{H}}_{\text{int}} \tag{3.143}$$

where P is idempotent ($P^2 = P$) and

$$P_{\bar{\mu}} = \frac{|\mathcal{H}_{\bar{\mu}})(\mathcal{H}_{\bar{\mu}}|}{(\mathcal{H}_{\bar{\mu}}|\mathcal{H}_{\bar{\mu}})}. \tag{3.144}$$

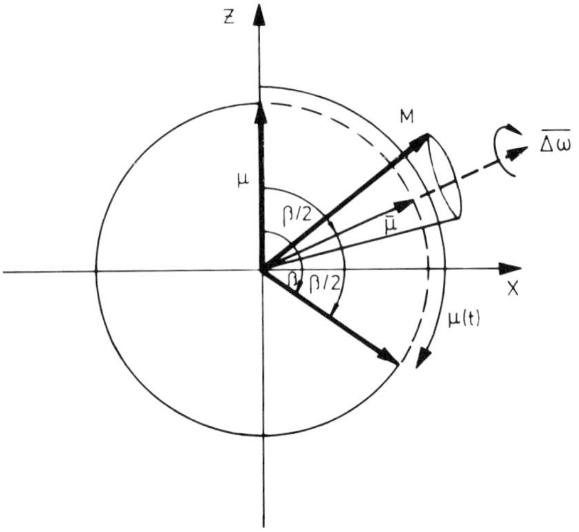

Fig. 3.26. Pictorial description of resonance offset averaging (second averaging) in the phase-alternated sequence, employing rf pulses in the y direction with a rotation angle β [67]. The unit magnetization vector μ is flipped by an angle β and $-\beta$ alternatively, leading to an average unit vector $\bar{\mu}$. The average offset frequency $\overline{\Delta\omega} = \cos(\beta/2)\,\Delta\omega$ points along $\bar{\mu}$ and any magnetization M processes on the average about $\bar{\mu}$ with frequency $\overline{\Delta\omega}$

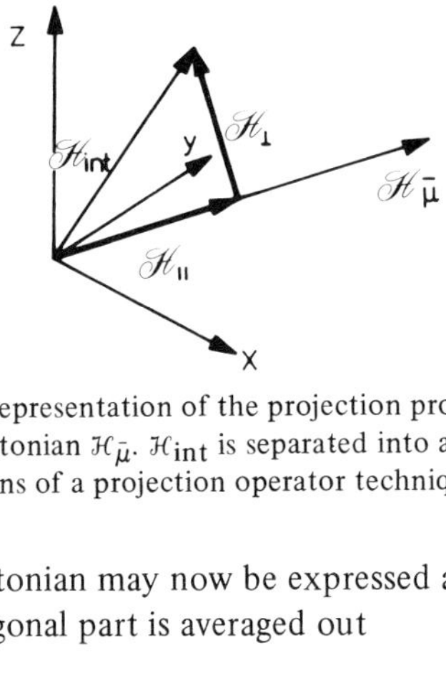

Fig. 3.27. Mnemonic vectorial representation of the projection procedure of an interaction Hamiltonian $\mathcal{H}_{int}$ onto a Hamiltonian $\mathcal{H}_{\bar{\mu}}$. $\mathcal{H}_{int}$ is separated into a parallel and an orthogonal part with respect to $\mathcal{H}_{\bar{\mu}}$ by means of a projection operator technique as defined in the text

The second averaged Hamiltonian may now be expressed as the "parallel" component of $\overline{\mathcal{H}}_{int}$, whereas, the orthogonal part is averaged out

$$\overline{\overline{\mathcal{H}}}{}^{(0)}_{int} = \mathcal{H}_{\parallel} = P_{\bar{\mu}}\overline{\mathcal{H}}_{int}. \tag{3.145}$$

Using Eqs. (3.142–3.145) we can write

$$\overline{\overline{\mathcal{H}}}{}^{(0)}_{int} = p\,\mathcal{H}_{\bar{\mu}} \tag{3.146}$$

where p is a scalar with

$$p = \frac{(\mathcal{H}_{\bar{\mu}}\,|\,\overline{\mathcal{H}}_{int.})}{(\mathcal{H}_{\bar{\mu}}\,|\,\mathcal{H}_{\bar{\mu}})}. \tag{3.147}$$

If we restrict ourselves to the zeroth order average Hamiltonian $\overline{\mathcal{H}}_{\text{int}}^{(0)}$ and remember, that

$$|\overline{\mathcal{H}}_{\text{int}}^{(0)}) = \frac{1}{t_c}\int_0^{t_c} dt\,\hat{\mathbf{L}}_1^{-1}(t)|\mathcal{H}_z)$$

we may write, using Eq. (3.141)

$$p = \frac{1}{t_c}\int_0^{t_c} dt\,\frac{(\mathcal{H}_z|\hat{\mathbf{R}}^{-1}\hat{\mathbf{L}}_1^{-1}(t)|\mathcal{H}_z)}{(\mathcal{H}_z|\mathcal{H}_z)}. \tag{3.148}$$

A very convenient expression for p can be obtained as shown by Pines and Waugh [67], with

$$\frac{(\mathcal{H}_z|\hat{\mathbf{R}}^{-1}\hat{\mathbf{L}}_1^{-1}(t)|\mathcal{H}_z)}{(\mathcal{H}_z|\mathcal{H}_z)} = P_k(\vec{\bar{\mu}}\cdot\vec{\mu}(t)) \tag{3.149}$$

where P_k is the Legendre polynomial of order k, and $\vec{\bar{\mu}}$ and $\vec{\mu}(t)$ are given by Eqs. (3.136) and (3.137). The order k depends on the rank of $\mathcal{H}_z$ e.g. k=2 for dipolar interaction and k = 1 for shift interaction.

Following Eqs. (3.148) and (3.149) p may be expressed as

$$p = \overline{P_k(\vec{\bar{\mu}}\cdot\vec{\mu}(t))} \tag{3.150a}$$

where

$$\overline{P_k(\vec{\bar{\mu}}\cdot\vec{\mu}(t))} = \frac{1}{t_c}\int_0^{t_c} dt\,P_k(\vec{\bar{\mu}}\cdot\vec{\mu}(t)). \tag{3.150b}$$

Combining Eqs. (3.146) and (3.150) leads to

$$\overline{\overline{\mathcal{H}}}_{\text{int}}^{(0)} = \overline{P_k(\vec{\bar{\mu}}\cdot\vec{\mu}(t))}\cdot\mathcal{H}_{\bar{\mu}} \tag{3.151}$$

where $\mathcal{H}_{\bar{\mu}}$ is according to Eq. (3.141) just the truncated or secular interaction Hamiltonian with respect to the axis $\vec{\bar{\mu}}$. The simple yet remarkable result obtained by Pines and Waugh [67] is, that no matter how complicated the multiple-pulse cycle is, as long as condition Eq. (3.135) is fulfilled, the average interaction Hamiltonian is just the secular interaction Hamiltonian itself with respect to some average axis in the rotating frame, scaled by the factor p. This factor p can be much more easily evaluated, than dealing with spin operators in a complicated multiple pulse sequence.

We shall treat now the phase-alternated sequence MW-2 as a simple example, to illustrate the usefulness of this approach. The phase-alternated sequence (MW-2) as shown in Fig. 3.19b consists of β pulses in the y and $\bar{y}$ direction alternatively. If we assume equal pulse spacing for all pulses (κ = 1) we find using Eq. (3.151) for the average dipolar Hamiltonian

$$\overline{\overline{\mathcal{H}}}_D^{(0)} = \mathcal{H}_{D\bar{\mu}}(3p_\delta(\beta) + 1)/4 \tag{3.152}$$

where

$$p_\delta(\beta) = (1 - \delta)\cos + \delta\,\frac{\sin\beta}{\beta}\,. \tag{3.153}$$

The $\bar{\mu}$ axis lies in the x-z plane. If there is a shift Hamiltonian $\bar{\mathcal{H}}_S^{(0)}$ present in the rotating frame Hamiltonian, the second average shift Hamiltonian is expressed as

$$\bar{\bar{\mathcal{H}}}_S^{(0)} = \mathcal{H}_{S\bar{\mu}} \cdot p_\delta(\beta/2) \tag{3.154}$$

where $p_\delta(\beta)$ is given by Eq. (3.153).

Figure 3.28 shows a plot of $(3\,p_\delta(\beta) + 1)/4$ as a function of β for different values of the duty factor δ. We see that $\bar{\bar{\mathcal{H}}}_D^{(0)}$ can be made to vanish for $\delta < 0.75$ opening the possibility for line narrowing experiments. If $\delta = 0$ it is evident from Fig. 3.28

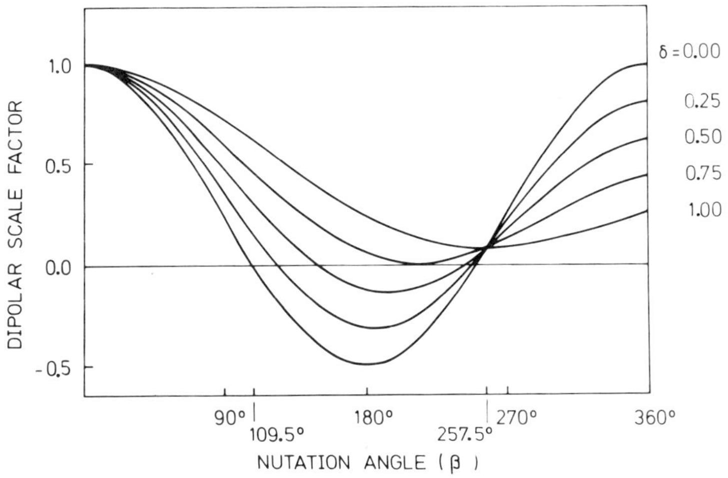

Fig. 3.28. Dipolar scaling factor $\overline{P_2(\bar{\mu}\cdot\mu(t))} = (3p\delta(\beta) + 1)/4$ [see Eq. (3.152)] for a phase-alternated sequence with a duty factor δ employing β pulses. (A. Pines and J. S. Waugh [67]). Line narrowing in solids is achieved with this sequence when $\bar{P}_2 = 0$, e.g. at 109.5° for $\delta = 0$ (phase-alternated tetrahedral experiment)

and Eq. (3.152) that $\bar{\bar{\mathcal{H}}}_D^{(0)} = 0$ for $\beta = \beta_t$, where β_t is the tetrahedral angle (109°28′). This sequence was named "phase-alternated tetrahedral" PAT sequence [*66*]. Line narrowing is still achieved if $\delta > 0$, however, the rotation angle β has to be larger than β_t in this case. The negative value of the dipolar scaling factor indicates, that "magic echoes" may be obtained in this region. For further details we refer the interested reader to Refs. [*66, 67*].

3.7 The Influence of Pulse Imperfection on Multiple-Pulse Experiments

One of the basic problems in multiple-pulse NMR concerns the question which resolution or line narrowing efficiency can be obtained. A number of derivates of the four-pulse cycle (WHH-4) have been proposed with the promise of better resolution. First, the second order correction term $\overline{\mathcal{H}}_D^{(2)}$ of the average dipolar Hamiltonian was felt to be the prime source for the limit in resolution, however, it was soon realized that pulse imperfections play a major role. The remedy usually prescribed is, to design cascaded cycles with different symmetry properties in order to cancel the effect of specific pulse imperfections [*26, 60, 68*].

Let us first start with the effect of magnetic field inhomogeneity. It is evident that inhomogeneity of the static magnetic field H_0 results in the same line broadening as in conventional NMR, however, scaled by the shift scaling factor S.

Inhomogeneity of the static magnetic field may be produced by the field itself of by the static magnetic susceptibility of the sample, if a non-spherical sample is used. The last effect, however, can be eliminated if a spherical sample is used. This is highly recommended, as is readily seen, when comparing the results in Table 3.1. The

Table 3.1. Influence of the static magnetic susceptibility of the sample on the linewidth (W. K. Rhim, D. D. Elleman and R. W. Vaughan [*33*])

Sample shape	Linewidth[a]
Sphere[b]	0.4 ppm
Cyclinder[c]	1.1 ppm
Rectangular parallelepiped[d]	1.4 ppm

[a] The chemical shift scaling factor has been taken into account for these values.
[b] 4 mm in diameter and spherical within 0.1%.
[c] 4 mm in diameter and 4.5 mm in length; cyclinder axis was perpendicular to the external field.
[d] 4 mm × 4 mm × 5 mm with cubic face perpendicular to field.

influence of the H_1 inhomogeneity, however, is more subtle and needs further consideration. A detailed discussion can be found in Refs. [*25, 60, 65, 68, 70*]. If a train of "inhomogenous" rf pulses is applied to a spin system, each individual spin I_i will experience a different rotation angle β_i, resulting in a distribution of rotation angles over the sample. An average rotation angle β_0 may be defined by

$$\beta_0 = \overline{\beta_i} \tag{3.155}$$

and the H_1 inhomogeneity by

$$\epsilon_i = (\beta_i - \beta_0)/\beta_0 \tag{3.156}$$

with

$$\sum_i \epsilon_i = 0$$

and the standard deviation

$$\epsilon = [\overline{\epsilon_i^2}]^{1/2}. \tag{3.157}$$

If a Gaussian distribution of the rf inhomogeneity is assumed, we obtain [*69*]

$$p(\epsilon_i) = \frac{1}{(2\pi M_\epsilon)^{1/2}} \exp(-\epsilon_i^2/2M_\epsilon) \tag{3.158a}$$

$$p(\beta_i) = \frac{N_0}{2\sqrt{\pi}} \exp[-(\frac{N_0}{2})^2 (\frac{\beta_i - \beta_0}{\beta_0})^2] \tag{3.158b}$$

where the second moment of the rf inhomogeneity distribution is given by

$$M_\epsilon = (\frac{4}{\pi})^2 \epsilon^2 \tag{3.159a}$$

and where

$$N_0 = \frac{\pi}{2\sqrt{2}\,\epsilon} \tag{3.159b}$$

There is a direct loss of coherence among the spins due to this rf inhomogeneity with a coherence time proportional to $M_\epsilon^{-1/2}$. This direct effect can be observed in a continuous rf field or a train of pulses with equal phase. A train of equally spaced $\pi/2$ pulses applied to a liquid sample is a suitable experiment for observing this direct effect and measuring the value M_ϵ, ϵ^2 or N_0 directly. The nuclear signal after the N-th $\pi/2$ pulse follows [*69*].

$$S(N) = \sin(N\pi/2) \exp[-(N/N_0)^2]. \tag{3.159c}$$

In phase-alternated cycles, however, this disorder is remedied due to the fact, that the initial state is recovered after each pair of phase-alternated pulses.

One should be aware, however, that the inhomogeneity effect will be transmitted to $\mathcal{H}_{int}(t)$, the "switched interaction Hamiltonian" under the condition $[\mathcal{H}_1(t), \mathcal{H}_{int}] \neq 0$. This indirect H_1 inhomogeneity effect is essential in all line narrowing sequences utilizing phase alternation. Since the average interaction Hamiltonian $\overline{\mathcal{H}}_0(\beta)$ depends on the rotation angle β of the rf pulses, the contribution of H_1 inhomogeneity to the average Hamiltonian may be conveniently expressed by

$$\overline{\mathcal{H}}_\epsilon = [\frac{d\overline{\mathcal{H}}_0(\beta)}{d\beta}]_{\beta_0} \cdot \epsilon_i \tag{3.160a}$$

if we limit ourselves to effects linear in ϵ_i.

Here β_0 denotes the value of β, for which maximum line narrowing efficiency is obtained.

Equation (3.160a) is applicable to any pulse sequence, employing arbitrary pulse width. The total average Hamiltonian can be expressed as

$$\overline{\mathcal{H}}_{tot} = \overline{\mathcal{H}}_0 + \overline{\mathcal{H}}_\epsilon \tag{3.160b}$$

where $\overline{\mathcal{H}}_0$ is the average Hamiltonian according to the average pulse rotation angle

β_0 of the H_1 inhomogeneity distribution. In order to introduce a measure for the influence the H_1 inhomogeneity has on the NMR signal, we consider the total scaling factor

$$S_{\text{tot}} = [\text{Tr}\{\overline{\mathcal{H}}^2_{\text{tot}}\}/\text{Tr}\{\mathcal{H}^2_{\text{int}}\}]^{1/2} \tag{3.160c}$$

where $\mathcal{H}_{\text{int}}$ is the interaction Hamiltonian in the rotating frame i.e. without application of any pulse sequence. Let us first discuss the effect of indirect

(i) off-resonance H_1 *inhomogeneity coupling*

The magnitude of this effect will be directly proportional to the H_1 inhomogeneity and the resonance offset and can become very large in multiple-pulse experiments, as can be easily demonstrated in a liquid sample.

Applying Eqs. (3.160a–c), we obtain ($\epsilon_i \ll 1$).

$$S_{\text{tot}} = S_0(1 + 2\Delta S\epsilon_i + \Delta S'\epsilon_i^2)^{1/2} \cong S_0(1 + \Delta S\epsilon_i) \tag{3.161a}$$

where

$$S_0 = [\text{Tr}\{\overline{\mathcal{H}}_0^2\}/\text{Tr}\{\mathcal{H}_S^2\}]^{1/2}$$

is the usual scaling factor of the multiple pulse experiment and

$$\Delta S = \text{Tr}\{\overline{\mathcal{H}}_0(\frac{d}{d_\beta}\overline{\mathcal{H}}_0)_{\beta_0}\}/\text{Tr}\{\overline{\mathcal{H}}_0^2\} \tag{3.161b}$$

$$\Delta S' = \text{Tr}\{(\frac{d}{d_\beta}\overline{\mathcal{H}}_0)^2_{\beta_0}\}/\text{Tr}\{\overline{\mathcal{H}}_0^2\}$$

are the scaling factors of the H_1 inhomogeneity.

Since ϵ_i is a small quantity the contribution of $\Delta S' \cdot \epsilon_i^2$ is negligible and the important parameter which describes the spectral distribution of the NMR line for a given H_1 inhomogeneity ϵ_i is ΔS, as far as shift interactions are concerned. In order to compare different pulse sequences we summarize in Table 3.2 the different expressions for $\overline{\mathcal{H}}_\epsilon$ and $\overline{\mathcal{H}}_0$ in the limit $\delta = 0$; $\beta_0 = \pi/2$ (see Tables 3.3 and 3.4) and include the parameter ΔS.

Table 3.2 Influence of H_1 inhomogeneity (ϵ_i) on shift interactions (resonance offset). ΔS is a measure of the spectral distribution due to H_1 inhomogeneity (see text)

	MW-2	WHH-4	HW-8	MREV-8
$\overline{\mathcal{H}}_0$	$\frac{1}{2}\Delta\omega \cdot (I_z - I_y)$	$\frac{1}{3}\Delta\omega \cdot (I_x + I_y + I_z)$	$\frac{1}{3}\Delta\omega \cdot I_z$	$\frac{1}{3}\Delta\omega \cdot (I_x + I_z)$
$\overline{\mathcal{H}}_\epsilon$	$-\frac{1}{2}\Delta\omega\epsilon I_z$	$-\frac{1}{3}\Delta\omega\epsilon(I_y + I_z)$	$-\frac{2}{3}\Delta\omega\epsilon I_z$	$-\frac{1}{3}\Delta\omega\epsilon I_z$
$\lvert\Delta S\rvert$	$\frac{1}{2}$	$\frac{2}{3}$	2	$\frac{1}{2}$

Notice, that the MW-2 and MREV-8 sequence have the smallest value of ΔS, whereas WHH-4 and especially HW-8 have larger ones. This means, e.g. that a HW-8 sequence shows four times stronger broadening due to H_1 inhomogeneity to off-resonance coupling than the MREV-8 sequence. Let us consider a simple example. Suppose the mean variation of the H_1 field in the sample coil is 2%, the MW-2 and MREV-8 sequences would show a line broadened by 1% with respect to the resonance offset, i.e. line width: 100 Hz for $\Delta\nu = 10$ kHz. Figure 3.29 demonstrates this effect in the case of a MREV-8 sequence applied to a liquid sample. The same procedure as outlined by Eqs. (3.160, 3.161) can be applied when arbitrary pulse width is assumed in the different multiple pulse sequences. As an example we shall treat the WHH-4 sequence, however, the extension to other pulse sequences is straightforward.

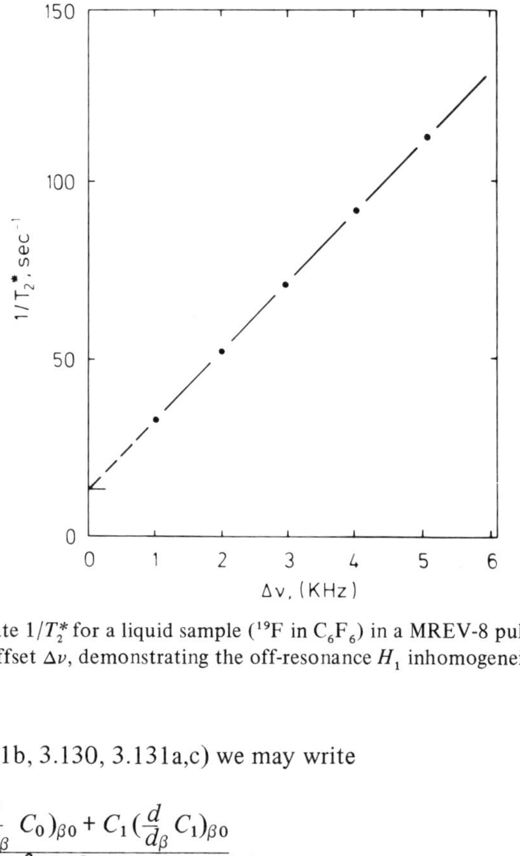

Fig. 3.29. Decay rate $1/T_2^*$ for a liquid sample (^{19}F in C_6F_6) in a MREV-8 pulse experiment versus resonance offset $\Delta\nu$, demonstrating the off-resonance H_1 inhomogeneity effect (Rhim *et al.* [*33*])

Using Eqs. (3.161b, 3.130, 3.131a,c) we may write

$$\Delta S = \frac{C_0\left(\frac{d}{d\beta}C_0\right)_{\beta 0} + C_1\left(\frac{d}{d\beta}C_1\right)_{\beta 0}}{C_0^2 + C_1^2} \quad .$$

Using the expressions for C_0 and C_1 as given by Eq. (3.130) for $\kappa = 2$, ΔS can be calculated for different duty factors δ or optimum pulse rotation angle β_0 in the WHH-4 experiment. Figure 3.30 represents such a calculation of $|\Delta S|$ versus δ or β_0 respectively. There is a slight change for ΔS although not a very drastic one with an increasing duty factor.

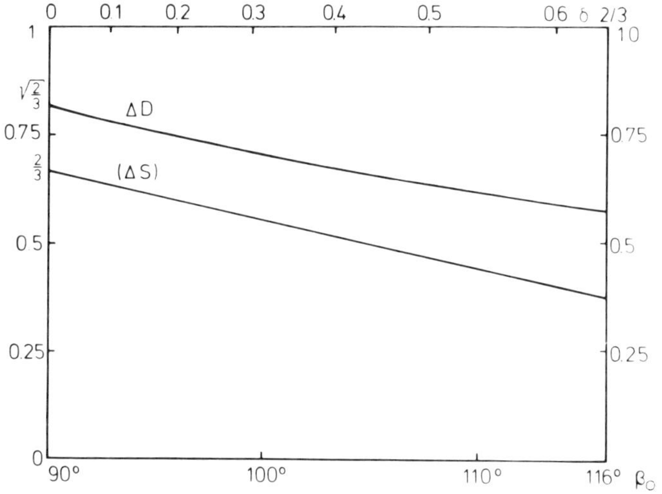

Fig. 3.30. Influence of the H_1 inhomogeneity coupling in a WHH-4 experiment versus duty factor δ or pulse rotation angle β_0 respectively. The coupling of H_1 inhomogeneity of the (a) dipolar interaction is represented by ΔD and (b) shift interaction and resonance offset is represented by ΔS (see text)

The H_1 inhomogeneity effect discussed so far vanishes at resonance ($\Delta\omega = 0$). In order to discuss

(ii) on-resonance H_1 *inhomogeneity effects*

we shall investigate how the H_1 inhomogoneity is transmitted to the average dipolar interactions.

This effect has been analyzed by Pfeifer and coworkers [*69*] and others [*60, 68, 70*]. The procedure is similar as before and Eq. (3.160) has to be applied. Let us again consider the WHH-4 sequence as an example. Since $\overline{\mathcal{H}}_D^{(0)} = 0$, if the pulse rotation angle $\beta = \beta_0$, we obtain for the total dipolar scaling factor according to Eq. (3.160)

$$S_{\text{tot}} = [\text{Tr}\{(\frac{d}{d\beta}\overline{\mathcal{H}}_D)^2\} / \text{Tr}\{\mathcal{H}_D^2\}]^{1/2}\epsilon_i = \Delta D\epsilon_i$$

where

$$\Delta D = [(\frac{d}{d\beta}D_0)_{\beta 0}^2 + (\frac{d}{d\beta}D_1)_{\beta 0}^2]^{1/2}. \qquad (3.162)$$

The coupling parameter ΔD of the dipolar interaction to the H_1 inhomogeneity has been calculated for the WHH-4 experiment ($\kappa = 2$) using Eqs. (3.129, 3.162) and is plotted versus β_0 in Fig. 3.30. Notice, that ΔD is of about the same size as ΔS, the coupling parameter for resonance-offset to H_1 inhomogeneity. Both effects have to be considered in the WHH-4 experiment and can severely limit the resolution.

It can be readily shown, that there is no H_1 inhomogeneity effect on the average dipolar Hamiltonian in the case of the HW-8 and the MREV-8 sequence [*66, 68, 70*] in the limit $\delta = 0, \beta = \pi/2$ (see Table 3.4). This fact combined with the small offset coupling is what makes the MREV-8 sequence superior to the others as claimed by its designers. However, if $\delta \neq 0$ H_1 inhomogeneity becomes effective.

The WHH-4 sequence shows the following contributions of H_1 inhomogeneity coupling to the dipolar interaction in the limit $\delta = 0, \beta_0 = \pi/2$:

$$\overline{\mathcal{H}}_D^{(0)}(\epsilon_i) = -i\sum_i \frac{1}{3}\epsilon_i[\mathcal{H}_{Dz}, (I_{xi} - I_{yi})] \tag{3.163a}$$

$$\overline{\mathcal{H}}_D^{(1)}(\epsilon_i) = \tau\sum_i \frac{1}{3}\,\epsilon_i[\mathcal{H}_{Dx}[(I_{xi} - I_{yi}), \mathcal{H}_{Dz}]]. \tag{3.163b}$$

We neglect the contribution of ϵ_i to $\overline{\mathcal{H}}_D^{(2)}$ and still use

$$\overline{\mathcal{H}}_D^{(2)} = \frac{\tau^2}{18}\,[(\mathcal{H}_{Dx} - \mathcal{H}_{Dy}), [\mathcal{H}_{Dy}, \mathcal{H}_{Dx}]] \tag{3.163c}$$

for the WHH-4, HW-8, and MREV-8 sequences.

Using the Gaussian H_1 field distribution [Eq. 3.158)] with the second moment M_ϵ due to the H_1 inhomogeneity, Pfeifer *et al.* [*69*] obtained for the average second moments of the different multiple-pulse experiments the following expressions

$$\overline{M}_2^{\text{WHH-4}} = \frac{2}{9}M_\epsilon[M_2 + \frac{(1+C_0)}{2}M_4\,\tau^2] + \frac{C_1}{324}M_6\tau^4 \tag{3.164}$$

and

$$\overline{M}_2^{\text{MREV-8}} = \overline{M}_2^{\text{HW-8}} = \frac{C_1}{324}\,M_6\tau^4 \tag{3.165}$$

where M_2, M_4, and M_6 are the corresponding moments of the dipolar interaction and where C_0 and C_1 are numbers defined by

$$C_0 = \frac{(I_z|\hat{\mathcal{H}}_{Dx}\hat{\mathcal{H}}_{Dz}^2\hat{\mathcal{H}}_{Dx}|I_z)}{(I_x|\hat{\mathcal{H}}_{Dz}^4|I_x)} \tag{3.166a}$$

and

$$C_1 = \frac{(I_z|\hat{\overline{\mathcal{H}}}{}_D^{(2)2}|I_z)}{(I_x|\hat{\mathcal{H}}_{Dz}^6|I_x)}. \tag{3.166b}$$

Using the following values for CaF_2 with H_0 parallel to the (111)-direction [*69*]

$M_2 = 1.23K^2$; $M_4 = 3.56K^4$; $M_6 = 15.7K^6$

$C_0 = 0.668$; $C_1 = 2.523$; $K = \gamma^2\hbar d^{-3}$

where d is the lattice constant, we obtain [69]

$$\overline{M}_2^{\text{WHH-4}} = M_\epsilon \cdot (0.22 M_2 + 0.217\, M_2^2\, \tau^2) + 0.065 M_2^3\, \tau^4 \tag{3.167a}$$

$$\overline{M}_2^{\text{MREV-8}} = 0.065\, M_2^3\, \tau^4 . \tag{3.167b}$$

Figure 3.31 confirms this dependence of the WHH-4 sequence on the on-resonance effect of the H_1 inhomogeneity in the case of CaF_2 in the (111)-direction.

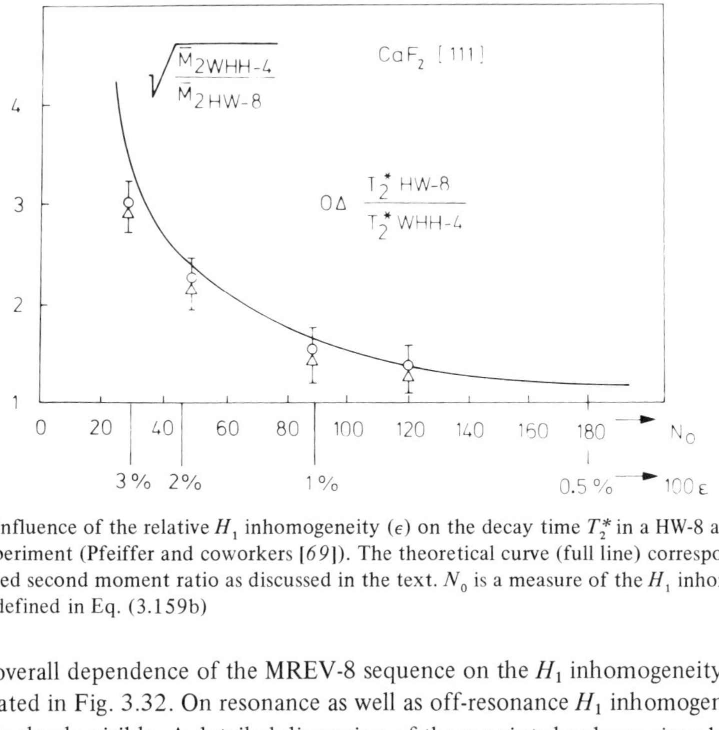

Fig. 3.31. Influence of the relative H_1 inhomogeneity (ϵ) on the decay time T_2^* in a HW-8 and WHH-4 experiment (Pfeiffer and coworkers [69]). The theoretical curve (full line) corresponds to a modified second moment ratio as discussed in the text. N_0 is a measure of the H_1 inhomogeneity as defined in Eq. (3.159b)

The overall dependence of the MREV-8 sequence on the H_1 inhomogeneity is demonstrated in Fig. 3.32. On resonance as well as off-resonance H_1 inhomogeneity coupling is clearly visible. A detailed discussion of these points has been given by Garroway *et al.* [70].

Other pulse imperfections may be investigated in a similar fashion. Indeed, this has been done by several authors [60, 68, 70, 71]. We shall follow here the meticulous work of Rhim, Elleman, Schreiber, and Vaughan [68]. Although H_1 inhomogeneity preserves the cyclic condition in all phase-alternated experiments, this may not be true for other pulse imperfections, such as phase transients, phase misadjustment, and errors in pulse length. In the following it is assumed that the violation of the cyclic condition is weak in the sense that the rf Hamiltonian $\mathcal{H}_1(t)$ can be split into a major part $\mathcal{H}_1^0(t)$, which still satisfies the cyclic condition and represents the ideal part of the rf pulse and into a small part $\mathcal{H}_1^1(t)$, which represents the non-ideal part of $\mathcal{H}_1(t)$

$$\mathcal{H}_1(t) = \mathcal{H}_1^0(t) + \mathcal{H}_1^1(t). \tag{3.168}$$

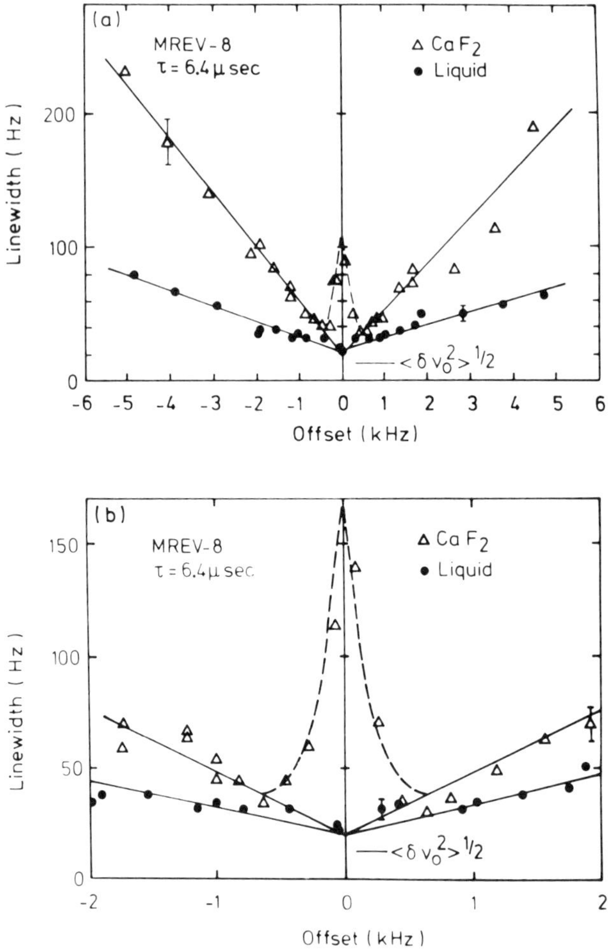

Fig. 3.32. (a) ^{19}F linewidth in CaF_2 $H_0 \parallel (111)$ and in CF_3COOH versus resonance offset for the MREV-8 sequence (Mansfield and co-workers [58]). (b) Behavior near resonance. The linewidth is defined here as the full width at half height as observed in the multiple-pulse experiment. $\langle \delta \nu_0^2 \rangle^{1/2}$ is the linewidth due to the static magnetic field inhomogeneity

We assume that

$$\| \mathcal{H}_1^0(t) \| \gg \| \mathcal{H}_1^1(t) \|$$

and combine $\mathcal{H}_1^1(t)$ with the interaction Hamiltonian $\mathcal{H}_{int}$. The ideal pulses now operate on the combined Hamiltonian $\mathcal{H}_{int} + \mathcal{H}_1^1(t)$ and coherent averaging theory may be applied to $\widetilde{\mathcal{H}}_{int}(t) + \widetilde{\mathcal{H}}_1^1(t)$. In general there are different imperfections present in multiple-pulse experiments and $\mathcal{H}_1^1(t)$ may be expressed by a sum as

$$\mathcal{H}_1^1(t) = \sum_k \mathcal{H}_k(t)$$

where k represents in the following:

P : phase misadjustment,
T : phase transients,
δ : pulse length misadjustments,
ϵ : H_1 inhomogeneity.
O : resonance offset

Then the average Hamiltonian takes the following form:

$$\bar{\mathcal{H}}_{\text{int}}^{(0)} = \bar{\mathcal{H}}_O^{(0)} + \bar{\mathcal{H}}_D^{(0)} + \sum_k \bar{\mathcal{H}}_k^{(0)}, \tag{3.169a}$$

$$\bar{\mathcal{H}}_{\text{int}}^{(1)} = \bar{\mathcal{H}}_O^{(1)} + \bar{\mathcal{H}}_D^{(1)} + \bar{\mathcal{H}}_{OD}^{(1)} + \sum_k (\bar{\mathcal{H}}_{Ok}^{(1)} + \bar{\mathcal{H}}_{Dk}^{(1)}), \tag{3.169b}$$

$$\bar{\mathcal{H}}_{\text{int}}^{(2)} = \bar{\mathcal{H}}_O^{(2)} + \bar{\mathcal{H}}_D^{(2)} + \bar{\mathcal{H}}_{OD}^{(2)} + \sum_k (\bar{\mathcal{H}}_{Ok}^{(2)} + \bar{\mathcal{H}}_{Dk}^{(2)}). \tag{3.169c}$$

The notation here is self-explaining with, for instance, $\mathcal{H}_{Ok}^{(1)}$ representing the first-order coupling between resonance offset $\mathcal{H}_O$ and the k-th imperfection $\mathcal{H}_k(t)$. Since the pulse imperfections $\mathcal{H}_1^1(t)$ will be in practice much smaller than the spin interactions $\mathcal{H}_{\text{int}}$, we have neglected cross terms between two imperfections. However, this can be incorporated if it is felt to be significant. Table 3.3 and Table 3.4, according to Rhim *et al.* [*68*] summarize the average interaction Hamiltonian due to different pulse imperfections for the following multiple-pulse sequences:

$(-\tau - P_x - 2\tau - P_x - 2\tau - P_x - 2\tau - P_x - \tau)_n$ MW-4
$(-\tau - P_x - 2\tau - P_{\bar{x}} - \tau -)_n$ MW-2
$(-\tau - P_{\bar{x}} - \tau - P_y - 2\tau - P_{\bar{y}} - \tau - P_x - \tau)_n$ WHH-4
$(-\tau - P_x - \tau - P_y - 2\tau - P_{\bar{y}} - \tau - P_{\bar{x}} - 2\tau - P_{\bar{x}} - \tau - P_y - 2\tau$
$- P_{\bar{y}} - \tau - P_x - \tau)_n$. MREV-8

These four sequences are the most important ones in multiple-pulse NMR. The MW-4 sequence is usually applied to a liquid sample in order ot adjust the pulse length precisely and to measure the H_1 inhomogeneity, whereas, the MW-2 sequence allows the precise adjustment of the 180° phase difference between two pulses. The WHH-4 and the MREV-8 sequences are the best known and most efficient line narrowing sequences in solids which suppress dipolar and quadrupolar interaction.

Due to the different pulse imperfections such as H_1 inhomogeneity (ϵ), pulse shape and pulse size (δ), phase errors (Φ), and effects of phase transients ($\omega_I(t)$, $\omega_T(t)$) the x pulse Hamiltonian for example becomes [*68*]

$$\begin{aligned}\mathcal{H}_1(t) = &-\omega_1 I_x - \sum_i \frac{\epsilon_i}{t_w} I_{xi} - \frac{\delta_x}{t_w} I_x \\ &- \omega_1 \Phi_x I_y - \omega_I(t) I_x + \omega_T(t) I_y\end{aligned} \tag{3.170}$$

where the following conditions are assumed to be satisfied:

$$\left|\frac{\epsilon_i}{t_w}\right|; \left|\frac{\delta_x}{t_w}\right|; |\Phi_x \omega_1|; |\omega_I(t)|; |\omega_T(t)| \ll |\omega_1|. \tag{3.171}$$

Table 3.3. Average interaction Hamiltonians for tune-up cycles, assuming the pulse rotation angle β_0 close to 90° (W. K. Rhim *et al.* [68]). (O, offset; P, phase error; T, phase transient; δ, pulse width error; ϵ, rf inhomogeneity; d, powder droop)

Average Hamiltonian	$(\frac{\pi}{2})_x - (\frac{\pi}{2})_x - (\frac{\pi}{2})_x - (\frac{\pi}{2})_x$ (MW-4)	Flip-flop cycle (MW-2)
$\bar{\mathcal{H}}_O^{(0)}$ [a]	0	$\frac{1}{2}\Delta\omega(1 + \frac{2}{3}a)(I_z - I_y)$
$\bar{\mathcal{H}}_O^{(1)}$	$-\frac{1}{16}t_c(\Delta\omega)^2 I_x$	0
$\bar{\mathcal{H}}_P^{(0)}$	0	$(1/t_c)(-\phi_x + \phi_{-x})(I_y + I_z)$
$\bar{\mathcal{H}}_{PO}^{(1)}$	0	$\frac{1}{4}\Delta\omega(\phi_x + \phi_{-x})I_x$
$\bar{\mathcal{H}}_T^{(0)}$ [b]	0	$(1/t_c)J_1(I_z - I_y)$
$\bar{\mathcal{H}}_{TO}^{(1)}$	$-\frac{1}{2}\Delta\omega J_1 I_x$	$-\frac{1}{4}\Delta\omega J_2 I_x$
$\bar{\mathcal{H}}_\delta^{(0)}$	$-(4/t_c)\delta_x I_x$	$(1/t_c)(-\delta_x + \delta_{-x})I_x$
$\bar{\mathcal{H}}_{\delta O}^{(1)}$	$\frac{1}{2}\Delta\omega\delta_x I_z$	$-\frac{1}{4}\Delta\omega(\delta_x + \delta_{-x})I_z$
$\bar{\mathcal{H}}_\epsilon^{(0)}$	$-(4/t_c)\sum_i \epsilon_i I_{xi}$	0
$\bar{\mathcal{H}}_{\epsilon O}^{(1)}$	$\frac{1}{2}\Delta\omega\sum_i \epsilon_i I_{zi}$	$-\frac{1}{2}\Delta\omega\sum_i \epsilon_i I_{zi}$
$\bar{\mathcal{H}}_d^{(0)}$ [c] (nth cycle)	$-\frac{\omega_s t_w}{t_c}4(1 - e^{-8n/b})I_x$	$-\frac{\omega_s t_w}{t_c}\frac{2}{b}e^{-4n/b}I_x$

[a] $a = 3t_w/t_c(4/\pi - 1)$.

[b] $J_1 = \int_0^{t_w}\omega_T(t)(\sin\omega_1 t - \cos\omega_1 t)dt$; $J_2 = \int_0^{t_w}\omega_T(t)(\sin\omega_1 t + \cos\omega_1 t)dt$.

[c] $\bar{\mathcal{H}}_d^{(0)}$(nth cycle) $= \bar{\mathcal{H}}_d^{(0)}[nt_c \rightarrow (n+1)t_c]$ Exponential decay of $\omega_{rf}(t)$ with time constant $b\tau$; $b \gg 1$ is assumed.

Table 3.4. Average interaction Hamiltonians for line narrowing multiple-pulse experiments, assuming the pulse rotation angle β_0 close to 90° (W. K. Rhim *et al.* [68]). (O, offset; D, dipolar; P, phase error; T, phase transient; δ, pulse width error; ϵ, rf inhomogeneity; d, powder droop)

	Four pulse cycle (WHH-4); $t_c = 6\tau$	Eight pulse cycle (MREV-8); $t_c = 12\tau$
$\bar{\mathcal{H}}_O^{(0)}$ a	$\frac{1}{3}\sum_i(\Delta\omega + \omega_0\sigma_{zzi})(1+a)(I_{xi}+I_{yi}+I_{zi})$	$\frac{1}{3}\sum_i(\Delta\omega + \omega_0\sigma_{zzi})(1+2a)(I_{xi}+I_{zi})$
$\bar{\mathcal{H}}_O^{(1)}$	0	$\frac{\tau}{3}\sum_i(\Delta\omega + \omega_0\sigma_{zzi})^2(I_{xi}-I_{zi})$
$\bar{\mathcal{H}}_D^{(0)}$	0^{b}	0
$\bar{\mathcal{H}}_D^{(2)}$	$(\tau^2/18)[\mathcal{H}_D^{(x)} - \mathcal{H}_D^{(z)}, [\mathcal{H}_D^{(x)}, \mathcal{H}_D^{(y)}]]$	$(\tau^2/18)[\mathcal{H}_D^{(x)} - \mathcal{H}_D^{(z)}, [\mathcal{H}_D^{(x)}, \mathcal{H}_D^{(y)}]]$
$\bar{\mathcal{H}}_{DO}^{(1)}$	0	0
$\bar{\mathcal{H}}_{DO}^{(2)}$ c	d	$(\tau^2/6)(\Delta\omega)[I_x, [\mathcal{H}_D^{(x)}, \mathcal{H}_D^{(y)}]] + (\tau^2/3)(\Delta\omega)^2(\mathcal{H}_D^{(z)} - \mathcal{H}_D^{(x)})$
$\bar{\mathcal{H}}_P^{(0)}$	$\frac{1}{6\tau}[(\phi_y - \phi_{-y})I_x + (-\phi_x + \phi_{-x} - \phi_y + \phi_{-y})I_y + (\phi_x - \phi_{-x})I_z]$	$\frac{1}{6\tau}[(\phi_y - \phi_{-y})I_x + (-\phi_x + \phi_{-x})I_y]$
$\bar{\mathcal{H}}_{PO}^{(1)}$	$\frac{1}{6}\sum_i(\Delta\omega + \omega_0\sigma_{zzi})[-(\phi_x + \phi_{-x})(I_{xi} - I_{yi})$ $+ (\phi_x + \phi_{-x} - \phi_y - \phi_{-y})I_{zi}]$	$\frac{1}{6}\sum_i(\Delta\omega + \omega_0\sigma_{zzi})[(-\phi_y + \phi_{-y})I_{xi} + 2\phi_x I_{yi}$
$\bar{\mathcal{H}}_{PD}^{(1)}$	$\frac{1}{2}(-\phi_x - \phi_{-x} + \phi_y + \phi_{-y})\sum_{i<j}A_{ij}(I_{xi}I_{zj} + I_{zi}I_{xj})$	0
$\bar{\mathcal{H}}_T^{(0)}$ e	$\frac{1}{6\tau}J_1(I_x + 2I_y + I_z)$	$\frac{1}{6\tau}J_1(I_x + I_z)$
$\bar{\mathcal{H}}_{TO}^{(1)}$ e	$\frac{1}{6}J_2\sum_i(\Delta\omega + \omega_0\sigma_{zzi})(I_{xi} - I_{yi})$	$\frac{1}{6}\sum_i(\Delta\omega + \omega_0\sigma_{zzi})[3J_1(I_{xi} - I_{zi}) - J_2 I_{yi}]$
$\bar{\mathcal{H}}_{TD}^{(1)}$	0	0

Table 3.4 (continued)

	Four pulse cycle (WHH-4); $t_c = 6\tau$	Eight pulse cycle (MREV-8); $t_c = 12\tau$
$\overline{\mathcal{H}}_{\delta}^{(0)}$	$\frac{1}{6\tau}[(-\delta_x + \delta_{-x})I_x + (\delta_y - \delta_{-y})I_z$	$\frac{1}{6\tau}(-\delta_x + \delta_{-x})I_x$
$\overline{\mathcal{H}}_{\delta D}^{(1)}$	$\frac{1}{2}\sum_{i<j} A_{ij}[(\delta_x + \delta_{-x})(I_{yi}I_{zj} + I_{zi}I_{yj})$ $+ (\delta_y + \delta_{-y})(I_{xi}I_{yj} + I_{yi}I_{xj})]$	0
$\overline{\mathcal{H}}_{\delta O}^{(1)}$	$-\frac{1}{6}\sum_i(\Delta\omega + \omega_0\sigma_{zzi})[(\delta_x + \delta_{-x})I_{zi} + (\delta_y + \delta_{-y})I_{yi}]$	$\frac{1}{6}\sum_i(\Delta\omega + \omega_0\sigma_{zzi})[-2\delta_{-x}I_{zi} + (\delta_y - \delta_{-y})I_{yi}]$
$\overline{\mathcal{H}}_{\epsilon}^{(0)}$	0	0
$\overline{\mathcal{H}}_{\epsilon O}^{(1)}$	$-\frac{1}{3}\sum_i \epsilon_i(\Delta\omega + \omega_0\sigma_{zzi})(I_{yi} + I_{zi})$	$-\frac{1}{3}\sum_i \epsilon_i(\Delta\omega + \omega_0\sigma_{zzi})I_{zi}$
$\overline{\mathcal{H}}_{\epsilon D}^{(1)}$ [f]	$\sum_{i<j} \epsilon_i A_{ij}[I_{yi}(I_{xj} + I_{zj}) + (I_{xi} + I_{zi})I_{yj}]$	0
$\overline{\mathcal{H}}_{d}^{(0)}$ (nth cycle)[g]	$-\frac{\omega_s t_w}{3\tau}\frac{1}{b}e^{-6n/b}(2I_x + I_z)$	$\frac{\omega_s t_w}{\tau}\frac{1}{b^2}e^{-12n/b}(2I_x + I_z)$
$\mathcal{H}_{dD}^{(1)}$ (nth cycle)[g]	$\omega_s t_w[1+(-1+3/b)e^{-6n/b}\cdot\sum_{i<j} A_{ij}[I_{yi}(I_{xj}+I_{zj})+(I_{xi}+I_{zi})I_{yj}]$	$(3\omega_s t_w/b)e^{-12n/b}\cdot\sum_{i<j} A_{ij}[I_{yi}(I_{xj} + I_{zj})+(I_{xi}+I_{zi})I_{yj}]$

[a] $a = (3t_w/t_c)(4/\pi - 1)$, assuming 90° pulses (see however Sect. 3.5).

[b] Pulse rotation angle β_0 has to be adjusted according to duty factor (see Sect. 3.5).

[c] This calculation assumes all nuclei are chemically and geometrically equivalent.

[d] $\overline{\mathcal{H}}_{DO}^{(2)} = (t_c^2/648)\{-3\Delta\omega[\mathcal{H}_D^{(x)},[\mathcal{H}_D^{(y)}, I_x]] - 3\Delta\omega[\mathcal{H}_D^{(z)}, [\mathcal{H}_D^{(x)}, I_y]] + (\Delta\omega)^2[I_y + I_z, [\mathcal{H}_D^{(y)}, I_x]] + (\Delta\omega)^2[4I_x + 3I_y + I_z, [\mathcal{H}_D^{(x)}, I_y]] + i(\Delta\omega)^2[\mathcal{H}_D^{(y)}, I_x] - i(\Delta\omega)^2[\mathcal{H}_D^{(x)}, I_y] + i(\Delta\omega)^2[\mathcal{H}_D^{(x)}, I_z]\}.$

[e] $J_1 = \int_0^{t_w} \omega_T(t)(\sin\omega_1 t - \cos\omega_1 t)dt;\ J_2 = \int_0^{t_w} \omega_T(t)(\sin\omega_1 + \cos\omega_1 t)dt.$

[f] It is assumed that ω_1 is constant over a scale of molecular dimensions.

[g] $\overline{\mathcal{H}}_d^{(0)}$ (nth cycle) $= \overline{\mathcal{H}}_d^{(0)}$ $[nt_c \rightarrow (n+1)t_c]$ and similarly for $\overline{\mathcal{H}}_{dD}^{(1)}$. Exponential decay of $\omega_{rf}(t)$ with time constant $b\tau$; $b \gg 1$ is assumed.

We note again, that we are dealing here with rf pulses which are close to δ pulses. The whole procedure, however, may be readily applied to pulses of arbitrary width. In Eq. (3.170) the ideal x pulse is represented by $\omega_1 I_x$, whereas ϵ_i takes into account the deviation of the rotation angle at the i-th nucleus caused by H_1 inhomogeneity, with

$$\sum_i \epsilon_i = 0.$$

δ_x represents the pulse size misadjustment; Φ_x is a phase angle misadjustment; $\omega_I(t)$ is the difference between the in-phase component of the real rf pulse and the idealized pulse with the condition

$$\int_0^{t_w} dt\, \omega_I(t) = 0.$$

Rhim *et al.* [*68*] have investigated the effect of these different pulse imperfections on the above mentioned multiple-pulse sequences in great detail, and the interested reader is referred to Ref. [*68*]. A detailed treatment of different pulse imperfections has also been given by Haeberlen [*25*].

3.8 Resolution of Multiple-Pulse Experiments

One of the most important questions is concerned with the resolution obtainable in a given multiple-pulse experiment [*51, 60, 70, 72, 73*]. As shown in the preceeding section, the resolution of line narrowing sequences may be considerably reduced by pulse imperfections. Although all line narrowing type multiple-pulse sequences are designed to achieve vanishing average dipolar (and quadrupolar) interaction, leftover higher order correction terms, such as $\bar{\mathcal{H}}_D^{(2)}$ and other higher order terms may limit the resolution even in the case of ideal rf pulses. Which of the many possible Hamiltonian, listed in Tables 3.3 and 3.4 of the preceeding section gives the major contribution, has to be investigated separately in each case. We shall present here a general approach to the problem of resolution and discuss some representative examples. The average interaction Hamiltonian in a multiple-pulse experiment can be expressed in general as [*51*]

$$\bar{\mathcal{H}}_{\text{int}} = \mathbf{F}_0 + t_c \mathbf{F}_1 + t_c^2\ \mathbf{F}_2 + \ldots \tag{3.172}$$

where $\mathbf{F}_0$, $\mathbf{F}_1$, $\mathbf{F}_2$ etc. are the different contributions due to zeroth, first and second order correction terms according to the Magnus expansion.

The nuclear response under the action of the average interaction Hamiltonian $\bar{\mathcal{H}}_{\text{int}}$ will be a decay of magnetization $G(t)$ and an oscillation. Shift Hamiltonian do not contribute to the decay time constant T_2^*, whereas dipolar interaction and cross coupling of dipolar interactions with shift interactions including H_1 inhomogeneity are directly responsible for a decay in magnetization. $\bar{\mathcal{H}}_{\text{int}}$ in Eq. (3.172) may represent those interactions. In an ideal line narrowing experiment we usually have $F_0 = F_1 = 0$, leaving

$$\overline{\mathcal{H}}_{\text{int}} = t_c^2\ \mathbf{F}_2 \tag{3.173}$$

as the leading correction term.

According to Eq. (3.173) we could argue, that simply the time scale of the decay function is scaled by t_c^2, leading to a decay time $T_2^* \sim t_c^{-2}$. The last conclusion is wrong, although the beginning of the decay may depend on t_c^2.

In order to describe the total decay we apply the line shape theory as discussed in Section 3 and Appendix E. There we have shown that an exact derivation starting from the Liouville-v. Neumann Equation leads under certain conditions to

$$\frac{d}{dt} G(t) = -\int_0^t dt'\, K(t-t')\, G(t')$$

where the "memory function" $K(t-t')$ may be replaced by an approximate expression. Since the decay time constant T_2^* in multiple-pulse experiments is usually long compared with the cycle time t_c, we invoke the "short correlation" limit, i.e. we expect the correlation time of $K(t-t')$ to be on the order of $t_c \ll T_2^*$. In this case an exponential decay results or a Lorentzian line shape with [see Eq. (3.32a)]

$$\frac{1}{T_2^*} = \int_0^\infty d\tau K(\tau).$$

This is confirmed by experiments, i.e. the line shapes observed in multiple-pulse experiments are usually close to a Lorentzian. As shown in Section 3 Eq. (3.38) this leads to a line width δ of

$$\delta = \frac{1}{T_2^*} = \sqrt{\frac{\pi}{2}}\left[\frac{M_2}{\mu-1}\right]^{1/2} \tag{3.174}$$

where in multiple-pulse experiments, we usually find

$$\mu = M_4/M_2^2 \gg 1.$$

Using the average interaction Hamiltonian as given by Eq. (3.172) we define the following "multiple-pulse moments"

$$M_2^* = \frac{(I_\mu | \hat{\mathcal{H}}_{\text{int}}^2 | I_\mu)}{(I_z | I_z)} \tag{3.175}$$

$$M_4^* = \frac{(I_\mu | \hat{\mathcal{H}}_{\text{int}}^4 | I_\mu)}{(I_z | I_z)}$$

where the appropriate component I_μ has to be chosen, accordingly. Now we obtain the general expression for the multiple-pulse decay time T_2^* by using Eqs. (3.174) and (3.173) as

$$\frac{1}{T_2^*} = \sqrt{\frac{\pi}{2}} \left[\frac{M_2^{*3}}{M_4^*}\right]^{1/2} \tag{3.176}$$

where $\mu - 1$ has been replaced by μ because of $\mu \gg 1$.

Let us now discuss several cases:

(i) $\overline{\mathcal{H}}_D = t_c^2 \, \mathbf{F}_2$

which is the case for example in the WHH-4, HW-8 and MREV-8 sequence on resonance.

It follows immediately that

$$M_2^* = a t_c^4 \quad \text{and} \; M_4^* = b t_c^8 \tag{3.177}$$

where

$$a = \frac{(I_\mu | \hat{\mathbf{F}}_2^2 | I_\mu)}{(I_z | I_z)}$$

and

$$b = \frac{(I_\mu | \hat{\mathbf{F}}_2^4 | I_\mu)}{(I_z | I_z)} .$$

The decay time T_2^* according to Eq. (3.176) equals in this case

$$\frac{1}{T_2^*} = \sqrt{\frac{\pi}{2}} \left[\frac{a^3}{b}\right]^{1/2} \cdot t_c^2 . \tag{3.178}$$

It is, however, well known that the t_c dependence of $1/T_2^*$ in multiple-pulse experiments ranges from t_c^2 to t_c^5 as observed experimentally (see Fig. 3.33).

It will be shown in the following that this is to be expected, if different correction terms of the average Hamiltonian are governing the decay.

The next case we are going to discuss is represented by a strong zeroth order contribution as is the case in the MW-2 and MW-4 experiment

(ii) $\overline{\mathcal{H}}_D = \mathbf{F}_0 + t_c^2 \, \mathbf{F}_2$

with $\mathbf{F}_0 = -\frac{1}{2} \mathcal{H}_{Dx}$ in the case of MW-2 and MW-4.

The $\mathbf{F}_2$ term will be usually represented by the $\overline{\mathcal{H}}_D^{(2)}$ part of the average dipolar Hamiltonian. In multiple-pulse line narrowing experiments, however, the above condition is also applicable, if cross terms between dipolar interaction and H_1

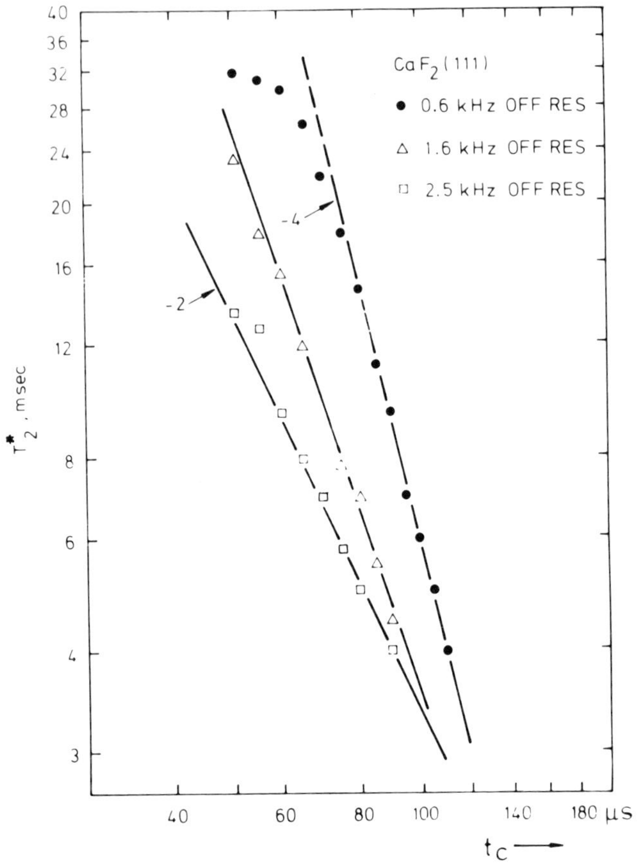

Fig. 3.33. Decay time constant T_2^* in CaF_2 observed in a MREV-8 pulse experiment versus cycle time t_c for three different resonance offsets according to W. K. Rhim, D. D. Elleman and R. W. Vaughan [*33*]

inhomogeneity, resonance-offset etc. are taken into account. Schmiedel [*72*], Ernst *et al.* [*73*] and others [*60, 70*], have discussed these effects in detail.

In general, we obtain in this case

$$M_2^* = a_0 + a_1 t_c^2 + a_2 t_c^4 \tag{3.179}$$

where

$$a_0 = (I_\mu | \hat{\mathbf{F}}_0^2 | I_\mu)/(I_z | I_z)$$

$$a_1 = (I_\mu | \hat{\mathbf{F}}_0 \hat{\mathbf{F}}_2 | I_\mu)/(I_z | I_z)$$

$$a_2 = (I_\mu | \hat{\mathbf{F}}_2^2 | I_\mu)/(I_z / I_z)$$

and

$$M_4^* = C_0 + C_1 t_c^2 + C_2 t_c^4 + C_3 t_c^6 + C_4 t_c^8 \tag{3.180}$$

where

$C_0 = \mathrm{Tr}\{k_0^2\}/\mathrm{Tr}\{I_z^2\}\,;\; C_1 = 2\,\mathrm{Tr}\{k_0 k_1\}/\mathrm{Tr}\{I_z^2\}$

$C_2 = \mathrm{Tr}\{(2\,k_0 k_2 + k_1^2)\}/\mathrm{Tr}\{I_z^2\}$

$C_3 = 2\,\mathrm{Tr}\{k_1 k_2\}/\mathrm{Tr}\{I_z^2\}$

$C_4 = \mathrm{Tr}\{k_2^2\}/\mathrm{Tr}\{I_z^2\}$

with the commutators

$k_0 = \hat{\mathbf{F}}_0^2 I_\mu$

$k_1 = (\hat{\mathbf{F}}_0 \hat{\mathbf{F}}_2 + \hat{\mathbf{F}}_2 \hat{\mathbf{F}}_0)\, I_\mu$

$k_2 = \hat{\mathbf{F}}_2^2\, I_\mu .$

Inserting M_2^* and M_4^* into Eq. (3.176) yields any t_c dependence of T_2^* up to t_c^6, depending on the relative size of the coefficients a_i and c_i.

Let us analyse in more detail the phase-alternated sequence (MW-2) for which we have

$$\mathbf{F}_0 = -1/2\,\mathcal{H}_{Dx}\,;\; \mathbf{F}_2 = \frac{1}{192}\,[(\mathcal{H}_{Dy} - \mathcal{H}_{Dx}), [\mathcal{H}_{Dy}, \mathcal{H}_{Dz}]]. \tag{3.181}$$

Because of $[\mathcal{H}_{Dx}, I_x] = 0$ we obtain $a_0 = a_1 = 0$ resulting in $M_2^* = a_2 t_c^4$.

For the same reason we have $k_0 = 0$ and $C_0 = C_1 = 0$, leaving as the leading term

$$C_2 = \mathrm{Tr}\{[\mathbf{F}_0,[\mathbf{F}_2, I_x]]^2\}\,/\mathrm{Tr}\,\{I_z^2\}$$

and

$$M_4^* = C_2 t_c^4 . \tag{3.182}$$

Inserting M_2^* and M_4^* into Eqs. (3.176) result in

$$\frac{1}{T_2^*} = \sqrt{\frac{\pi}{2}}\,[\frac{a_2^3}{c_2}]^{1/2} \cdot t_c^4 . \tag{3.183}$$

This is verified experimentally as can be seen in Figs. 3.34 and 3.35. In line narrowing multiple-pulse experiments, however, it is difficult to disentangle the different contributions of the pulse imperfections to line broadening in order to make theoretical predictions of the t_c dependence of T_2^*. The t_c dependence of T_2^* on the other hand may give some hints as to which terms contribute significantly.

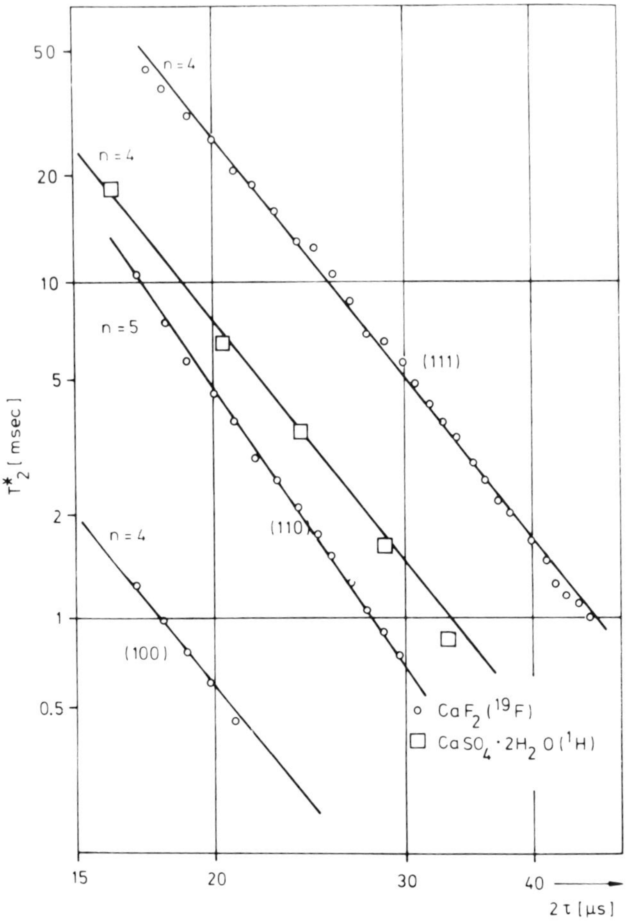

Fig. 3.34. Decay time T_2^* versus pulse spacing 2τ in different MW-4 multiple pulse experiments according to H. Ernst *et al.* [65]

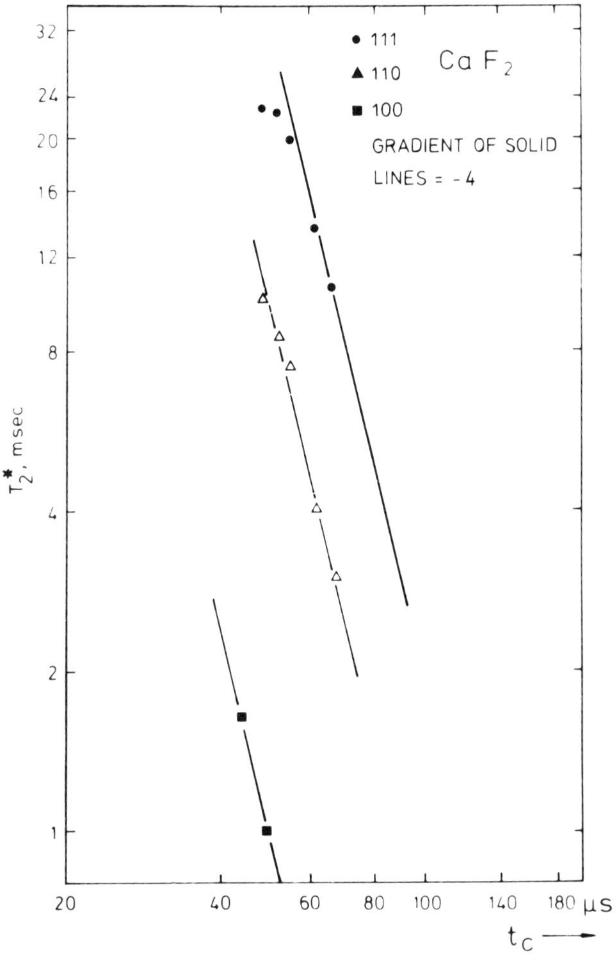

Fig. 3.35. Decay time T_2^* versus cycle time t_c in CaF_2 for different orientations of the magnetic field, observed in a MREV-8 pulse experiment (Rhim *et al.* [*33*])

3.9 Magic Angle Rotation Frame Line Narrowing Experiments

For completeness we shall briefly touch on a second class of line narrowing experiments. These experiments are equivalent to the "magic angle" specimen rotation method discussed in Section 2.5, besides that, rotation is performed in spin space. This has the advantage of leaving all anisotropies in the shift Hamiltonian unchanged and allowing, on the other hand, a very rapid "rotation" without any moving part simply by applying magnetic fields. Fig. 3.36 shows how this may be achieved. The external field Hamiltonian $\mathcal{H}_1(t)$ in the rotating frame must be proportional to

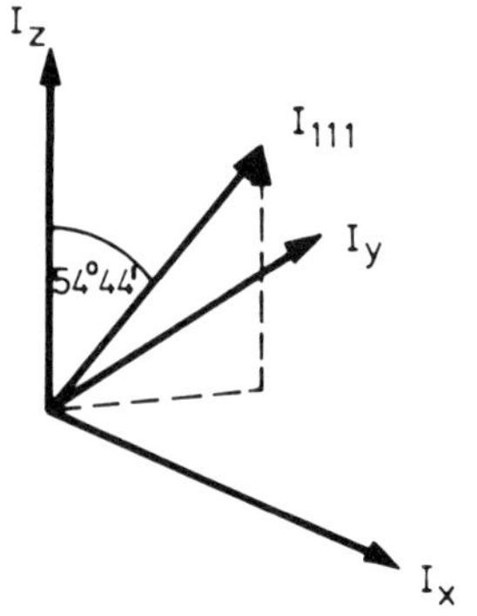

Fig. 3.36. Pictorial representation of the operator I_{111} in spin space

$I_{111} = (I_x + I_y + I_z)/\sqrt{3}$. One way of producing this field, would be to apply the H_1 field off-resonance by $\Delta\omega$ i.e.

$$\mathcal{H}_1 = \Delta\omega\, I_z + \omega_1 I_x = \omega_e(I_z \cos\vartheta + I_x \sin\vartheta) \tag{3.184}$$

where

$$\omega_e^2 = \Delta\omega^2 + \omega_1^2$$

and

$$\tan\vartheta = \frac{\omega_1}{\Delta\omega}.$$

The angle ϑ can be adjusted easily to fulfill the magic angle condition $\tan\vartheta = \sqrt{2}$.

This technique has been applied by Lee and Goldburg (LG) [*74*] to obtain a lengthened decay in a solid. Notice, however, that no phase alternation can be performed in this case, since a phase reversal of ω_1 does not reverse $\mathcal{H}_1$. Even application of the H_1 field below resonance ($-\Delta\omega$) does not correspond to a phase reversal, since the two different reference frames involved are in general not coherent. One way around this is a dc "video" field in the z direction which may be switched to $+\Delta\omega$. A convenient means of achieving this is the tilted coil arrangement as shown in Fig. 3.37. The coil makes an angle ϑ with the direction of H_0 and is excited at the same time by a "video" field H_v and a rf field H_1.

The corresponding fields in the z and x, y direction are

$$H_z = H_v \cos\vartheta; H_{x,y} = H_1 \sin\vartheta.$$

The magic angle condition can now easily be satisfied by

$$\tan\vartheta_m = \frac{H_{x,y}}{H_z} = \frac{H_1}{H_v} \tan\vartheta = \sqrt{2}.$$

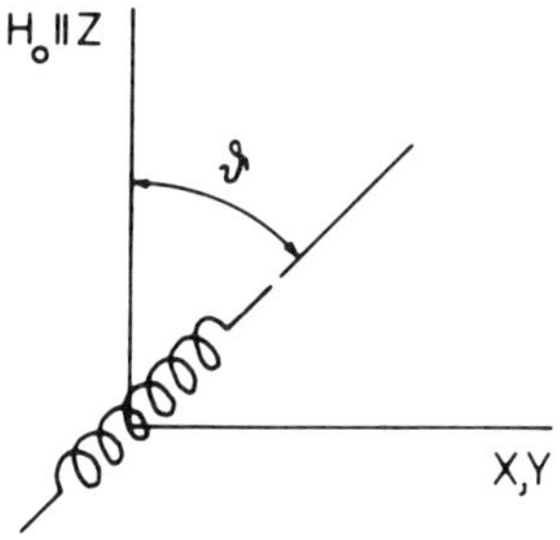

Fig. 3.37. Tilted coil arrangement for producing rf field and video field in the same coil in order to perform magic angle experiments in the rotating frame

In practice ϑ is conveniently chosen to be $\vartheta = 45^\circ$. Haeberlen and Waugh [*12*] have carried out a modification of the LG experiment by applying a train of 120° pulses under the magic angle to CaF_2, as shown in the schematic diagram in Fig. 3.38a. The one cycle propagator can be written down readily as

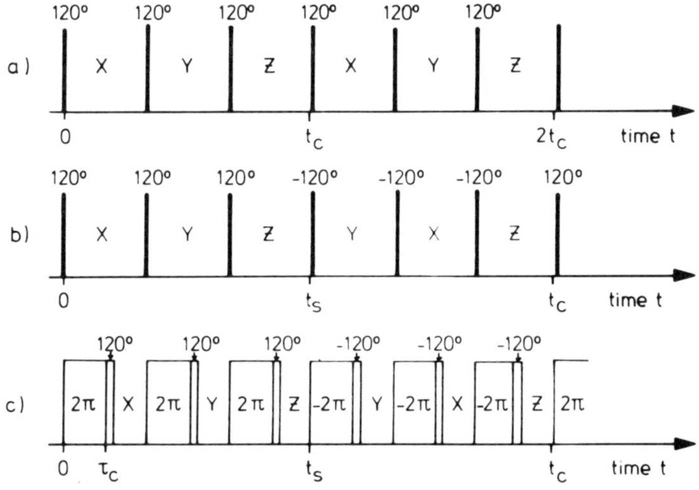

Fig. 3.38. Pulse timing for three different magic angle experiments in spin space. (a) Three pulse cycle (b) Six pulse cycle (c) Nested cycles (see Ref. [*56*])

$$\mathbf{L}(t_c) = [\mathbf{X}, \mathbf{Y}, \mathbf{Z}] \qquad (3.185)$$

with

$$\overline{\mathcal{H}}^{(0)} = \frac{1}{3}(\mathbf{X} + \mathbf{Y} + \mathbf{Z})$$

as in any line narrowing multiple-pulse experiment, i.e. $\overline{\mathcal{H}}_D^{(0)} = 0$. Thus line narrowing in solids is obtained in such a sequence and is demonstrated in Fig. 3.39. It is evident,

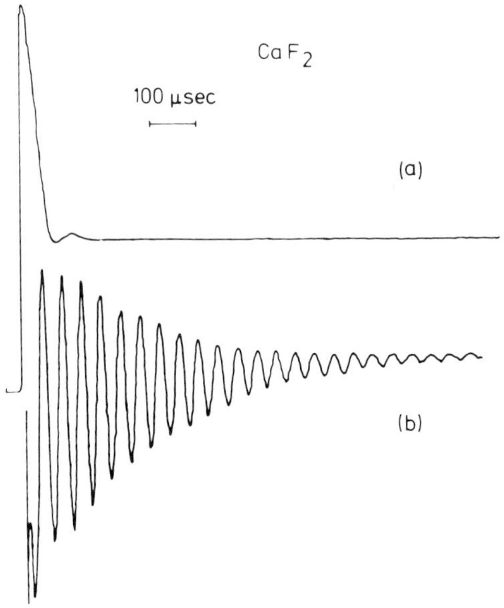

Fig. 3.39. (a) Bloch decay and (b) three pulse magic angle decay (see Fig. 3.38a) of ^{19}F in CaF_2 (U. Haeberlen and J. S. Waugh [*12*])

however, that this cycle possesses no reflection symmetry and indeed the first correction term $\bar{\mathcal{H}}^{(1)} \neq 0$ and especially $\bar{\mathcal{H}}_D^{(1)} \neq 0$, explaining the destruction of the decay in Fig. 3.39. Therefore, it has been proposed by HW [*12*] to apply phase reversed LG cycles alternatively in order to reduce the first order correction term $\bar{\mathcal{H}}^{(1)}$. This is achieved in the second sequence in Fig. 3.38b. The leading correction term to the average dipolar interaction is then $\bar{\mathcal{H}}_D^{(2)}$, which may be readily obtained like in the other multiple-pulse experiments as

$$\bar{\mathcal{H}}_D^{(2)} = \frac{\tau^2}{18} [(\mathcal{H}_{Dy} - \mathcal{H}_{Dx}), [\mathcal{H}_{Dx}, \mathcal{H}_{Dy}]]. \tag{3.186}$$

The scaling factor in these experiments is evidently $S = 1/\sqrt{3}$. Several modifications of these pulsed LG experiments have been applied to solids [*56*]. One of these sequences is schematically drawn in Fig. 3.38c. The idea behind these sequences is to implement a fast LG cycle, which corresponds to a 2π rotation around the (111)direction in the rotating frame, in front of each 120° pulse in order to reduce second order correction terms more effectively. The observation windows (X, Y, Z) are "precessing" in the sense that averaging of the dipolar interaction is obtained in the windows alone. Nesting full LG cycles between "precessing windows" in this way has been successfully applied to solids and considerable line narrowing has been achieved. A representative example is given in Fig. 3.40. For further discussion consult Ref. [*12, 56*].

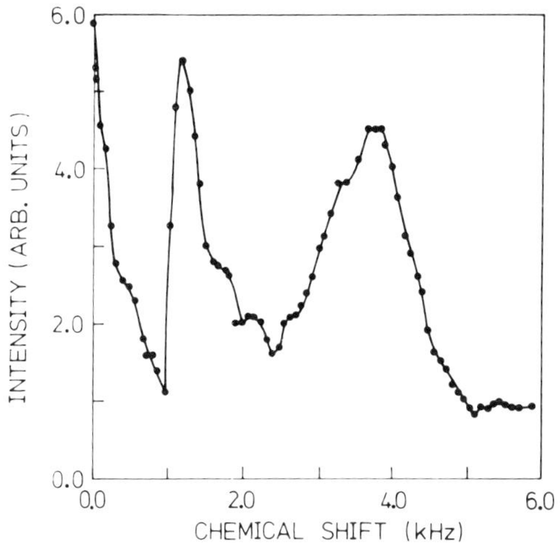

Fig. 3.40. ^{19}F spectrum observed in solid TFE/PFMVE copolymer at 54 MHz by applying a nested cycle magic angle experiment according to Fig. 3.38c (see Ref. [*56*]). The sharp peak corresponds to the rapidly rotating OCF_3-group

4. Double Resonance Experiments

There are many nuclei, namely, ^{13}C, ^{15}N etc. which can give valuable information about the electronic structure of molecules. Because of their low natural abundance their homonuclear dipolar coupling is very weak and the coupling to abundant spins such as protons can be eliminated by decoupling techniques, resulting in a high resolution spectrum.

The NMR signals to be observed, however, are extremely weak, because of (i) the low natural abundance, (ii) the small gyromagnetic ratio and (iii) usually long spin lattice relaxation times. To make an estimate on the detectability of those spins let us make the following assumptions, which are appropriate in NMR pulse experiments:

a) Signals are to be compared in a constant static magnetic field H_0.

b) The quality factor Q of the probe coil and the filling factor are constant at different frequencies.

c) The detector bandwidth is constant and less than the bandwidth of the resonance circuit.

d) The measuring time is constant.

Under these conditions the observed signal to noise ratio S/N depends on the spin quantum number I, the gyromagnetic ratio γ and the number of spins as [*4a*]

$$S/N \sim I(I+1)N\gamma^{5/2}\,.$$

We can increase S/N by relaxing assumption d) and by accumulating more and more spectra i.e. by increasing the measuring time. Let us define a normalized measuring time T_N, which is the time a measurement has to be performed to give the same signal to noise ratio as an equivalent proton sample. If we assume in addition that e) the spin lattice relaxation time T_1 is constant for all our samples, (which would be a favourable case, if we refer to the T_1 of protons), we may write

$$T_N \sim (S/N)^{-2}\,.$$

Some representative values of S/N and T_N are given in Table 4.1. There is clear evidence that it is a formidable task to detect those rare spins.

There is another class of rare spins which are neighbours to impurities, such as: (i) point defects in alkali halides (ii) paramagnetic impurities in metals (Kondo effect) (iii) paramagnetic impurities in diamagnetic crystals etc. The number of these spins depends on the number of the impurities which may be very small.

In order to detect these different types of rare spins, a wealth of double resonance techniques have been developed. The basic ideas were proposed by Hartmann and Hahn [*1*] from which different schemes are derived. Since the NMR signal of the rare spins is very weak, a gain in sensitivity can be obtained by utilizing the reservoir of abundant spins. This can be done in two ways: (i) by the indirect method [*1, 2*],

Table 4.1. Normalized signal to noise ratio S/N and normalized measuring time T_N with respect to protons for different natural abundant nuclei in the same static magnetic field

Spin	Natural abundance %	S/N	Measuring time T_N
^{1}H ($I = 1/2$)	100	1	1
^{19}F ($I = 1/2$)	100	0.858	1.36
^{31}P ($I = 1/2$)	100	0.104	92
^{39}K ($I = 3/2$)	93.08	$2.2.10^{-3}$	$2.1.10^{5}$
^{13}C ($I = 1/2$)	1.108	$3.5.10^{-4}$	$8.1.10^{6}$
^{109}Ag ($I = 1/2$)	48.65	$2.27.10^{-4}$	$2.0.10^{7}$
^{43}Ca ($I = 7/2$)	0.13	$3.2.10^{-5}$	$9.7.10^{8}$
^{15}N ($I = 1/2$)	0.365	$1.2.10^{-5}$	$7.0.10^{9}$
^{2}H ($I = 1$)	0.0156	$5.8.10^{-6}$	$3.0.10^{10}$

where the rare spins are detected via the abundant spins and (ii) by the direct method [*3*], where the rare spins are polarized by the abundant spins. Although the indirect method is of higher sensitivity in principle in the case where the rare spin spectrum contains little spectral information, the direct method has some practical advantages and has so far furnished almost all the high resolution double resonance NMR spectra in solids of rare spins.

In Section 4.1 we give a basic account on double resonance experiments, whereas, in Section 4.2 we describe the cross-polarization technique of rare spins, which has furnished most of the high resolution spectra in solids to date. In Section 4.3 we shall deal with the cross-polarization dynamics and in Section 4.4 with the spin-decoupling dynamics, which are both closely interrelated.

4.1 Basic Principles of Double Resonance Experiments

Suppose we have a system of abundant I and rare S spins i.e. $N_I \gg N_S$ where N is the number of spins. Each spin system is coupled to the lattice and approaches the lattice temperature with the spin lattice relaxation time T_{1I} and T_{1S} respectively, as shown schematically in Fig. 4.1. The I and S spins may be coupled by some inter-

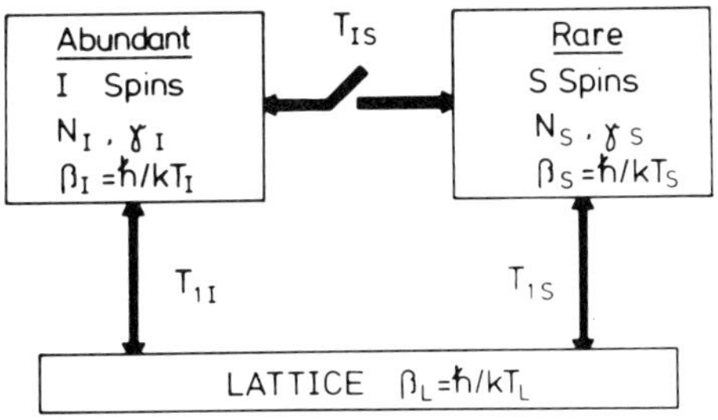

Fig. 4.1. Schematic representation of an abundant I spin reservoir and a rare S spin reservoir, which are coupled to the lattice as expressed by their spin lattice relaxation times T_{1I} and T_{1S}. The coupling between the two reservoirs as represented by the cross relaxation time T_{IS} can be varied at the experimenters discretion by a suitable application of rf fields

action represented by the cross relaxation time T_{IS}. For the basic understanding let us apply the spin temperature [*4, 5*] concept, which is discussed in great detail in the standard book by Goldman [*6*].

In the high temperature appoximation we write for the spin density matrix [*4, 6*].

$$\rho = Z^{-1}(1 - \beta \mathcal{H}) \tag{4.1}$$

where

$$Z = \mathrm{Tr}\{\mathbf{1}\} = (2I + 1)^{N_I}(2S + 1)^{N_S}$$

and with the inverse temperature

$$\beta = \hbar/kT.$$

Let us define the quantities: [*4, 6*]

$$\text{magnetization} \quad M_i = \hbar\gamma\,\mathrm{Tr}\{\rho I_i\} \quad i = x, y, z \tag{4.2a}$$

$$\text{energy} \quad E = \hbar \cdot \mathrm{Tr}\{\rho\mathcal{H}\} = -\beta\hbar \mathrm{Tr}\, \mathcal{H}^2 \tag{4.2b}$$

$$\text{entropy} \quad S = -k\,\mathrm{Tr}\{\rho \log \rho\}. \tag{4.2c}$$

In the case of Zeeman interaction, with

$$\mathcal{H} = -\gamma H \cdot I_i$$

these quantities reduce to

$$M_i = \beta C \cdot H, \tag{4.3a}$$

$$E = -\beta \cdot C \cdot H^2, \tag{4.3b}$$

$$S = \mathrm{const} - k\beta^2 C H^2, \tag{4.3c}$$

where $C = \frac{1}{3} \cdot NI(I + 1)\gamma^2 \hbar$.

After waiting several time T_{1I}, T_{1S} when placing a sample into the static magnetic field H_0, the I and S spins reach the magnetization

$$M_{0I} = \beta_L C_I H_0 \quad \text{and} \quad M_{0S} = \beta_L\, C_S\, H_0$$

where β_L is the inverse lattice temperature.

The first step in double resonance consists now in

(i) Cooling of the abundant I *spin system*

This can be achieved e.g. by locking the spins in a field $H_I \ll H_0$. In this case

$$M_{0I} = \beta_L \cdot C_I \cdot H_0 \quad = \beta_I \cdot C_I \cdot H_I \tag{4.4}$$

with

$$\beta_I/\beta_L = H_0/H_I \gg 1$$

or

$$T_I \ll T_L .$$

There are three basic schemes for achieving a cooling of a spin system, which are represented in the schematical diagram Fig. 4.2.

These are

a) Spin locking [1–7] in the rotating frame with a field $H_{1x} = 2H_{1I}\cos\omega_{0I} \cdot t$, which results with ($H_{1I} \gg H'_{LI}$) in

$$\beta_I = \frac{H_0}{H_{1I}} \beta_L ; M_I = \beta_I \cdot C_I \cdot H_{1I}; E = -\beta_I\, C_I\, H_{1I}^2 . \tag{4.5}$$

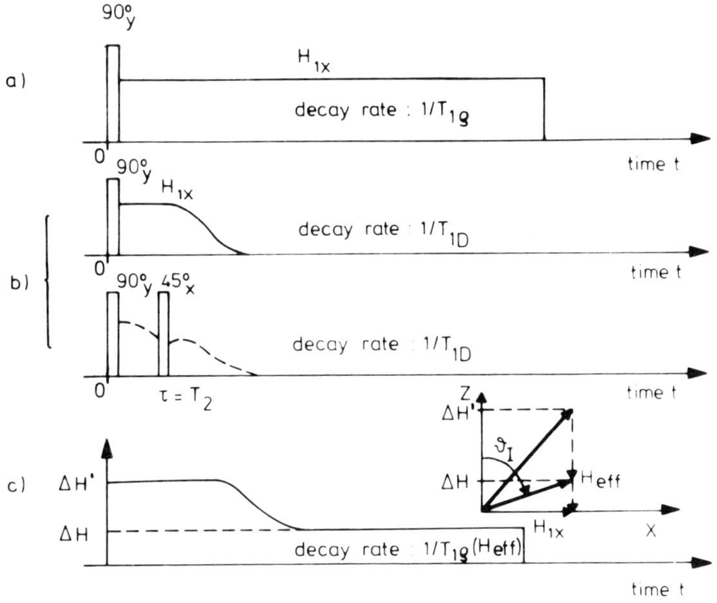

Fig. 4.2 a–c. Schematic representation of different means for reducing the spin temperature in the rotating frame. (a) Spin locking, (b) adiabatic demagnetization in the rotating frame (ADRF) and Jeener-Broekaert [9] two pulse experiment, (c) demagnetization in an effective field by sweeping on resonance with a small rf field

The inverse spin temperature β_I will approach the inverse lattice temperature β_L with the time constant $T_{1\rho}$, the spin lattice relaxation time in the rotating frame.

b) Adiabatic demagnetization in the rotating frame (ADRF) [8] by turning off the H_{1X} field adiabatically, leaving the spins in the "dipolar field" H'_{LI}, where

$$\mathrm{Tr}\{\mathcal{H}'^2_D\} = \gamma^2 H'^2_L \,\mathrm{Tr}\{I_z^2\}$$

$$\beta_I = \frac{H_0}{H'_{LI}}\,\beta_L;\; E_I = -\beta_I\, C_I\, H'^2_{LI}. \tag{4.6}$$

The inverse spin temperature β_I will approach the inverse lattice temperature β_L with the time constant T_{1D}, the spin lattice relaxation time of the dipolar state.

A different way of achieving a dipolar state was proposed by Jeener and Brokaert, [9], applying a $90^\circ_y - \tau - 45^\circ_x$ pulse sequence. Since this process is of course not adiabatic, a somewhat smaller inverse temperature β_I is achieved, therefore, [6]

$$\beta_I \cong 0.525\,\frac{H_0}{H'_{LI}}\,\beta_L. \tag{4.7}$$

Nevertheless, this is a very convenient method for cooling the abundant spin system.

c) Adiabatic passage and stop at ΔH. In this case the spins are locked in an effective field $H^2_{\mathrm{eff}} = \Delta H^2 + H^2_{1x}$ which makes an angle $\vartheta_I = \arctan(H_{1x}/\Delta H)$ with the z direction.

$$\beta_I = \frac{H_0}{H_{\mathrm{eff}}}\,\beta_L;\, M_I = \beta_I \cdot C_I \cdot H_{\mathrm{eff}};\, E_I = -\beta_I \cdot C_I \cdot H^2_{\mathrm{eff}}. \tag{4.8}$$

The next step ist

(ii) Bring I *and* S *spins into contact*

Since the I spins are cold and the S spins are hot, there will be a calorimetric effect [2] and energy exchange may proceed with the time constant T_{IS}. Only if $T_{IS} \ll T_{1DI}, T_{1\rho I}$ this energy exchange is considerable and can be utilized for a double resonance effect. Rapid energy transfer is possible only under total energy conservation. No such transfer is possible in the laboratory frame. However, in the rotating frame a matching of the energy levels is possible as shown schematically in Fig. 4.3, allowing rapid transfer under energy conservation (in the rotating frame) if the Hartmann-Hahn [1] condition

$$\gamma_S H_{1S} = \gamma_I H_{1I} \text{ for } I = S = 1/2 \tag{4.9}$$

or

$$\omega_{1S} = \omega_{1I}$$

is fulfilled, where H_{1S} and H_{1I} are the rf fields in the rotating frame of the I and S

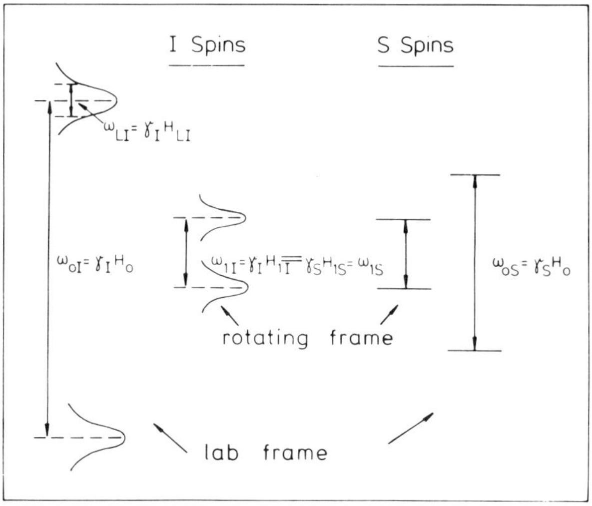

Fig. 4.3. Pictorial representation of level matching in the rotating frame (Hartmann-Hahn condition) for spin 1/2 systems

spins respectively. Of course, these fields may be effective fields, or correspondingly "effective frequencies" ω_{eS} and ω_{eI}.

As will be shown in more detail in Section 4.3, the transfer rate can be expressed as [*10,11*]

$$\text{ADRF case: } \left(\frac{1}{T_{IS}}\right)_{\text{ADRF}} = \sin^2 \vartheta_s M_2^{IS}\, J_{\text{ADRF}}(\omega_{eS}) \tag{4.10}$$

and in the

$$\text{spin locking case [11] } \left(\frac{1}{T_{IS}}\right)_{SL} = \frac{1}{2} \sin^2 \vartheta_S \sin^2 \vartheta_I M_2^{IS}\, J_{SL}(\Delta\omega_e)$$

$$\omega_{eS}, \omega_{eI} \gg \omega_{LI} \tag{4.11}$$

where

$$\tan \vartheta_S = H_{1S}/\Delta H_S \text{ and } \tan \vartheta_I = H_{1I}/\Delta H_S$$

$$\Delta\omega_e = \omega_{eS} - \omega_{eI}$$

and where M_2^{IS} is the second moment of the *I-S* coupling Hamiltonian. The spectral distribution function for the cross relaxation process $J(\omega)$ decreases monotonically to zero for increasing ω, i.e. with increasing mismatch of the Hartmann-Hahn condition. This results in a drastic decrease of the transfer-rate $1/T_{IS}$, as is expected. McArthur, Hahn and Walstedt [*10*] have used an intuitive approach, based on the experimental data to express the functional form of $J_{\text{ADRF}}^{(\omega_{eS})}$ as

$$J_{\text{ADRF}}(\omega_{eS}) = \frac{1}{2}\tau_c \exp(-\omega_{eS}\tau_c) \tag{4.12}$$

where τ_c is the correlation time. A more general approach using the memory function technique was applied by Demco, Tegenfeldt and Waugh [*11*] and will be discussed in more detail in Section 4.3.

The spin temperature exchange occuring in a single *I-S* contact is shown schematically in Fig. 4.4. We suppose again a high inverse spin temperature β_I at the begin-

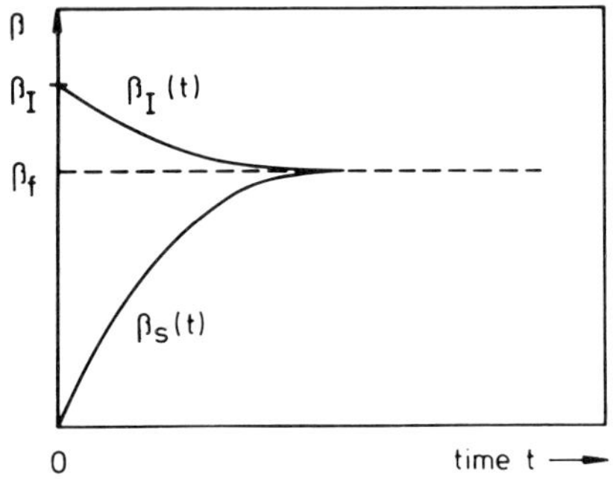

Fig. 4.4. Time evolution of *I* and *S* spin inverse temperatures β_I and β_S, when the spin systems are brought into contact at $t = 0$. Initial condition: $\beta_I \neq 0$; $\beta_S = 0$. A final inverse spin temperature β_f is reached after several T_{IS}

ning of the contact ($t = 0$) and a zero inverse temperature β_S at $t = 0$. If we neglect spin lattice relaxation, both spin temperatures will finally reach the same value β_f, assuming exponential relaxation as

$$\beta_I(t) = (\beta_I - \beta_f)\, e^{-t/T_{IS}} + \beta_f, \tag{4.13}$$

$$\beta_S(t) = \beta_f(1 - e^{-t/T_{IS}}). \tag{4.14}$$

Assuming energy conservation

$$\beta_I \cdot C_I \cdot H_I^2 + \beta_S\, C_S \cdot H_S^2 = \beta_f[C_I H_I^2 + C_S H_S^2] \tag{4.15}$$

and with $\beta_S = 0$, Eq. (4.15) leads to

$$\frac{\beta_f}{\beta_I} = \frac{1}{1 + \epsilon'}, \tag{4.16}$$

where

$$\epsilon' = \frac{C_S H_S^2}{C_I H_I^2} \tag{4.17}$$

is the ratio of the heat capacities of the *S* and *I* spins.

If the Hartmann-Hahn condition $\gamma_S H_S = \gamma_I H_I$ is fulfilled, we obtain

$$\epsilon = \frac{N_S S(S+1)}{N_I I(I+1)} \ll 1. \tag{4.18}$$

With $M_I^{(f)} = \beta_f \cdot C_I \cdot H_I$ and $M_S^{(f)} = \beta_f \cdot C_S \cdot H_S$ the final I and S spin magnetization reaches

$$M_I^{(f)} = \frac{1}{1+\epsilon} \; M_I^{(i)} \tag{4.19}$$

and

$$M_S^{(f)} = \frac{\gamma_I}{\gamma_S} \cdot \frac{1}{1+\epsilon} \cdot M_{0S} \tag{4.20}$$

where $M_I^{(i)}$ is the initial magnetization of the I spins and M_{0S} is the Zeeman magnetization of the S spins. Since ϵ is a very small number of the order of N_S/N_I we may write $1/(1+\epsilon) \cong 1 - \epsilon$, i.e. according to Eqs. (4.19) and (4.20) the I spin magnetization does not decrease very much in a single contact, however, the S spin magnetization may have been increased if $\gamma_I/\gamma_S > 1$.

In order to achieve a noticable destruction of the I spin magnetization, multiple contacts [*2, 3*] have to be performed as demonstrated in Fig. 4.5 The I spins are spin-locked in the field H_{1I}, whereas, the S spins become polarized in the field H_{1S} which may be an effective field in the rotating frame. The pulsed H_{1S} field is of duration t_w with a spacing τ_i. Coupling between the I and S spins is achieved, when the H_{1S} field is turned on and I and S are decoupled consecutively, then H_{1S} is turned off. A free precession of the S spins can be observed during this time.

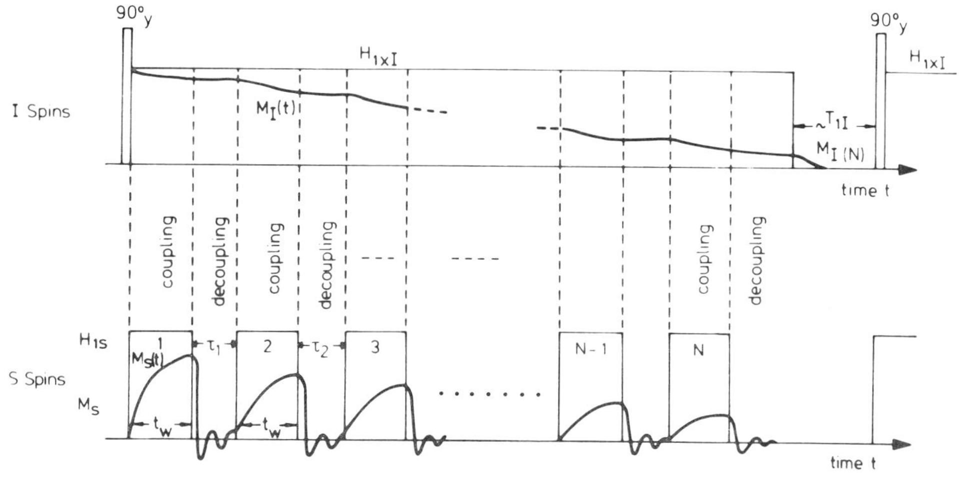

Fig. 4.5. Pulse timing of a typical double resonance experiment in the rotating frame

After the k-th contact we may write for the I spin magnetization [1–3, 12]

$$M_I(k) = \left(\frac{1}{1+\epsilon}\right)^k M_{0I} \cong (1-\epsilon)^k M_{0I} \tag{4.21}$$

or

$$M_I(k) \cong \exp(-k\epsilon)\, M_{0I} \tag{4.22}$$

and for the S spin magnetization

$$M_S(k) = \frac{\gamma_I}{\gamma_S}\,(1-\epsilon)^k\, M_{0S} \cong \frac{\gamma_I}{\gamma_S}\,\exp(-k\epsilon) M_{0S} \tag{4.23}$$

under the condition that the S spin inverse temperature $\beta_S = 0$ at the beginning of each contact. If we sum up all S signals we may write after N contacts

$$M_{SN} = \sum_{k=1}^{N} M_S(k) = \frac{\gamma_I}{\gamma_S}\, M_{0S} \sum_{k=1}^{N} (1-\epsilon)^k \tag{4.24}$$

or to a good approximation [3, 12]

$$M_{SN} \cong \frac{\gamma_I}{\gamma_S}\, M_{0S} \sum_{k=1}^{N} e^{-k\epsilon}. \tag{4.25}$$

Thus we distinguish between the indirect method, where the I spin magnetization is observed after N contacts and the direct method, where the S spin signals are accumulated.

We are now going to relax the condition $\beta_S = 0$ at the beginning of each contact and discuss the general case. In the direct method, we are interested in $M_I(N)/M_{0I}$ after N contacts i.e.

$$\frac{M_I(N)}{M_{0I}} = \frac{\beta_{NI}}{\beta_{0I}} \tag{4.26}$$

where

$$\beta_{0I} = \beta_L \cdot \frac{H_0}{H_{1x}}.$$

Several authors have discussed the behavior of I and S spin magnetization in a multiple contact double resonance experiment [2, 3, 12–14]. We shall follow here the discussion given by H. Ernst [13] which we find particularly useful for a unified description of all rotating frame double resonance experiments. We shall use the following notation

$\beta_{kI}^{(i)}$: initial inverse temperature of the I spins at the beginning of the k-th H_{1S} pulse,

$\beta_{kI}^{(f)}$: final inverse temperature of the I spins after the k-th H_{1S} pulse.

The expression "temperature" is not to be taken too seriously in this context. We simply use it as a thermodynamic parameter which describes the evolution of the energy as the only constant of the motion [6] (see also Section 4.3).

We may write:

$$\beta_{(k+1)I}^{(i)} = \beta_{kI}^{(f)} \text{ no contact between } I \text{ and } S \text{ spins during } \tau_k \tag{4.27}$$

$$\beta_{(k+1)S}^{(i)} = \beta_{kS}^{(f)} G(\tau_k) \tag{4.28}$$

where $G(\tau)$ is the f.i.d. of the S spins. We shall write in short form notation $G(\tau_k)$ $G(\tau_k) = g_k$ with $|g_k| \leq 1$.

Energy conservation demands

$$\beta_I^{(i)} + \epsilon\beta_S^{(i)} = \beta_I^{(f)} + \epsilon\beta_S^{(f)}. \tag{4.29}$$

If we further assume complete spin temperature exchange within each contact ($t_w \gg T_{IS}$) i.e.

$$\beta_{kI}^{(f)} = \beta_{kS}^{(f)} \tag{4.30}$$

we can combine Eqs. (4.27)–(4.30) to obtain

$$\beta_{(k+1)I}^{(f)} = \beta_{kI}^{(f)} \frac{1 + \epsilon g_k}{1 + \epsilon} \tag{4.31}$$

or

$$\beta_{NI}^{(f)} = \beta_{1I}^{(f)} \cdot \prod_{k=1}^{N-1} \left(\frac{1 + \epsilon g_k}{1 + \epsilon}\right). \tag{4.32}$$

Let us keep in mind, that we are looking for $\beta_{NI}^{(f)} / \beta_{0I}$ according to Eq. (4.26) in order to obtain $M_I(N)/M_{0I}$. Using total energy conservation [Eq. (4.29)] and complete spin temperature exchange [Eq. (4.30)], we arrive at

$$\frac{\beta_{1I}^{(f)}}{\beta_{1I}^{(i)}} = \frac{1 + \lambda\epsilon}{1 + \epsilon} \tag{4.33}$$

where

$$\lambda = \beta_{1S}^{(i)} / \beta_{1I}^{(i)} = \frac{H_{1xI}}{H_{\text{eff}S}} \cos\vartheta_S. \tag{4.34}$$

Combining Eqs. (4.26, 4.32, 4.33) leads to

$$\frac{M_I(N)}{M_{0I}} = \frac{1 + \lambda\epsilon}{(1 + \epsilon)^N} \prod_{k=1}^{N-1} (1 + \epsilon g_k). \tag{4.35}$$

With $|g_k| \leqslant 1; \epsilon \ll 1; \lambda \leqslant 1$ and $(1+\epsilon)^{-N} \cong \exp(-N\epsilon)$ we may write

$$\frac{M_I(N)}{M_{0I}} = e^{-N\epsilon} \prod_{k=1}^{N-1} (1 + \epsilon g_k). \tag{4.36}$$

A further approximation is obtained by

$$\ln(M_I(N)/M_{0I}) = -N\epsilon + \sum_{k=1}^{N-1} \ln(1+\epsilon g_k) \cong -N\epsilon + \sum_{k=1}^{N-1} \epsilon g_k$$

or

$$M_I(N) = M_{0I}\, e^{-\delta} \tag{4.37}$$

where the damping constant δ for N different I-S contacts equals

$$\delta = \epsilon\left[1 + \sum_{k=1}^{N-1} (1 - g_k)\right]. \tag{4.38}$$

The quantity, still to be determined, is g_k.

In order to get a better understanding of the motion of the S spin magnetization verctor, we follow here the vector representation according to H. Ernst [*13*], as drawn schematically in Fig. 4.6 The notation is self-explanatory and we realize that the magnetization $M_{kS}^{(f)}$ at the end of the k-th contact pulse is polarized, along the effective field $H_{\mathrm{eff}S}$ and has the component $M_{kS}^{\perp}$ in the direction of H_{1xS} and the component $M_{kS}^{\parallel}$ parallel to ΔH_S. If the H_{1xS} field is turned off for the time τ_k, only the component $M_S^{\perp}$ is changed to the new value $M_{(k+1)S}^{\perp}$ at the beginning of the $(k+1)$-th contact pulse.

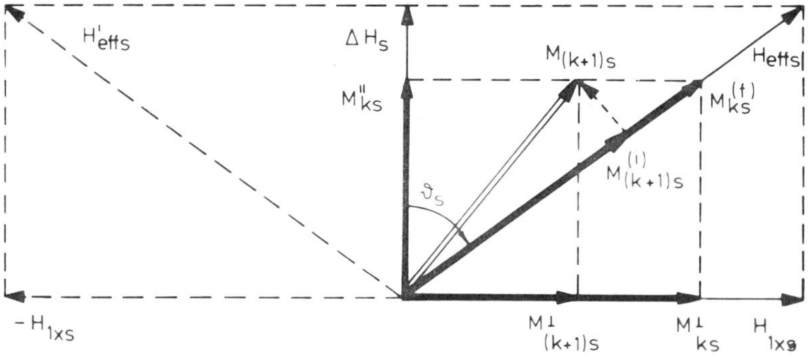

Fig. 4.6. Vector representation of the S spin magnetization at different times during a double resonance experiment according to Fig. 4.5 (H. Ernst [*13*])

The total S spin magnetization $M_{(k+1)S}$ is no longer parallel to the effective field $H_{\text{eff}S}$ and the initial magnetization $M^{(i)}_{(k+1)S}$ is given by the projection of $M_{(k+1)S}$ onto the effective field $H_{\text{eff}S}$. We readily realize, that

$$M^{\perp}_{(k+1)S} = M^{\perp}_{kS}\, g_{xk} \tag{4.39}$$

where

$$g_{xk} = G(\tau_k)\cos\Delta\omega_S\tau_k . \tag{4.40}$$

The quantity wanted is

$$g_k = M^{(i)}_{(k+1)S}/M^{(f)}_{kS}$$

which can be written according to Fig. 4.6 as

$$g_k = \cos^2\vartheta_S \pm \sin^2\vartheta_S g_{xk} \tag{4.41}$$

where the minus sign stands for phase alternation of H_{1xS}. We may now write the damping constant according to Eqs. (4.37) and (4.38) as [*13*]

$$\delta = \epsilon[1 + \sin^2\vartheta_S \sum_{k=1}^{N-1} (1 \mp g_{xk})]. \quad \begin{matrix}\text{“–” no phase alternation}\\ \text{“+” phase alternation}\end{matrix} \tag{4.42}$$

Equation (4.42) is the basic equation which describes virtually all indirect double resonance methods as will be discussed in the following:

a) $\tau_k = 0$ i.e. $g_{xk} = 1$

$$\delta = \begin{cases} \epsilon & \text{no phase alternation, practically no destruction of } I \text{ magnetization, since } \epsilon \ll 1. \\ 2\epsilon N\sin^2\vartheta_S + \underbrace{\epsilon\cos^2\vartheta_S}_{\cong 0}. & \text{phase alternation.} \end{cases} \tag{4.43}$$

With phase alternation of H_{1xS} we arrive at the Hartmann-Hahn experiment [*1*] where $\delta = 2\epsilon N$ if irradiation is applied on resonance. The observed line width is determined by the influence of $\sin^2\vartheta_S$ and the increase of T_{IS} by mismatching the Hartmann-Hahn condition. The line width is typically a few kHz, whereas, the natural line width of the S spins may be a few Hertz. This method has been extensively applied to the investigation of point defects in cubic crystals [*14*].

b) $\tau_k \gg T^*_{2S}$ i.e. $g_{xk} = 0$

$$\delta = \epsilon N\sin^2\vartheta_S + \underbrace{\epsilon\cos^2\vartheta_S}_{\cong 0}. \tag{4.44}$$

On resonance ($\vartheta_S = 90^\circ$) this corresponds to the Lurie-Slichter [*2*] experiment with $\delta = \epsilon N$. The line width is the same as in a).

c) $\tau_k = \tau = \text{const}$ i.e. $g_{xk} = G(\tau)$ if $\Delta\omega_S = 0$.

It follows

$$\delta = \epsilon N(1 - G(\tau)) - \underbrace{\epsilon G(\tau)}_{\cong 0} \tag{4.45}$$

which corresponds to the experiment of McArthur, Hahn and Walstedt [*10*], who mapped out the S spin f.i.d. $G(\tau)$ by varying τ with $0 \leqslant \tau \leqslant T_{2S}^*$.

McArthur [*10*] *et al.*, however, applied ADRF instead of spin locking to the I spins, that is why no decoupling of the I spins was obtained. The S spin f.i.d. is determined by the I-S dipolar coupling mainly in this case.

d) same as c) but with a small deviation from resonance

$\Delta\omega_S \neq 0$.

It follows

$$\delta = \epsilon N(1 - G(\tau)\cos\Delta\omega_S\tau) \tag{4.46}$$

where δ is now a function of $\Delta\omega_S$. This represents the experiment performed by Mansfield [*12*] and co-workers, which allows high resolution spectroscopy to be performed by the indirect method.

In the special case

$$n \cdot 2\pi + \pi/2 < \Delta\omega_S\tau \leqslant n2\pi + \frac{3}{2}\pi$$

it follows $|\delta| > \epsilon N$ and the double resonance process becomes very effective. The spectrum $I(\Delta\omega_S)$ of the S spins can be obtained by summing the I signal for different values of τ as

$$\sum_{\tau=\tau_0}^{\tau=n\tau_0}(\delta + \epsilon N) \cong \frac{N_\epsilon}{\tau_0}\int_0^\infty d\tau G(\tau)\cos\Delta\omega_S\tau = \frac{N_\epsilon}{\tau_0} I(\Delta\omega_S).$$

A high resolution spectrum of ^{13}C in adamantane was obtained by Mansfield [*12*] and co-workers with these techniques, as shown in Fig. 4.7. Figure 1.5 in the introduction shows the same spectrum obtained by the direct method of Pines [*3*] *et al.*, to be discussed in Section 4.2.

For completeness we want to discuss briefly some *steady state methods.* Suppose the I spins are in a spin locked state, i.e., they are kept at a low spin temperature. The S spins approach the magnetization

$$M_{0S} = \beta_I \cdot C_S \cdot H_{\text{eff}S}$$

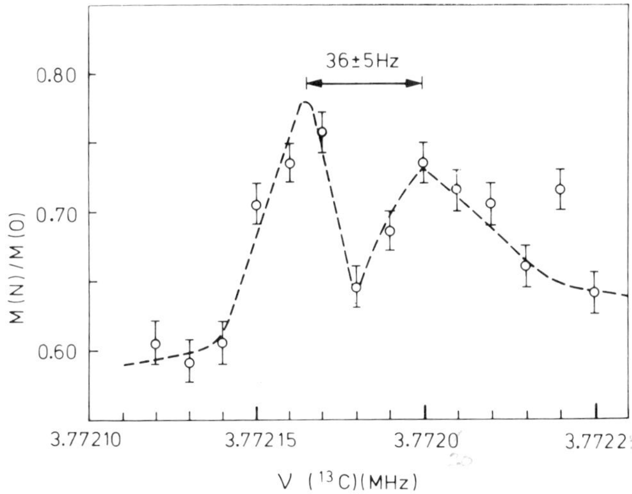

Fig. 4.7. ^{13}C spectrum of solid adamantane obtained by Mansfield and co-workers [*12*] by a high resolution proton destruction double resonance technique as described in the text

parallel to H_{effS} with the rate $1/T_{IS}$. In addition another S spin saturation field $H_{1SS}^{(t)} = 2 \cdot H_{SS} \cos \omega_{SS} \cdot t$ is applied to the S spins as shown vectorially in Fig. 4.8.

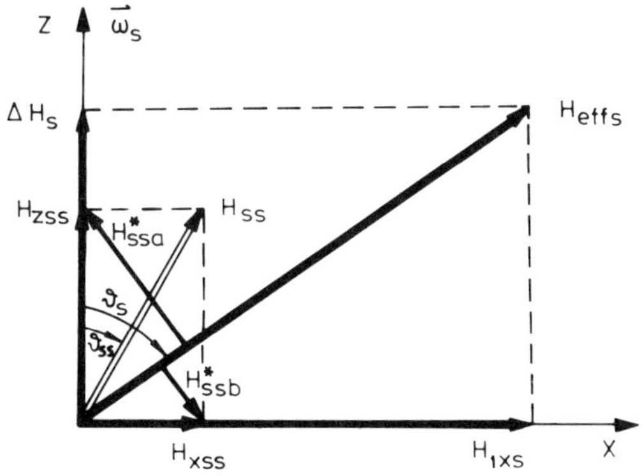

Fig. 4.8. Vector representation of rf field applied in *cw* double resonance experiments (H. Ernst [*13*])

The angle between the z axis and H_{1SS} may be labelled ϑ_{SS}, whereas the angle between the z axis and H_{effS} may be called ϑ_S as usually. The power absorbed from the saturating field may be expressed as (H. Ernst [*13*])

$$P_S = -\frac{\omega_{es} M_{0S}}{\gamma_S T_{IS}[1 + (\frac{\Delta\omega_{SS}}{\gamma_S H_{SS}^*})^2]} \tag{4.47}$$

where $\Delta\omega_{SS}$ is the deviation from resonance and H_{SS} an effective field according to Fig. 4.8 to be explained later.

The power P_S absorbed by the S spins is transferred to the I spins and changes their magnetization as

$$P_S = \frac{dM_I}{dt} \cdot H_{1xI}. \tag{4.48}$$

With

$$M_I = \beta_I \cdot C_I \cdot H_{1xI}$$

we obtain [*13*]

$$\delta(\Delta\omega_{SS}) = \ln M_I/M_{0I} = \epsilon' \frac{T_{1I}}{T_{IS}} \frac{1}{[1 + (\frac{\Delta\omega_{SS}}{\gamma_S H_{SS}^*})^2]} \tag{4.49}$$

where we remind the reader of

$$\epsilon' = \frac{C_S H_{\text{eff}S}^2}{C_I H_{1xI}^2}.$$

$\delta(\Delta\omega_{SS})$ is a Lorentzian with the half width at half height

$$\Delta\omega = \gamma_S H_{SS}^* < 1/T_{IS}.$$

$H_{1SS}(t)$ may be expressed in the ω_S rotating frame by the vector

$$H_{1SS}(t) = H_{SS}\{\sin\vartheta_{SS}\cos(\omega_{SS} - \omega_S)t, \sin\vartheta_{SS}\sin(\omega_{SS} - \omega_S)t,$$

$$2\cos\vartheta_{SS}\cos\omega_{SS}\cdot t\}. \tag{4.50}$$

Two possible resonances can occur:

(i) $\omega_{SS} = \omega_{eS}$ i.e. $H_{SSa}^* = H_{zSS}\sin\vartheta_S = H_{SS}\cos\vartheta_{SS}\sin\vartheta_S$.

The saturating field H_{SS} induces transitions in the effective field $H_{\text{eff}S}$ i.e., it is applied in the audio frequency range. Such an experiment was performed by Walstedt *et al.* [*15*] and McArthur *et al.* [*10*] by applying an audio-field in the z-direction $\vartheta_{SS} = 0$; $H_{SSa}^* = H_{SS}\sin\vartheta_S$ at the frequency $\omega_{SS} = \omega_{1S} = \gamma_S H_{1S}$.

For $\omega_{SS} > \omega_{1S}$, but $\omega_{SS} = \omega_{eS}$ two resonance minima are observed with varying $\Delta\omega_S$. The condition for the two minima is readily obtained as

$$\Delta\omega_S = \pm(\omega_{SS}^2 - \omega_{1S}^2)^{1/2}. \tag{4.51}$$

Because the Hartmann-Hahn condition is fulfilled, T_{IS} is short ($\sim$1 m s) and the resulting line width is about 1 kHz.

(ii) $\omega_{SS} = \omega_S + \omega_{eS}$ i.e. $H^*_{SSb} = H_{xSS}\cos\vartheta_S = H_{SS}\sin\vartheta_{SS}\cos\vartheta_S$.

Such an experiment was performed by Bleich and Redfield [*16, 17*] where H_{SS} lies in the x-direction ($\vartheta_{SS} = 90°$) and $H^*_{SSb} = H_{SS}\cos\vartheta_S$. Bleich and Redfield used $\vartheta_S \cong 5°$ to $10°$ by applying the H_{1xS}-field far off resonance. Already a relative small H_{1xS} produces a large $H_{\text{eff}S}$ in order to fulfil the Hartmann-Hahn condition. T_{IS} on the other hand becomes very long, because of the factor $\sin^2\vartheta_S$, which in this case is $\sin^2\vartheta_S \cong 10^{-2}$.

This leads to a high resolution spectrum, as demonstrated in Fig. 4.9. However, the observed signal is proportional to the quantity $I(\omega)/T_{IS}$ i.e., a strong deformation of the spectrum results, since T_{IS} varies with ϑ_{IS} which is the angle of the I-S internuclear vector with respect to the z axis. If ϑ_{IS} equals the magic angle, the next nearest neighbour coupling of ^{13}C to ^{1}H in benzene vanishes and T_{IS} becomes very large, leading to a hole in the spectrum at the corresponding frequency (see Fig. 4.9). These disturbing effects are absent in the direct method of Pines *et al.* [*3*] unless they are produced on purpose.

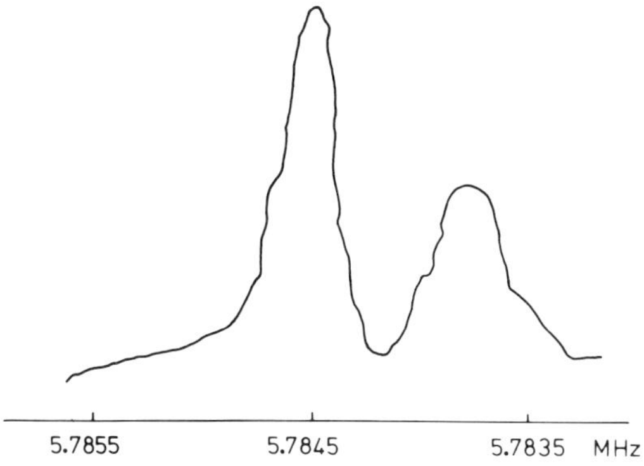

Fig. 4.9. ^{13}C double resonance spectrum of polycrystalline C_6H_6 at $-40°$ C, obtained by C. S. Yannoni and H. E. Bleich [*17a*] with a *cw* double resonance technique of the Bleich-Redfield [*16*] type

4.2 Cross-Polarization of Dilute Spins

As we have seen in the preceeding section, the I spin magnetization decreases, whereas, the S spin magnetization increases during an I-S contact, given the I spins are prepared in a low spin temperature state. Pines, Gibby and Waugh [*3*] have designed a technique to utilize this fact, in order to obtain high resolution spectra of rare spins in solids. They have named their method: "Proton Enhanced Nuclear Induction Spectroscopy" (PENIS). The timing of this technique is the same as schematically shown in Fig. 4.5, where the pulse spacing τ is large to allow the S spin magnetization to decrease fully to zero. The S spin magnetization between the pulses is observed and successive S spin f.i.d.'s are accumulated in an online computer and Fourier transformed to obtain the S spin spectrum. Since the I-field H_{1xI} is kept on during the S spin f.i.d. the abundant I spins are decoupled, resulting in a high resolution spectrum of the S spins. Figure 4.10 shows some representative ^{13}C spectra obtained by this method.

The S spin magnetization after the k-th contact can be expressed according to Eq. (4.23) as [3]

$$M_S(k) = \frac{\gamma_I}{\gamma_S} e^{-k\epsilon} M_{0S}$$

where

$$\epsilon = \frac{N_S(S+1)}{N_I I(I+1)}$$

since the Hartmann-Hahn condition is fulfilled in this experiment. For the same reason the contact time t_w is on the order of milliseconds, whereas, τ may be on the order of a few hundred milliseconds, depending on the wanted spectral resolution. The decrease of I spin magnetization due to $T_{1\rho}$ is not taken into account in this discussion, it may, however, spoil the whole beauty of this experiment, when very high resolution is demanded and also slow motions are present. The total coadded magnetization after N contacts may be expressed according to Eq. (4.25) as

$$M_{SN} = \frac{\gamma_I}{\gamma_S} M_{0S} \sum_{k=1}^{N} e^{-k\epsilon}.$$

In order to estimate the sensitivity of this method by comparing the enhanced S spin signal with the ordinary Bloch decay, we assume a given quality factor Q of the probe coil, a constant filling factor and a constant detector bandwidth, producing a constant detector noise, all represented by the constant K. The signal to noise ratio equals

$$S/N = K \cdot \omega_0^{1/2} \cdot M$$

under these conditions as outlined in Section 4.

The signal to noise ratio of the f.i.d. of the S spins may now be expressed by

$$(S/N)_{\text{fid}} = K\omega_{0S}^{1/2} \cdot M_{0S} \tag{4.52}$$

whereas, the signal to noise ratio of N accumulated multiple contact spectra in a cross-polarization (cp) experiment is given by

$$(S/N)_{cp} = \frac{1}{\sqrt{N}} K \cdot \omega_{0S}^{1/2} \cdot M_{SN} \tag{4.53}$$

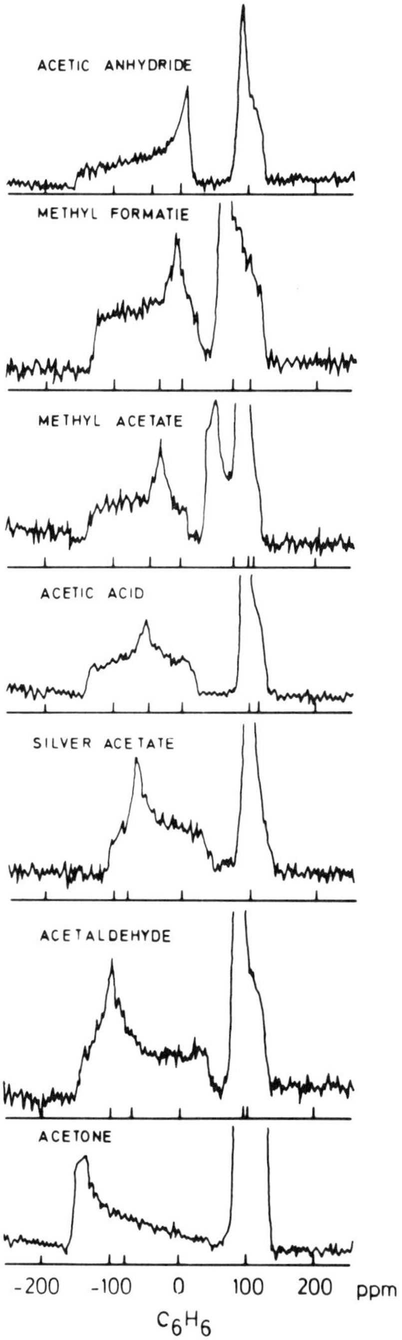

Fig. 4.10. Proton enhanced ^{13}C spectra of polycrystalline compounds containing carbonyl groups (large shielding anisotropy, left side) and methyl groups (small anisotropy, right side) obtained by A. Pines, M. G. Gibby and J. S. Waugh [*3b*] at liquid nitrogen temperature using PENIS technique

which results in a gain factor G_{cp} of the multiple contact cross-polarization experiment over an ordinary f.i.d. as [*3, 12*]

$$G_{cp} = \frac{\gamma_I}{\gamma_S} \cdot \frac{1}{\sqrt{N}} \sum_{k=1}^{N} e^{-k\epsilon}. \tag{4.54}$$

This is maximized for $N\epsilon \cong 1.3$ from which follows

$$G_{cp} = 0.64 \frac{\gamma_I}{\gamma_S} \left[\frac{N_I I(I+1)}{N_S S(S+1)}\right]^{1/2} \tag{4.55}$$

or

$$(S/N)_{cp} = 0.64 \frac{\gamma_I}{\gamma_S} \left[\frac{N_I I(I+1)}{N_S S(S+1)}\right]^{1/2} K \cdot \omega_{0S}^{1/2} M_{0S}. \tag{4.56}$$

For ^{13}C at natural isotopic abundance in organic solids, where $N_I/N_S \cong 150$, we find $G_{cp} \cong 30$. Notice, that it takes $G_{cp}^2 \cong 10^3$ accumulations of the f.i.d. to reach the same signal to-noise ratio as in the cross-polarization experiment. Another point to be mentioned is the fact that it takes T_{1S} to repeat the f.i.d. measurements, whereas, it takes T_{1I} for replicating the cross-polarization experiment. Since usually $T_{1S} > T_{1I}$, a factor T_{1S}/T_{1I} in front of Eq. (4.55) increases further the "gain" of the cross-polarization experiment.

If for technical reasons, or because of $T_{1\rho}$ effects, only a few contacts can be performed, the gain G_{cp} is of course drastically reduced. In these cases it is favourable to reduce T_{1I}. This has been discussed in more detail by Pines *et al.* [*3*].

In order to compare the sensitivity of the direct multiple-contact cross-polarization experiment with the indirect method, we have to calculate $(S/N)_{\text{ind}}/(S/N)_{cp}$, where $(S/N)_{cp}$ is given by Eq. (4.56). Whereas, in the cross-polarization experiment the total spectrum is obtained in one shot, each spectral element had to be measured step by step in the indirect method. If we assume that the I magnetization always decreases to $1/e$ of its initial value, $(S/N)_{\text{ind}}$ for each spectral element is given by

$$(S/N)_{\text{ind}} = \left(1 - \frac{1}{e}\right) \cdot K \cdot \omega_{0I}^{1/2} M_{0I}. \tag{4.57}$$

Suppose the S spin spectrum contains n spectral elements, we then define the signal-to-noise ratio of the whole spectrum as

$$(S/N)_{\text{ind}} = 0.632 \frac{1}{\sqrt{n}} K \cdot \omega_{0I}^{1/2} M_{0I}. \tag{4.58}$$

Remember, that in the cross-polarization experiment also the whole spectrum is obtained, however in one shot.

If we assume the same quality factor Q of the coil, the same detector bandwidth and noise characteristic in all cases, the factor K is the same in Eqs. (4.56) and (4.58) and we can write [*3, 12*]

$$\frac{(S/N)_{\text{ind}}}{(S/N)_{cp}} \cong \frac{1}{\sqrt{n}} \left(\frac{\gamma_I}{\gamma_S}\right)^{3/2} \left[\frac{N_I I(I+1)}{N_S S(S+1)}\right]^{1/2} \tag{4.59}$$

for the signal-to-noise ratio of a spectrum, recorded with the indirect method as compared with the cross-polarization method.

This is about 2.85 in the case of ^{13}C and ^{1}H in organic compounds ($N_I/N_S \cong 150$) if $n = 1024$.

The advantage in sensitivity of the indirect method over the cross-polarization experiment is even more pronounced if fewer spectral elements are needed to represent the S spin spectrum. The break-even point is reached in the case of ^{13}C and ^{1}H when about 8300 spectral elements are needed which has so far rarely been observed.

Although the indirect method is of greater sensitivity, almost all of the high resolution spectra of rare spins in solids have been observed by the cross-polarization technique to date. This is because of distortions which may result in the indirect method, and which have been discussed in the last section. On the other hand, very often the indirect method does not produce the total spectrum as fast as the cross-polarization technique.

We are now going to discuss briefly the different steps in the direct detection method of rare spins, following closely the paper of Pines *et al.* [*3*]. Four major steps have to be performed:

(1) *Preparation* of the I spins implies the polarization of the I spins in a static magnetic field.

(2) The "*hold*" period keeps the I spin order in some suitable reference frame.

(3) "*Mix*" constitutes the transfer of spin order form the I to the S spin system.

(4) *Observation* of the S spin signal is the final step, where the I spins may be decoupled if high resolution S spectra are desired.

Some suitable procedures to perform the four essential steps are summarized in the following:

(1) *Preparation:*
  (a) polarize I by the spin-lattice relaxation in H_0
  (b) dynamically polarize I by optical or microwave polarization [*18*, *19*]
  (c) polarize I using quadrupolar nucleus with short spin-lattice relaxation time [*20–24*].

(2) *Hold*
  (a) hold M_I along H_0 in laboratory frame,
  (b) spin lock M_I along H_{1I} in rotating frame,
  (c) hold I order in dipolar state in laboratory or rotating frame [*6*, *8*].

(3) *Mix*
  (a) Hartmann-Hahn [*1*], matched or unmatched,
  (b) solid effect in laboratory or rotating frames [*6*],
  (c) adiabatic crossover [*3b*, *6*].

(4) *Observation*
  (a) undecoupled I
  (b) continous I spin decoupling [*25*, *26*]
  (c) pulsed I spin decoupling [*27*]

We want to discuss now some examples of high resolution cross-polarization experiments using an I dipolar state in the rotating frame. Figure 4.11 explains different feasable steps [*3b*]. The essential part in these experiments is the ADRF of the I spins, followed by turning on the H_{1S}-field and mixing of the I and S spin reservoir. Before the mixing sets in the inverse "dipolar-spin-temperature" β_{DI} of the I spins and the inverse spin-temperature β_S of the S spins are

$$\beta_{DI} = \beta_L \cdot \frac{H_0}{H'_{LI}}; \qquad \beta_S = 0$$

where the "local field" of the I spins is defined by

$$\mathrm{Tr}\{\mathcal{H}'^2_{II}\} = \gamma^2 H'_{LI}\, \mathrm{Tr}\,\{I_z^2\}.$$

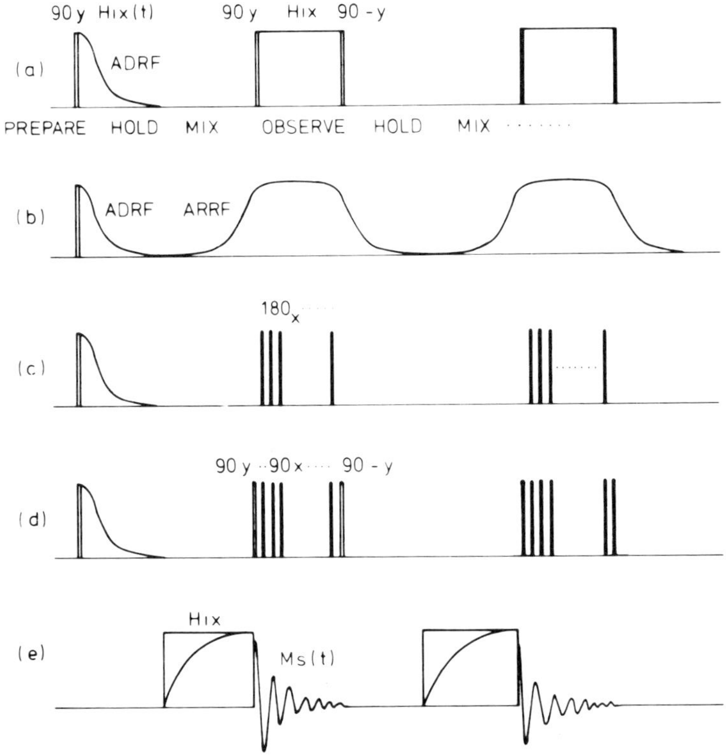

Fig. 4.11 a–e. Different double resonance schemes proposed by A. Pines, M. G. Gibby and J. S. Waugh [*3b*] using an I dipolar state created by ADRF. The following I spin decoupling schemes may be employed: (a) Magic sandwich decoupling. The 90° pulses cause a "spin locking" of the I dipolar state. (b) Adiabatic remagnetization in the rotating frame (ADRF) spin locks the I magnetization along H_{1X} and decouples the S spins. (c) The I spins are decoupled by 180° pulses and maintained in the dipolar state, since $\dot{\mathcal{H}}'_{II}$ is invariant under 180° rotation (only in the case of δ pulses, see Fig. 3.23). (d) Pulsed version of (a) applying 90° pulses for decoupling. (e) S spin irradiation and magnetization during polarization and observation

The prime labels the secular part of the dipolar Hamiltonian. We assume the mixing to proceed at constant energy i.e., the same final spin-Temperature β_f for I and S spins is reached

$$\beta_{DI} \cdot C_I H_{LI}'^2 = \beta_f [C_I H_{LI}'^2 + C_S H_{1S}^2]$$

resulting in

$$\frac{\beta_f}{\beta_{DI}} = \frac{1}{1+\epsilon'} \tag{4.60}$$

where

$$\epsilon' = \frac{\mathrm{Tr}\{\mathcal{H}_{1S}^2\}}{\mathrm{Tr}\{\mathcal{H}_{II}'^2\}}$$

or

$$\epsilon' = \frac{N_S S(S+1)}{N_I I(I+1)} \cdot \alpha^2 = \epsilon\alpha^2 \tag{4.61}$$

with

$$\alpha = \frac{\gamma_S H_{1S}}{\gamma_I H_{LI}'}. \tag{4.62}$$

If the Hartmann-Hahn condition is matched i.e., $\alpha = 1$ it follows $\epsilon' = \epsilon$. The final magnetization of the S spins is readily obtained using the above equation to be [*28*]

$$\frac{M_{S\infty}}{M_{0S}} = \frac{\beta_f H_{1S}}{\beta_L H_0} = \frac{\gamma_I}{\gamma_S} \frac{\alpha}{1+\epsilon\alpha^2}. \tag{4.63}$$

Let us discuss two different cases:
(i) $\alpha = 1$, matched Hartmann-Hahn conditon

$$\frac{M_{S\infty}}{M_{0S}} = \frac{\gamma_I}{\gamma_S} \frac{1}{1+\epsilon}$$

which is the same as in Eq. (4.20) in the case of spin locking of the I spins. The gain in magnetization equals about the ratio of the gyromagnetic moments, however, multiple contacts may be achieved. The transfer time corresponds to milliseconds according to Eq. (4.10).
(ii) $\alpha \gg 1$, unmatched Hartmann-Hahn condition.
The S spin magnetization reaches a maximum according to Eq. (4.63) for $\alpha \cdot \sqrt{\epsilon} = 1$, with

$$\left.\frac{M_S}{M_{0S}}\right|_{\max} = \frac{1}{2} \frac{\gamma_I}{\gamma_S} \left[\frac{N_I I(I+1)}{N_S S(S+1)}\right]^{1/2}. \tag{4.64}$$

This maximum is reached when the heat capacities of the I and S spins are equal. By comparing Eq. (4.64) with (4.55) we realize that the same total S spin magnetization as in a multiple-contact cross-polarization experiment can be obtained in one shot [*19b*]. Such a one-shot polarization experiment with unmatched Hartmann-Hahn condition was first performed by A. Pines [*29*] on adamantane. Figure 4.12 shows

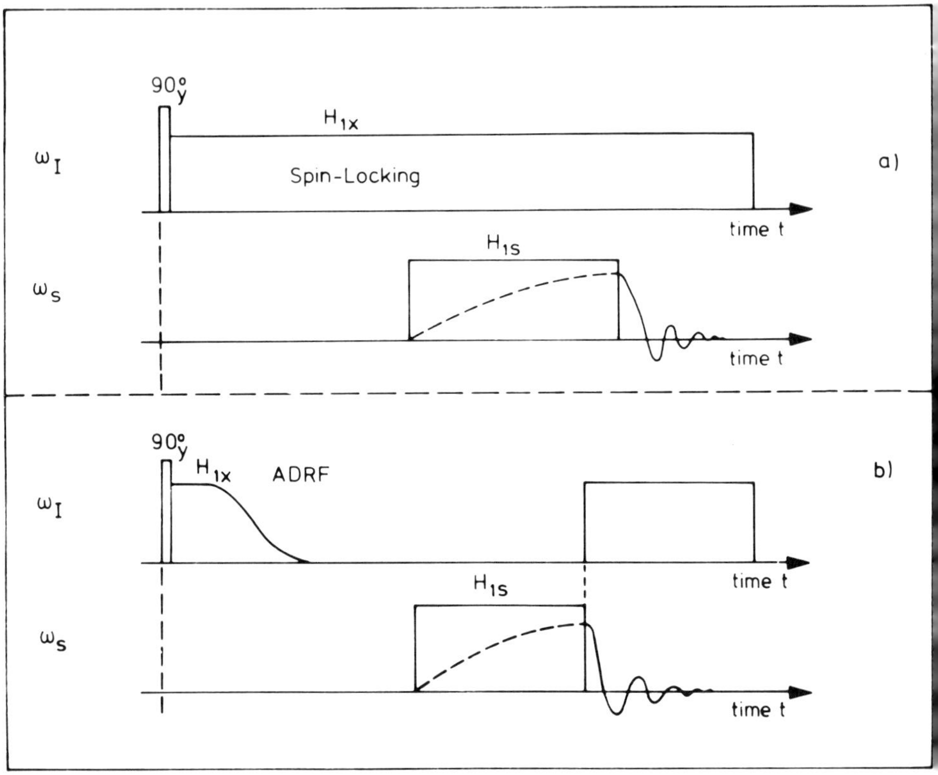

Fig. 4.12. Schematic representation of the one shot cross-polarization experiment according to A. Pines [*29*]

schematically how this one-shot cross-polarization experiment can be performed [*3b*]. The interested reader may readily see by himself that the factor 1/2 vanishes in Eq. (4.64) if the H_{1S} field is turned on adiabytically. In this case entropy conservation according Eq. (4.3c) has to be invoked instead of energy conservation, as used above. Figure 4.13 demonstrates the dependence of M_S on α in the case of the methyl ^{13}C in CF_3COOAg [*28*]. The experimental data are obtained in a one shot cross-polarization experiment which includes T_{1D} effects to be discussed in the following. T_{1D} effects of the I spin dipole reservoir can severely limit the beauty of the one-shot cross-polarization experiment. Since $\alpha \gg 1$ the transfer time T_{IS} increases exponentially with α, according to Eqs. (4.10, 4.12) and may exceed T_{1D} way before the optimum value of α is reached. This is why in practice maximum S spin magnetization is barely observed.

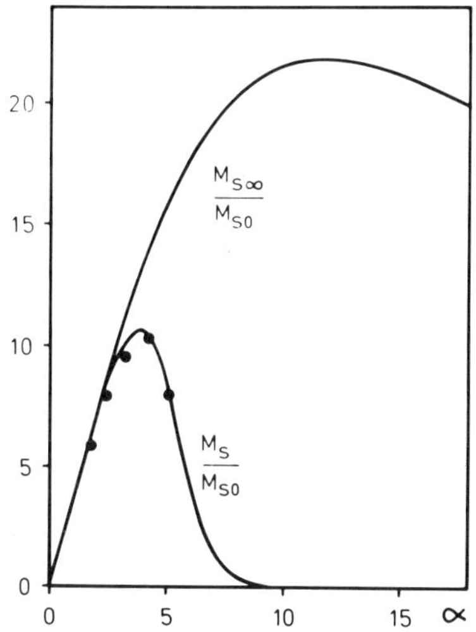

Fig. 4.13. S spin magnetization versus the mismatch parameter α in a one shot cross-polarization experiment employing ADRF. M_{S_0} is the Boltzman magnetization and $M_{S\infty}$ corresponds to the equilibrium magnetization for infinite relaxation times [Eq. (4.63)]. In the case of finite relaxation time T_{1D} of the I spins the S spin magnetizations M_S does not reach its full maximum and its dependence on α is modified by T_{1D} and the cross-relaxation time T_{IS} (α) according to Eq. (4.68c). The data points correspond to the maximum magnetization of methyl ^{13}C in a single crystal of CF_3COOAg oriented with methyl group C_3-axis at the magic angle with respect to the magnetic field, leaving only intermolecular dipolar interactions effective. (Courtesy of G. Sinning)

The dynamics of the cross-polarization process will be discussed in the next section. Chemical shift tensors which have been determined by this method are summarized in Chapter 5. Applications of the one-shot cross-polarization to relaxation studies will be presented in Chapter 6.

4.3 Cross-Polarization Dynamics

Let us start with a phenomenological approach to describe the variation of the inverse spin temperatures β_I and β_S of the I and S spins respectively in the "mixing" part of the two basic cross-polarization experiments as depicted in Fig. 4.4.

For the initial S spin-temperature, we assume $\beta_S = 0$ and recall for the I spin-temperature according to Section 4.1.

$$\text{(a)}\quad \beta_I = \beta_L \frac{H_0}{H_{1I}} \qquad \text{spin locking } H_{1I} \gg H'_{LI}$$

and

$$\text{(b)}\quad \beta_I = \beta_L \frac{H_0}{H'_{LI}} \qquad \text{ADRF.}$$

If we invoke energy conservation in the rotating frame we can write

$$\frac{d}{dt}\beta_I + \epsilon' \frac{d}{dt}\beta_S = 0 \qquad (4.65)$$

where $\epsilon' = \epsilon\alpha^2$ is given by Eq. (4.61) of the preceeding section, if H'_{LI} is replaced by H_{1I} in the case of spin locking. The S spin temperature β_S is relaxed with the time constant T_{IS} towards the instantanous I spin-temperature which results in the following coupled differential equations [*10*]

$$\frac{d}{dt}\beta_S = -\frac{1}{T_{IS}}(\beta_S - \beta_I) \qquad (4.66a)$$

$$\frac{d}{dt}\beta_I = -\frac{\epsilon'}{T_{IS}}(\beta_I - \beta_S) - \frac{1}{T_{1x}}\beta_I \qquad (4.66b)$$

where the last term in Eq. (4.66b) has been added to account for relaxation of the I spins such as $T_{1\rho}$ (spin locking; $x = \rho$) or T_{1D} (ADRF; $x = D$) respectively.

The nature of T_{IS} and its dependence on H_{1S} will be discussed in more detail later in this section. Notice, however, that expressions for $1/T_{IS}$ have been quoted already in Eqs. (4.10) and (4.11). These coupled differential equations (4.66) are straightforwardly solved under the initial conditions:

$\beta_S(0) = 0$ and $\beta_I(0) = \beta_{I0}$.

$$\beta_S(t) = \beta_{I0}\frac{1}{a_+ - a_-}(e^{-a_- t/T_{IS}} - e^{-a_+ t/T_{IS}}) \qquad (4.67a)$$

$$\beta_I(t) = \beta_{I0}\frac{1}{a_+ - a_-}[(1 - a_-)\, e^{-a_- t/T_{IS}} - (1 - a_+)\, e^{-a_+ t/T_{IS}}] \qquad (4.67b)$$

where

$$a_\pm = \frac{1}{2}(1 + \epsilon\alpha^2 + T_{IS}/T_{1x})[1 \pm (1 - \frac{4\, T_{IS}/T_{1x}}{1 + \epsilon\alpha^2 + T_{IS}/T_{1x}})^{1/2}]$$

and with $\alpha = \omega_{eS}/\omega_{LI}$; $\qquad \epsilon = \dfrac{N_S S(S+1)}{N_I I(I+1)}$.

The variation of the magnetization $M_S(t)$ of the S spins with the coupling time is

$$\frac{M_S(t)}{M_{S0}} = \frac{\beta_S(t) H_{1S}}{\beta_L H_0} = \alpha\frac{\gamma_I}{\gamma_S}\frac{\beta_S(t)}{\beta_{I0}} \qquad (4.68a)$$

where $\beta_S(t)/\beta_{I0}$ must be inserted according to Eq. (4.67a).

The S spin magnetization reaches a maximum at time t_m (see Fig. 4.14 and 4.15):

$$t_m = \frac{T_{IS}}{a_+ - a_-}\ln\left(\frac{a_+}{a_-}\right). \qquad (4.68b)$$

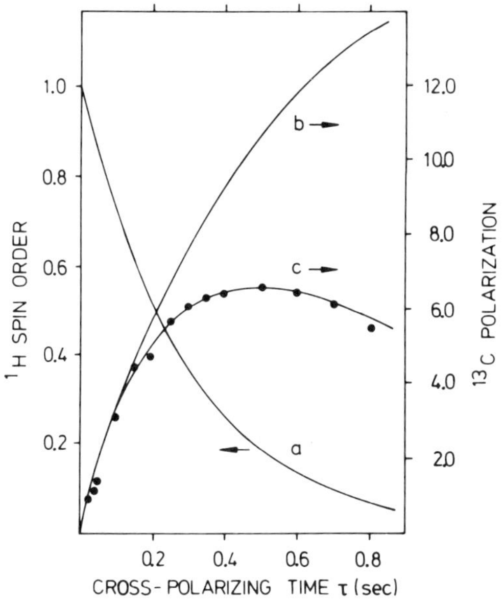

Fig. 4.14. Time evolution of ^{13}C polarization in adamantane at room temperature in a one shot cross-polarization experiment (Pines *et al.* [*30*]). The loss of ^{1}H spin order due to T_{1D} has been considered

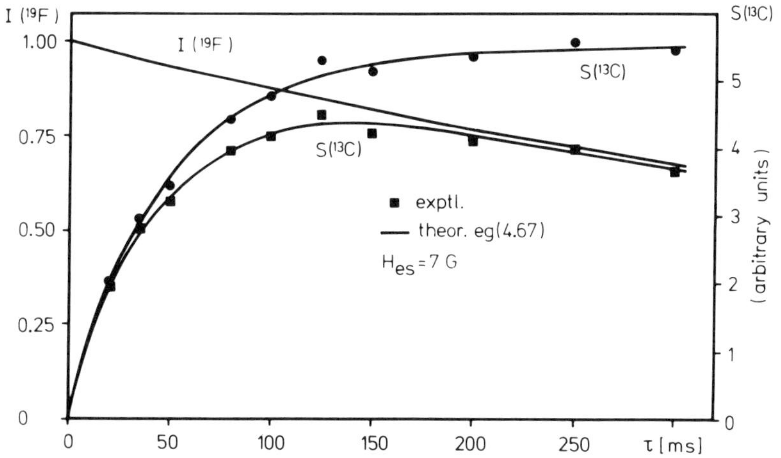

Fig. 4.15. Time evolution of methyl group ^{13}C polarization and ^{19}F order in a CF_3COOAg single crystal oriented as described in Fig. 4.13. The cross relaxation time T_{IS} (see Fig. 4.16a) and the maximum S spin magnetization M_S (see Fig. 4.13) are obtained from this graph, by fitting the data to Eq. (4.67a) (see Ref. [*28*])

Notice, that for large values of T_{1x}/T_{IS} one has to wait a long time compared with T_{IS} for the S spin signal to reach its maximum. However, remenber that already 95% of the maximum signal is reached after 3 T_{IS}. Other characteristic cases may be discussed by the interested reader. We turn now to the maximum S spin magnetization M_S obtainable at time t_m for a given value of the "mismatch paramter" α. Applying Eqs. (4.68a, b) we obtain

$$t = t_m : \frac{M_S}{M_{S0}} = \alpha \frac{\gamma_I}{\gamma_S} \frac{1}{a_+ - a_-} [(\frac{a_-}{a_+})^{\frac{a_-}{a_+ - a_-}} - (\frac{a_-}{a_+})^{\frac{a_-}{a_+ - a_-}}]. \quad (4.68c)$$

In the case of negligible I spin relaxation ($T_{IS}/T_{1x} \ll 1$, $t_m = \infty$) we obtain the same result as for $M_{S\infty}/M_{S0}$ [see Eq. (4.63)]. This has been schematically drawn in Fig. 4.13. However, in practice, for large rf fields at the S spin resonance ($\alpha \gg 1$), the cross relaxation time T_{IS} becomes comparable to the I spin relaxation time in the rotating frame T_{1x} (whether T_{1D} or $T_{1\rho}$) and Eqs. (4.68a, c) have to be applied. M_S/M_{S0} has been plotted in Fig. 4.13 versus the mismatch parameter α for the case of the methyl ^{13}C in CF_3COOAg.

The data included in Fig. 4.13 are obtained from a series of experiments analogous to the one represented in Fig. 4.15. Similarly one can discuss the I spin order represented by $\beta_I(t)/\beta_{I0}$, which may be deduced straightforwardly for different experimental situations by applying Eq. (4.67b).

Let us now discuss some special cases of Eq. (4.67)

(i) $\epsilon = 0$; $T_{IS}/T_{1x} = 0$ i.e. vanishing heat capacity of the S spins; negligible T_{1x} relaxation of the I spins

$$\beta_S(t) = (1 - e^{-t/T_{IS}})\, \beta_{I0} \quad (4.69a)$$

$$\beta_I(t) = \beta_{I0}.$$

(ii) $\epsilon = 0$; $T_{IS}/T_{1x} \neq 0$ i.e. same as (i), but T_{1x} relaxation of the I spins.

$$\beta_S(t) = \frac{1}{1-\lambda} (1 - e^{-(1-\lambda)t/T_{IS}})\, e^{-t/T_{1x}}\, \beta_{I0} \quad (4.69b)$$

$$\beta_I(t) = e^{-t/T_{1x}} \cdot \beta_{I0}$$

where $\lambda = T_{IS}/T_{1x}$.

This case is usually met under extreme dilution of S spins, and with matched Hartmann-Hahn condition ($\alpha = 1$), but with short T_{1x} relaxation time.

(iii) $\epsilon' \neq 0$; $\lambda = 0$ i.e., non-negligible heat capacity of the S spins, but negligible T_{1x} relaxation.

$$\beta_S(t) = \frac{1}{1+\epsilon'} (1 - e^{-(1+\epsilon')t/T_{IS}})\, \beta_{I0} \quad (4.70a)$$

$$\beta_I(t) = \frac{1}{1+\epsilon'} (1 + \epsilon' e^{-(1+\epsilon')t/T_{IS}})\, \beta_{I0}. \quad (4.70b)$$

This case is usually met, no matter if the Hartmann-Hahn condition is matched or unmatched, as long as $T_{IS} \ll T_{1X}$.

In the one shot cross-polarization experiment, however, ϵ' usually becomes large and since the Hartmann-Hahn condition is not matched ($\alpha \gg 1$) T_{IS} is comparable with T_{1x} or $\lambda \cong 1$ i.e., Eq. (4.67) has to be used. This case is demonstrated in Fig. 4.14 for ^{13}C in adamantane und in Fig. 4.15 for the methyl ^{13}C in a single crystal of CF_3COOAg [*28*]. By fitting the experimental data to Eq. (4.67) the equilibrium magnetization M_S and the cross-relaxation time T_{IS} can be determined for different coupling fields H_{1S}, or values of ω_{eS} respectively. The cross-relaxation rate $W_{IS} = 1/T_{IS}$ depends strongly on H_{1S} more or less exponentially in the ADRF case, as suggested by McArthur *et al.* [*10*] [see Eq. (4.12)]. In Fig. 4.16 and 4.17 the modified cross-polarization rate $1/(T_{IS} \sin^2 \vartheta_S)$ is plotted versus H_{1I}, H_{eS}, ν_{eS} according to Eqs. (4.10, 4.12) together with the experimental data, for those completely different systems as ^{13}C in adamantane, methyl ^{13}C in CF_3COOAg and ^{43}Ca in CaF_2. In all cases good agreement between the experimental data and an exponential cross-polarization spectrum is obtained. Demco *et al.* [*11*], however, have shown, that this behavior is purely accidental and they have provided a more subtle discussion of cross-polarization spectra. We shall follow briefly the discussion of cross-polarization spectra as initiated by McArthur *et al.* [*10*] and improved by Demco, Tegenfeld and Waugh [*11*].

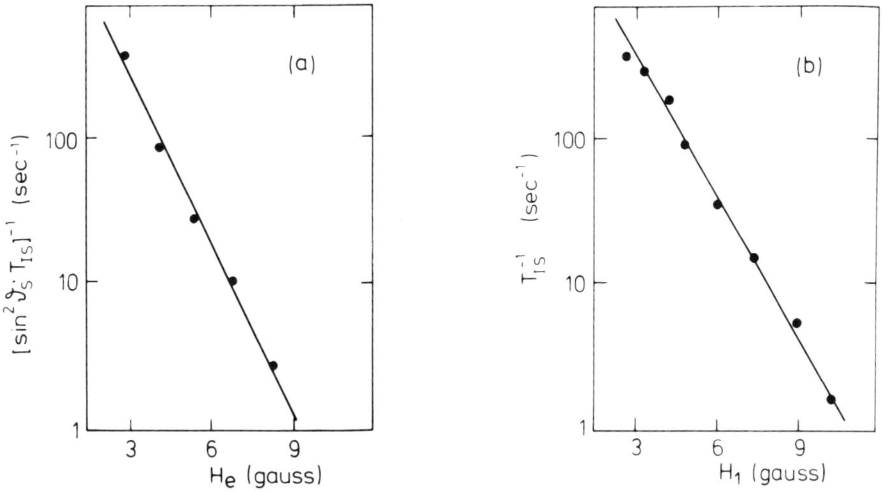

Fig. 4.16. Cross relaxation rate $W_{IS} = 1/T_{IS}$ obtained in an ADRF one shot cross-polarization experiment versus the effective field H_e, H_1 at S spin frequency for (a) the methyl ^{13}C in a CF_3COOAg single crystal [*28*] and (b) ^{13}C in adamantane (A. Pines *et al.* [*30*]) as obtained from different sets of Figs. 4.14 and 4.15 with varying rf fields close to the ^{13}C resonance frequency. The straight lines correspond to an exponential cross relaxation spectrum Eq. (4.12) with correlation times $\tau_c = 150\ \mu s$ in the case of CF_3COOAg and $\tau_c = 140\ \mu s$ in the case of adamantane. (Fig. 4.16b courtesy of A. Pines)

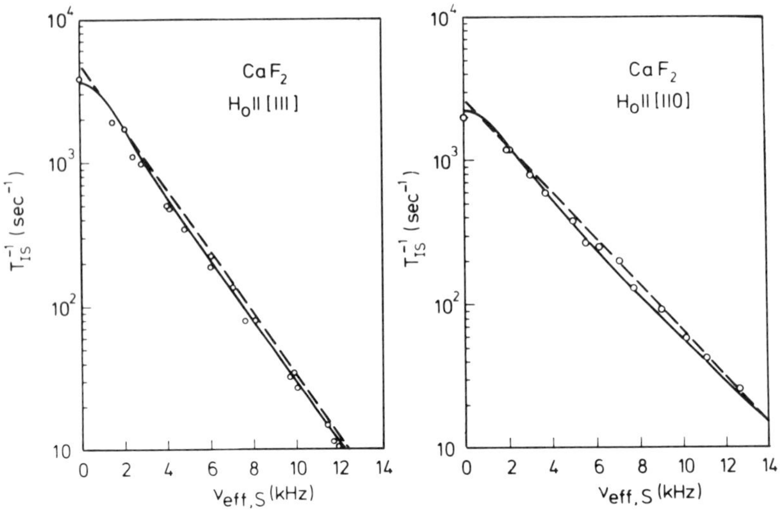

Fig. 4.17. Cross relaxation rate $1/T_{IS}$ versus effective field at S spin resonance (^{43}Ca) under ADRF conditions in $^{43}Ca^{19}F_2$ for two different orientations (a) $H_0 \| (110)$, (b) $H_0 \| (111)$. The data points are obtained by McArthur *et al.* [*10*] and the dashed line is their fit to an exponential spectrum Eq. (4.12). The solid lines were calculated by D. Demco, J. Tegenfeldt and J. S. Waugh [*11*] by using a memory function approach

Let us apply an rf field, with

$$\mathcal{H}_{\mathrm{rf}}(t) = -2\omega_{1I} I_x \cos\omega_I t - 2\omega_{1S} S_x \cos\omega_S t \tag{4.71}$$

to the coupled *I-S* spin system. The total Hamiltonian, neglecting spin lattice relaxation is conveniently expressed in the "tilted rotating frame" (TR), by using the following transformations [*11*]:

$$\mathbf{R} = \mathbf{R}_I\,\mathbf{R}_S;\ \mathbf{R}_I = e^{\,i\omega_I I_z t};\ \mathbf{R}_S = e^{i\omega_S S_z t} \tag{4.72a}$$

$$\mathbf{T} = \mathbf{T}_I \mathbf{T}_S;\ \mathbf{T}_I = e^{i\vartheta_I I_y};\ \mathbf{T}_S = e^{i\vartheta_S S_y}; \tag{4.72b}$$

with $\tan\vartheta_I = \dfrac{\omega_{1I}}{\Delta\omega_I}$; $\tan\vartheta_S = \dfrac{\omega_{1S}}{\Delta\omega_S}$

where $\Delta\omega = \omega_0 - \omega$; $\omega_e^2 = \Delta\omega^2 + \omega_1^2$.

The total Hamiltonian in the "tilted rotating frame" may now be expressed as

$$\mathcal{H} = \mathcal{H}_I + \mathcal{H}_S + \mathcal{H}_{IS} \tag{4.73}$$

where

(i) spin locking case [*11*]

$$\mathcal{H}_I = -\omega_{eI} I_z + P_2(\cos\vartheta_I)\mathcal{H}_{II}^{(0)} + \mathcal{H}_{II}^{(ns)} \tag{4.74}$$

$$\mathcal{H}_S = -\omega_{eS} S_z \tag{4.75}$$

$$\mathcal{H}_{IS} = \cos\vartheta_I \cos\vartheta_S \, \mathcal{H}_{IzSz} + \sin\vartheta_I \sin\vartheta_S \, \mathcal{H}_{IxSx} - \sin\vartheta_I \cos\vartheta_S \, \mathcal{H}_{IxSz}$$
$$- \cos\vartheta_I \sin\vartheta_S \, \mathcal{H}_{IzSx}. \tag{4.76}$$

(ii) ADRF case [*11*]

$$\mathcal{H}_I = \mathcal{H}_{II}^{(0)}; \mathcal{H}_S = -\omega_{eS} S_z \tag{4.77}$$

$$\mathcal{H}_{IS} = \cos\vartheta_S \, \mathcal{H}_{IzSz} - \sin\vartheta_S \, \mathcal{H}_{IzSx} \tag{4.78a}$$

with

$$\mathcal{H}_{I\mu S\upsilon} = \sum_{i,j} B_{ij} I_{\mu i} S_{\upsilon j} \tag{4.78b}$$

where

$$B_{ij} = \gamma_I \gamma_S \hbar (1 - 3\cos^2\vartheta_{ij})/r_{ij}^3. \tag{4.78c}$$

Because of $N_I \gg N_S$ and $\gamma_I \gg \gamma_S$ in the cases considered in this context, we have $\|\mathcal{H}_I\|; \|\mathcal{H}_S\| \gg \|\mathcal{H}_{IS}\|$ and $\mathcal{H}_{IS}$ may be considered as a small perturbation. $\mathcal{H}_I$ and $\mathcal{H}_S$ are orthogonal operators in the sense:

$$(\mathcal{H}_I | \mathcal{H}_S) = 0$$

where the definition

$$(\mathbf{A} | \mathbf{B}) \underset{\text{def}}{=} \mathrm{Tr}\{\mathbf{A}^+ \mathbf{B}\}$$

as in Section 3 and Appendix E has been used.

Expressing the total Hamiltonian as

$$\mathcal{H} = \mathcal{H}_1 + \mathcal{H}_2 + \mathcal{H}_p \tag{4.79}$$

where $\mathcal{H}_1 = \mathcal{H}_I; \mathcal{H}_2 = \mathcal{H}_S; \mathcal{H}_p = \mathcal{H}_{IS}$ will be assumed to be Hermitian in our case. Expanding the density matrix into orthogonal operators leads to

$$\rho = Z^{-1}[1 - \beta_1 \mathcal{H}_1 - \beta_2 \mathcal{H}_2 - \mathbf{R}] \tag{4.80}$$

with

$$(\mathcal{H}_1 | \mathcal{H}_2) = (\mathcal{H}_1 | \mathbf{R}) = (\mathcal{H}_2 | \mathbf{R}) = 0. \tag{4.81}$$

An equation of motion may be derived for the expectation value of the operators $\mathcal{H}_i; i = 1, 2$.

$$\langle \mathcal{H}_i \rangle = (\mathcal{H}_i | \rho), \; i = 1, 2. \tag{4.82}$$

Using the first Born approximation in the time evolution operator

$$\mathbf{S}(t) = \exp\left[-i(1-P)(\hat{\mathcal{H}}_1 + \hat{\mathcal{H}}_2 + \hat{\mathcal{H}}_p)t\right] \tag{4.83a}$$

we arrive at

$$\mathbf{S}(t) \cong \mathbf{S}_0(t) = \exp\left[-i(1-P)(\hat{\mathcal{H}}_1 + \hat{\mathcal{H}}_2)t\right] \tag{4.83b}$$

where the projector

$$P = \frac{|\mathcal{H}_1)(\mathcal{H}_1|}{(\mathcal{H}_1|\mathcal{H}_1)} + \frac{|\mathcal{H}_2)(\mathcal{H}_2|}{(\mathcal{H}_2|\mathcal{H}_2)} \tag{4.84}$$

has been used and where $\hat{\mathcal{H}}$ denotes a Liouvillian. We arrive at the following integro-differential equation (see Appendix E)

$$\frac{d}{dt}\langle\mathcal{H}_i(t)\rangle = -\left\{\int_0^t dt' K_{ii}(t-t')\langle\mathcal{H}_i(t')\rangle\right.$$

$$\left. + \int_0^t dt' K_{ij}(t-t')\langle\mathcal{H}_j(t')\rangle\right\} \qquad i,j = 1,2 \quad i \neq j \tag{4.85}$$

where

$$K_{ii}(t-t') = \frac{(\mathcal{H}_i|\hat{\mathcal{H}}_p\, \mathbf{S}_0(t,t')\, \hat{\mathcal{H}}_p|\mathcal{H}_i)}{(\mathcal{H}_i|\mathcal{H}_i)} \tag{4.86}$$

$$K_{ij}(t-t') = \frac{(\mathcal{H}_i|\hat{\mathcal{H}}_p\, \mathbf{S}_0(t,t')\, \hat{\mathcal{H}}_p|\mathcal{H}_j)}{(\mathcal{H}_j|\mathcal{H}_j)}. \tag{4.87}$$

In the fast correlation limit ($\tau_c \ll t$, $t - t' = \tau$) this may be expressed as (see Appendix E)

$$\frac{d}{dt}\langle\mathcal{H}_i(t)\rangle = -\left\{\int_0^\infty d\tau\, K_{ii}(\tau)\langle\mathcal{H}_i(t)\rangle + \int_0^\infty d\tau\, K_{ij}(\tau)\langle\mathcal{H}_j(t)\rangle\right\}. \tag{4.88}$$

Introducing the new thermodynamic coordinates [*11*]

$$\beta_i = \frac{(\mathcal{H}_i|\,\rho(t))}{(\mathcal{H}_i|\mathcal{H}_i)} = \frac{\langle\mathcal{H}_i(t)\rangle}{(\mathcal{H}_i|\mathcal{H}_i)} \tag{4.89}$$

and the correlation function $C(\mathcal{H}_i, \mathcal{H}_j, \tau)$, this results readily in the same pair of differential equations as Eq. (4.66), which was obtained by a phenomenological approach:

$$\frac{d}{dt}\beta_i(t) = -\int_0^\infty d\tau\, C(\mathcal{H}_i;\mathcal{H}_i;\tau)\,\beta_i(t) - \int_0^\infty d\tau\, C(\mathcal{H}_i;\mathcal{H}_j;\tau)\,\beta_j(t) \tag{4.90}$$

$i, j = 1, 2$
$i \neq j$

where

$$C(\mathcal{H}_i; \mathcal{H}_i; \tau) = K_{ii}(\tau)$$

and

$$C(\mathcal{H}_i; \mathcal{H}_j; \tau) = \frac{(\mathcal{H}_j | \mathcal{H}_j)}{(\mathcal{H}_i | \mathcal{H}_i)} K_{ij}(\tau).$$

Using energy conservation

$$\frac{d}{dt} [\langle \mathcal{H}_1 \rangle + \langle \mathcal{H}_2 \rangle] = 0.$$

Equation (4.90) can be casted into the form of Eq. (4.66) for vanishing spin lattice relaxation by using

$$\epsilon' = \frac{(\mathcal{H}_2 | \mathcal{H}_2)}{(\mathcal{H}_1 | \mathcal{H}_1)} \tag{4.91}$$

and

$$\frac{1}{T_{IS}} = \int_0^\infty d\tau \, a(\tau) \tag{4.92}$$

where

$$a(\tau) = K_{22}(\tau) = \frac{(\mathcal{H}_2 | \hat{\mathcal{H}}_p \mathbf{S}_0(\tau) \hat{\mathcal{H}}_p | \mathcal{H}_2)}{(\mathcal{H}_2 | \mathcal{H}_2)} \tag{4.93}$$

The cross relaxation rate $1/T_{IS}$ is the issue of this discussion and we are setting out to calculate it by obtaining an expression for the correlation function $a(\tau)$. A tedious, but straightforward calculation, using Eqs. (4.74–4.78, 4.83) leads to

(i) spin locking case [*11*]

$$a(\tau) = \sin^2 \vartheta_I \sin^2 \vartheta_S \, a_x(\tau) \frac{1}{2} [\cos(\omega_{eS} - \omega_{eI})\tau + \cos(\omega_{eS} + \omega_{eI})\tau]$$
$$+ \cos^2 \vartheta_I \sin^2 \vartheta_S \, a_z(\tau) \cos \omega_{eS} \tau. \tag{4.94}$$

(ii) ADRF case [*11*]

$$a(\tau) = \sin^2 \vartheta_S \, a_z(\tau) \cos \omega_{eS} \tau \tag{4.95}$$

where in the two different cases (i) and (ii) we define

$$a_z(\tau) = (\sum_i B_i I_{zi} | e^{-iP_2(\cos\vartheta_I)\hat{\mathcal{H}}_{II}^{(0)}\tau} | \sum_i B_i I_{zi}) \tag{4.96a}$$

and

$$a_x(\tau) = (\sum_i B_i I_{xi} | e^{-iP_2(\cos\vartheta_I)\hat{\mathcal{H}}_{II}^{(0)}\tau} | \sum_i B_i I_{xi}). \tag{4.96b}$$

In the ADRF case (ii) we set especially $\cos\vartheta_I = 1$.

It may be convenient to introduce the second moment M_2^{IS} of the heteronuclear dipole coupling as

$$M_2^{IS} = \frac{(S_x | \hat{\mathcal{H}}_{IS}^2 | S_x)}{(S_x | S_x)} = \mathrm{Tr}\,\{(\sum_i B_i I_{zi})^2\}. \tag{4.97}$$

Let us introduce the correlation functions $C_x(\tau)$ and $C_z(\tau)$ [*11*]:

$$C_x(\tau) = a_x(\tau)/M_2^{IS}$$

$$C_z(\tau) = a_z(\tau)/M_2^{IS}.$$

i.e.

$$\underset{\mu = x, z}{C_\mu(\tau)} = \frac{(\sum_i B_i \cdot I_{\mu i} | e^{-iP_2(\cos\vartheta_I)\hat{\mathcal{H}}_{II}^{(0)}\tau} | \sum_i B_i I_{\mu i})}{(\sum_i B_i I_{\mu i} | \sum_i B_i I_{\mu i})}. \tag{4.98}$$

Combining Eqs. (4.92–4.98) we are now well equipped to express the cross-polarization rate $1/T_{IS}$ by the cross-polarization spectral densities

$$J_x(\omega) = \int_0^\infty d\tau \cos\omega\tau\, C_x(\tau), \tag{4.99a}$$

$$J_z(\omega) = \int_0^\infty d\tau \cos\omega\tau\, C_z(\tau) \tag{4.99b}$$

as

(i) spin locking case

$$\frac{1}{T_{IS}} = \sin^2\vartheta_S M_2^{IS} [\cos^2\vartheta_I J_z(\omega_{eS}) + \sin^2\vartheta_I \tfrac{1}{2}[J_x(\omega_{eS} - \omega_{eI}) + J_x(\omega_{eS} + \omega_{eI})]]. \tag{4.100}$$

Since $J_x(\omega); J_z(\omega)$ approach zero for $\omega\tau_c \gg 1$, we find it to be a good approximation in the case $\omega_{eS}, \omega_{eI} \cong 1/T_{2I}$ and $\tau_c \cong T_{2I}$ to write, following DTW [*11*]

$$\frac{1}{T_{IS}} = \frac{1}{2}\sin^2\vartheta_S \sin^2\vartheta_I M_2^{IS} J_x(\Delta\omega_e). \tag{4.101}$$

In the

(ii) ADRF case [*10, 11*] we obtain

$$\frac{1}{T_{IS}} = \sin^2\vartheta_S M_2^{IS} J_z(\omega_{eS}). \tag{4.102}$$

For a full description of the cross-polarization dynamics it is mandatory to calculate the cross-polarization spectral densities $J_x(\omega)$ and $J_z(\omega)$.

Before setting out to calculate $C_x(\tau)$ and $C_z(\tau)$ or $J_x(\omega)$ and $J_z(\omega)$ correspondingly, we would like to discuss briefly the

short time behavior:

Two dramatic transient effects have appeared in this regime which have to be accounted for theoretically. The first one to be discussed was observed by Strombotne and Hahn [*6, 31*] in the labaratory frame and by McArthur, Walstedt and Hahn [*10*] in the rotating frame (see Fig. 4.18). These effects have been treated by

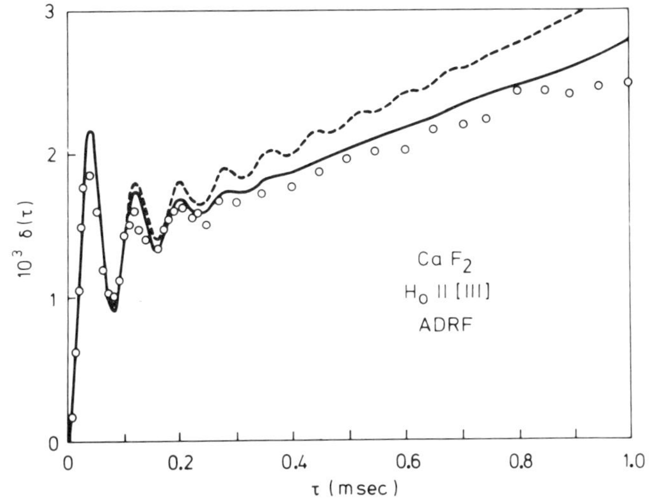

Fig. 4.18. Transient oscillations in the rotating frame in $^{43}Ca^{19}F_2$ under ADRF conditions as observed by McArthur *et al.* [*10*] (open circles) with ν_{effS} = 12.6 kHz. The fractional decrease $\delta(\tau)$ of *I* spin (^{19}F) order, when a mixing pulse is applied to the *S* spins (^{43}Ca) was calculated by McArthur *et al.* [*10*] by using a Lorentzian autocorrelation function with τ_c = 77 μs (dashed curve) and by Demco *et al.* [*11*] by using their autocorrelation function obtained by a memory function approach (solid curve)

the spin temperature concept in the past, describing it as an adjustment of the spin temperatures between the Zeeman reservoir and the non-secular dipole reservoir. The approach taken here (DTW [*11*]) is more rigorous, since no explicit spin temperature assumption is involved.

The second transient effect in the short time behavior was observed by R. Ernst [*32*] and co-workers (see Fig. 4.19) and is due to the discrete *I-S* dipole spectrum, damped by spin diffusion among the abundant *I* spins.

Recently J. S. Waugh [*33*] and co-workers have exploited this effect and designed a sequence which applies rf irradiation at the "magic angle" of the *I* spins in order to reduce the damping of these oscillations. This sequence is sketched in Fig. 4.20 together with the high-resolution *I-S* dipole spectrum obtained [*33*] (Fig. 4.21).

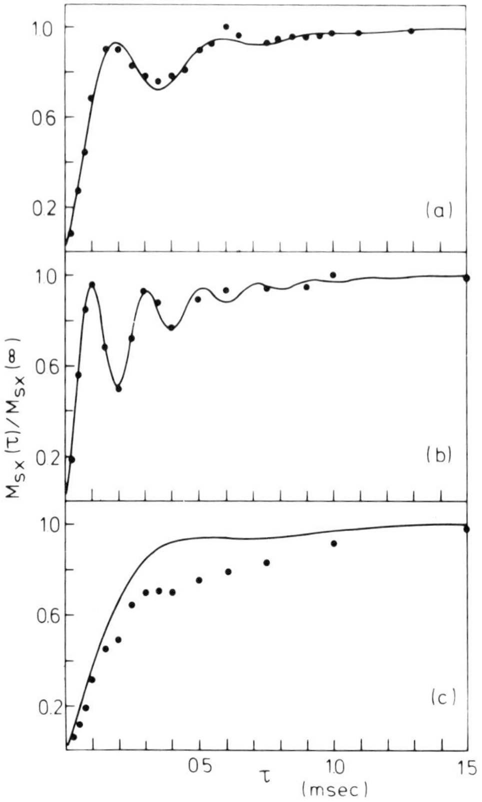

Fig. 4.19. Normalized S spin magnetization $M_{SX}(\tau)$ of ^{13}C in a ferrocene single crystal for three different orientations versus coupling time τ in a one shot cross-polarization experiment of the spin locking type, with $\nu_{1I} = \nu_{1S} = 16$ kHz. (R. R. Ernst and co-workers [*32*]). The transient oscillations of the ^{13}C magnetization correspond to the I-S dipolar interaction of next nearest protons. The damping of the oscillations is caused by spin diffusion. For further details see Ref. [*32*]

All these effects can be accounted for by the present theory, if we expand the density matrix for short times. We might as well use Eq. (4.85), where we replace in the right hand side

$\langle \mathcal{H}_{i,j}(t') \rangle$ by $\langle \mathcal{H}_{i,j}(0) \rangle$.

Integration leads to

$$\langle \mathcal{H}_i(t) \rangle - \langle \mathcal{H}_i(0) \rangle = -\int_0^t dt_2 \int_0^{t_2} dt_1 K_{ii}(t_2 - t_1) \langle \mathcal{H}_i(0)$$

$$-\int_0^t dt_2 \int_0^{t_2} dt_1 K_{ij}(t_2 - t_1)(i) \langle \mathcal{H}_j(0) \rangle. \tag{4.103}$$

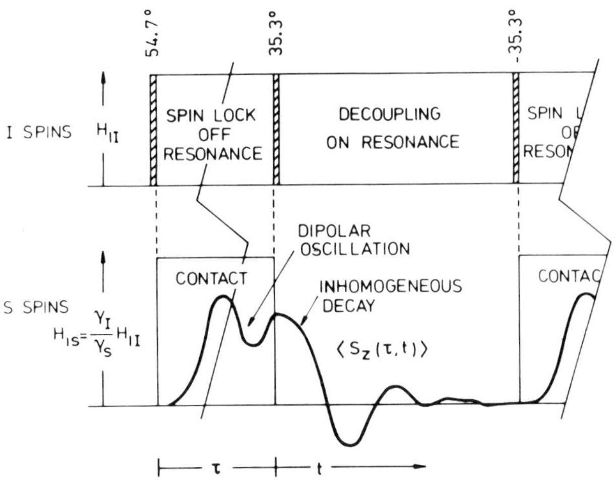

Fig. 4.20. A scheme proposed by J. S. Waugh and co-workers [*33*] for observing resolved dipolar coupling spectra of dilute nuclear spins in solids

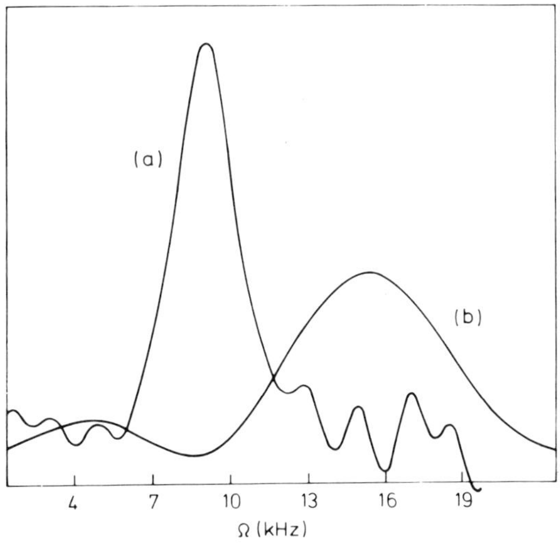

Fig. 4.21 a and b. ^{13}C-^{1}H dipolar spectra obtained by J. S. Waugh and co-workers [*33*] at a certain orientation of an ammonium tartrate single crystal, employing the scheme of Fig. 4.20. The observed line (a) at a dipolar frequency Ω = 9.7 kHz arises from the C–H coupling of the CHOH group with off-resonance spin-locking of the protons, whereas, in case (b) resonant spin-locking of the protons is applied

With the initial condition

$$\langle \mathcal{H}_2(0)\rangle = 0; \langle \mathcal{H}_1(0)\rangle \neq 0$$

we obtain

$$\frac{\langle \mathcal{H}_2(t)\rangle}{\langle \mathcal{H}_1(0)\rangle} = -\int_0^t dt_2 \int_0^{t_2} dt_1 K_{21}(t_2 - t_1).$$

Using energy conservation, the equation above is conveniently expressed as

$$\frac{\langle \mathcal{H}_2(t)\rangle}{\langle \mathcal{H}_1(0)\rangle} = \epsilon' \int_0^t dt_2 \int_0^{t_2} dt_1 a(\tau) \tag{4.104}$$

where $a(\tau)$ has been defined above for the two different cases (i) spin locking and (ii) ADRF [Eq. (4.94–4.98)]. It might be more convenient to express the energy $\langle \mathcal{H}_2(t)\rangle$ in terms of the equilibrium energy $\langle \mathcal{H}_{2,\mathrm{eq}}\rangle = \langle \mathcal{H}_2(t = \infty)\rangle$. It is easily recognized that

$$\langle \mathcal{H}_1(0)\rangle = \frac{1+\epsilon'}{\epsilon'} \langle \mathcal{H}_{2eq}\rangle$$

and with

$$\frac{\langle \mathcal{H}_2(t)\rangle}{\langle \mathcal{H}_{2,eq}\rangle} = \frac{M_S}{M_{S,eq}}$$

we rewrite Eq. (4.104) as [*11*]

$$\frac{M_S}{M_{S,eq}} = (1 + \epsilon') \int_0^t dt_2 \int_0^{t_2} dt_1 a(\tau). \tag{4.105}$$

As shown above [Eq. (4.94–4.98)] $a(\tau)$ is an oscillatory function, even when the correlation functions $a_x(\tau)$ and $a_z(\tau)$ or $C_x(\tau)$, $C_z(\tau)$ respectively, are monotonic functions due to a more or less isotropic I spin surrounding of the rare S spins, which is e.g. the case in a cubic crystal. The oscillation frequencies are $\omega_{eS} \pm \omega_{eI}$ and ω_{eS} as was demonstrated experimentally [*10*] (see Fig. 4.18). If the I-S interaction is more discrete i.e., a few next neighbour I spins have a distinct dipolar coupling to a single S spin of strength much larger than any other I-S interaction, oscillating correlation functions $C_x(\tau)$ and $C_z(\tau)$ are obtained, producing oscillations in the cross-polarization S spin magnetization. This last effect was observed by R. Ernst [*32*] and co-workers as was mentioned above and has been demonstrated in Fig. 4.19.

Only in exceptional cases it is possible to calculate $C_x(\tau)$ and $C_z(\tau)$ exactly. In order to describe experimental results by the current theory we will have to employ a certain degree of approximation.

Approximations of $C_x(\tau)$ *and* $C_z(\tau)$

For the convenience of the reader we recall Eq. (4.98), here in a slightly different notation

$$C_\mu(\tau) = \frac{\mathrm{Tr}\{(\sum_i B_i I_{\mu i}) e^{-iP_2(\cos\vartheta_I)\hat{\mathcal{H}}_{II}^{(0)}\tau} (\sum_i B_i I_{\mu i})\}}{\mathrm{Tr}\{(\sum_i B_i I_{\mu i})^2\}} \qquad (4.106)$$

$$\mu = x, z$$

where in the ARDF case $\cos\vartheta_I = 1$.

A moment expansion of $C_x(\tau)$ and $C_z(\tau)$ can, of course, be done exactly, leading, however, to more and more complicated expressions which become practically unmanagable when moments higher than the sixth moment are involved. Already the fourth moment of $C_{x,z}(\tau)$ contains lattice sums which correspond to the sixth moment of the free induction decay. Low order moments, however, like the second and fourth moment may be conveniently used to adjust approximations to the correct expressions.

A sort of zeroth order approximation would be to guess the functional form of $C_x(\tau)$ and $C_z(\tau)$. Although this may not be very appealing theorectically, it may, however, be very usefull, if this method succeeds in representing experimental data. The guessing, of course, has to be strongly supported by the experimental data.

Taking a closer look at the expressions for $C_x(\tau)$ and $C_z(\tau)$, we recognize, that in the case of B_i = const. the correlation function $C_x(\tau)$ would represent the free induction decay of the I spins, where the time scale is scaled by $P_2(\cos\vartheta_I)$. Since the B_i vary slightly in a cubic solid an additional "inhomogenous broadening" (*cum grano salis*) is introduced which inhibits slightly I spin flip-flop processes. Since free induction decays of abundant spins in cubic solids are often close to a Gaussian, it is the best guess to represent $C_x(\tau)$ by a Gaussian function. This is confirmed by the experimental data and also by higher order approximations to be discussed shortly. $C_z(\tau)$, however, depends merely on the local character of the I-S coupling; note, that $C_z(\tau) = 1$ if B_i = const. Since flip-flop processes among the I spins play a dominant role in $C_z(\tau)$ a sort of Anderson Weiss model would be appropriate, leading to a more or less Lorentzian character of $C_z(\tau)$, which was confirmed experimentally and is also displayed in higher order approximation [*11*]. Summarizing these arguments we are tempted to write:

$$C_x(\tau) = \exp(-t^2/\tau_c^2) \qquad (4.107a)$$

and

$$C_z(\tau) = \frac{1}{1 + (\tau/\tau_c)^2}. \qquad (4.107b)$$

In both cases the correlation time τ_c may be expressed by the second moment N_2 of $C_x(\tau)$ and $C_z(\tau)$, respectively, as

$$\frac{1}{\tau_c^2} = \frac{1}{2} \cdot N_2. \qquad (4.108)$$

Using Eq. (4.106) a straightforward calculation yields for

$$C_x(\tau) : N_2 = \frac{1}{3} I(I+1) P_2(\cos\vartheta_I)^2 \cdot \frac{5\sum_{i<j} A_{ij}^2 (B_i^2 + B_j^2) + 8\sum_{i<j} A_{ij}^2 B_i B_j}{\sum_i B_i^2} \tag{4.109}$$

and for

$$C_z(\tau) : N_2 = \frac{2}{3} I(I+1) P_2(\cos\vartheta_I)^2 \frac{\sum_{i<j} A_{ij}^2 (B_i - B_j)^2}{\sum_i B_i^2} \tag{4.110}$$

where

$$B_i = \gamma_I \gamma_S \hbar (1 - 3\cos^2\vartheta_i)/r_i^3 \tag{4.111a}$$

and

$$\mathcal{H}_{II} = \sum_{i<j} A_{ij} (3 I_{zi} I_{zj} - \mathbf{I}_i \cdot \mathbf{I}_j)$$

with

$$A_{ij} = \frac{1}{2} \gamma_I^2 \hbar (1 - 3\cos^2\vartheta_{ij})/r_{ij}^3. \tag{4.111b}$$

In the ADRF case we set $P_2(\cos\vartheta_I) = 1$.

Eq. (4.109) and (4.110) may be expressed in a more compact form, which is more suitable for a numerical computation as

$$C_x(\tau) : N_2 = \frac{1}{24} I(I+1) P_2(\cos\vartheta_I)^2 \frac{\gamma_I^4 \hbar^2}{a^6} \frac{5 S_4 + 18 S_3}{S_1} \tag{4.112a}$$

and

$$C_z(\tau) : N_2 = \frac{1}{12} I(I+1) P_2(\cos\vartheta_I)^2 \frac{\gamma_I^4 \hbar^2}{a^6} \cdot \frac{S_4}{S_1} \tag{4.112b}$$

using the lattice sums, defined as follows [*11*]:

$$\begin{aligned}
S_1 &= \sum_i b_i^2 & S_7 &= \sum_{i\neq j\neq k} b_{ij}^2 b_{ik}^2 b_i^2 \\
S_2 &= \sum_i b_i^4;\ S_2^* = \sum_i b_{ij}^4 & S_8 &= \sum_{i\neq j\neq k} b_{ij}^2 b_{ik}^2 b_i b_j \\
S_3 &= \sum_{i\neq j} b_{ij}^2 b_i b_j & S_9 &= \sum_{i\neq j\neq k} b_{ij}^2 b_{ik}^2 b_j b_k \\
S_4 &= \sum_{i\neq j} b_{ij}^2 (b_i - b_j)^2 & S_{10} &= \sum_{i\neq j\neq k} b_{ij}^2 b_{ik} b_{jk} b_k^2 \\
S_5 &= \sum_{i\neq j} b_{ij}^4 b_i b_j & S_{11} &= \sum_{i\neq j\neq k} b_{ij}^2 b_{ik} b_{jk} b_i b_j \\
S_6 &= \sum_{i\neq j} b_{ij}^4 (b_i - b_j)^2 & S_{12} &= \sum_{i\neq j\neq k} b_{ij}^2 b_{ik} b_{jk} b_j b_k
\end{aligned} \tag{4.113}$$

where

$$b_{ij} = (1 - 3\cos^2\vartheta_{ij})/r_{ij}^3 \tag{4.114}$$

$$b_i = (1 - 3\cos^2\vartheta_i)/r_i^3$$

with r_i being the distance between I and S spins and r_{ij} the distance between the spins I_i and I_j and where both distances are defined in units of the lattice constant a. The angle of the corresponding vector with the static magnetic field H_0 is denoted by $\cos\vartheta_i$ and $\cos\vartheta_{ij}$ respectively.

Different correlation times τ_c calculated according to Eqs. (4.108, 4.112) in the case of CaF_2 are compared with experimental values for different orientations of the magnetic field. The agreement is quite convincing (see Table 4.2).

Table 4.2. Correlation times τ_c for the dipolar fluctuation autocorrelation functions $(1 + \tau^2/\tau_c^2)^{-1}$ and $\exp(-\tau^2/\tau_c^2)$ for a $^{43}CaF_2$ crystal according to Demco, Tegenfeldt and Waugh (DTW) [11]. The correlation time $\tau_c = (2/N_2)^{1/2}$ is calculated, using Eq. (4.112) and compared with the experimental data of McArthur *et al.* [10]

Direction of H_0	$\tau_c(\mu\text{sec})$		
	exp (ADRF)[10]	DTW[11] (ADRF)	DTW[11] (SL, $\vartheta_I = \frac{\pi}{2}$)
(100)	–	45	37
(110)	57 ± 0.5	61	64
(111)	78 ± 1; 80 + 1	81	102

Nevertheless, Demco *et al.* [11] went a step further and applied the memory function approach to the calculation of the correlation functions $C_x(\tau)$ and $C_z(\tau)$. Recalling the discussion in Chapter 3 and Appendix E, we may write similar to Eq. (3.29)

$$\frac{d}{dt} C_\mu(t) = -\int_0^t dt' K(t - t')\, C_\mu(t') \tag{4.115}$$

or expressing the corresponding spectral distribution function $J(\omega)$ following Eq. (3.33) directly as

$$J(\omega) = \frac{K'(\omega)}{[\omega - K''(\omega)]^2 + [K'(\omega)]^2} \tag{4.116}$$

where

$$K'(\omega) = \int_0^\infty K(\tau)\cos\omega\tau\, d\tau$$

and

$$K''(\omega) = \int_0^\infty K(\tau)\sin\omega\tau\, d\tau.$$

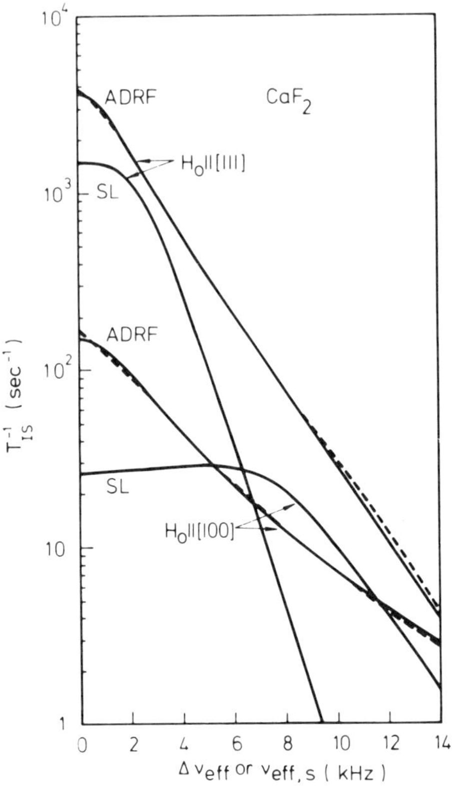

Fig. 4.22. Theoretical cross-polarization spectra for different orientations of the CaF_2 crystal as calculated by D. E. Demco *et al.* [*11*] for the spin-locking (SL) and ADRF case employing a memory function approach (solid lines) and an information theory approach (dashed line)

Approximation of $K(\tau)$ by a Gaussian function which can be rationalized by the calculation of the ratios of higher moments [*11*] leads to $K'(\omega)$ and $K''(\omega)$ equal to Eq. (3.38). Thus the second and the fourth moment N_2 and N_4 is all that is needed to calculate $J(\omega)$ in this degree of approximation. Expressions for the fourth moments of the correlation function can be obtained from Eq. (4.106) as

$$C_x(\tau) : N_4 = [P_2(\cos\vartheta_I)]^4 \frac{1}{3}\frac{I(I+1)}{16S_1} \left\{\frac{1}{5}[101I(I+1) - \frac{49}{2}]S_1S_2^* + \frac{1}{10}[176I(I+1) - 32]S_5 + \frac{1}{3}I(I+1)[77S_7 + 88S_8 + 24S_9 + 8S_{10} + 38S_{11} + 8S_{12}]\right\} \quad (4.117)$$

$$C_z(\tau) : N_4 = [P_2(\cos\vartheta_I]^4 \frac{1}{3}\frac{I(I+1)}{8S_1} \left\{\frac{1}{5}[16I(I+1) - 7] \cdot (S_1 \cdot S_2^* - S_5) + \frac{1}{3}I(I+1)[13S_7 - 16S_8 + 3S_9 - 8S_{10} + 16S_{11} - 8S_{12}]\right\} \quad (4.118)$$

Figure 4.17 compares those calculations with the experimental data of McArthur *et al.* [*10*]. Note, that no adjustable parameter of any kind was used in this theory. The agreement between theory and experiment is quite pleasing.

4.4 Spin-Decoupling Dynamics

Spin decoupling of abundant spins is a prerequisite in order to obtain high resolution NMR spectra in solids, containing different spin species [*3*]. Several NMR spectra of the cross-polarization type and also multiple-pulse experiments have been reported in the preceeding chapters which employed spin decoupling [*3, 25–28*]

A strong rf field H_{1I} is usually applied in a pulsed or continuous fashion at the resonance frequency of the non-observed abundant spins I which contribute to the coupling represented by $\mathcal{H}_{IS}$. If H_{1I} is strong such that I_z is flipped rapidly compared with the spin-spin interactions, the time average of $\mathcal{H}_{IS}$ vanishes and consequently the excess broadening due to the I spins is zero. The question arises: How strong does H_{1I} have to be in order to average out the heteronuclear coupling? On the first sight it would be sufficient, to make $\omega_{1I} = \gamma_I \cdot H_{1I}$ large compared with $\sqrt{M_2{}^{IS}}$ where M_2^{IS} is the second moment of the I-S coupling. This would be correct only, if the flip-flop term in the I-I coupling Hamiltonian would be zero. However, the flip-flop "motion" of the I-I spins modulates the I-S interaction which results in a narrowing of the resonance line of the S spins if γ_I/γ_S is large, as was first shown by Abragam and Winter [*4a, 34*]. In order to narrow the S spin resonance any further, ω_{1I} has to be comparable to the flip-flop rate of the I spins.

This leads to the interesting conclusion that a line broadening instead of a line narrowing will result when the "decoupling" field H_{1I} is applied at the "magic angle", i.e. off resonance by $\Delta\omega = \omega_{1I}/\sqrt{2}$. These aspects are borne out more quantitatively in the following:

We want to calculate the "excess broadening" of the S spin signal due to the I-S interaction.

Let us neglect homonuclear dipolar coupling of the S spins, i.e. we assume the following Hamiltonian in the rotating frame

$$\mathcal{H} = \mathcal{H}_{1I} + \mathcal{H}_{II} + \mathcal{H}_{IS} \tag{4.119}$$

where

$$\mathcal{H}_{1I} = -\omega_{eI} I_z ; \mathcal{H}_{II} = P_2(\cos\vartheta_I)\mathcal{H}_{II}^{(0)} + \mathcal{H}_{II}^{(ns)}$$

$$\mathcal{H}_{IS} = \cos\vartheta_I \mathcal{H}_{IzSz} - \sin\vartheta_I \mathcal{H}_{IxSz}.$$

We want to calculate $\langle S_x(t)\rangle$, after having the S spins prepared by an initializing 90° pulse in the y direction of the rotating frame. Using a similar orthogonal operator expansion of the density matrix as in Appendix E we arrive at

$$\frac{d}{dt}\langle S_x(t)\rangle = -\int_0^t dt' K(t-t')\ \langle S_x(t')\rangle \tag{4.120}$$

where $K(\tau)$ is obtained by a projection operator technique similar as in Appendix E as

$$K(\tau) = \frac{(S_x|\hat{\mathcal{H}}_{IS}\mathbf{S}(\tau)\,\hat{\mathcal{H}}_{IS}|S_x)}{(S_x|S_x)} \tag{4.121}$$

where

$$\mathbf{S}(\tau) = \exp\,[-i(1-P)\,\hat{\mathcal{H}}\tau]$$

with

$$P = \frac{|S_x)\,(S_x|}{(S_x|S_x)}$$

Notice, that

$$K(0) = \frac{(S_x|\hat{\mathcal{H}}_{IS}^2|S_x)}{(S_x|S_x)} = M_2^{IS}. \tag{4.122}$$

Recall from Appendix E that the line shape function

$$G(\omega) = \int_0^\infty dt \cos \omega t \,\langle S_x(t)\rangle$$

can be expressed as the Fourier transform of the Sspin free induction decay and may be obtained by Laplace inversion from Eq. (4.120) as

$$G(\omega) = \frac{K'(\omega)}{[\omega - K''(\omega)]^2 + [K'(\omega)]^2}$$

where $K'(\omega)$ and $K''(\omega)$ are the cosine and the sine transform of $K(\omega)$ respectively, according to Appendix E. If $K(\omega)$ is approximated by a Gaussian as before the line shape function $G(\omega)$ is straightforwardly calculated following along the lines of Appendix E.

Let us now discuss different limiting cases:

(i): $\omega_{1I} = 0$ i.e. $\Delta\omega = 0$ no decoupling

$$\mathbf{S}(\tau) = \exp\,[-i(1-P)\,(\mathcal{H}_{II} + \mathcal{H}_{IS})\tau]$$

from which follows for the second moment N_2 of $K(\tau)$

$$N_2 = [M_4^{ISIS} + M_4^{ISII} - (M_2^{IS})^2]/M_2^{IS} \tag{4.123}$$

where

$$M_4^{ISIS} = \frac{(S_x|\hat{\mathcal{H}}_{IS}^4|S_x)}{(S_x|S_x)}$$

and

$$M_4^{ISII} = \frac{(S_x | \hat{\mathcal{H}}_{IS} \hat{\mathcal{H}}_{II}^2 \hat{\mathcal{H}}_{IS} | S_x)}{(S_x | S_x)}$$

which may be expressed in terms of the lattice sums, as defined in the preceeding section by

$$M_4^{ISIS} = \frac{1}{3} I(I+1) \left(\frac{\gamma_I \gamma_S \hbar}{a^3}\right)^4 \left[\frac{1}{5}(3I^2 + 3I - 1) S_2 + I(I+1)(S_1^2 - S_2)\right], \quad (4.124)$$

$$M_4^{ISII} = \left[\frac{1}{3} I(I+1)\right]^2 \cdot \frac{1}{4} \left(\frac{\gamma_I}{\gamma_S}\right)^2 \left(\frac{\gamma_I \gamma_S \hbar}{a^3}\right)^4 S_4, \quad (4.125)$$

$$M_2^{IS} = \frac{1}{3} I(I+1) \left(\frac{\gamma_I \gamma_S \hbar}{a^3}\right) \cdot S_1. \quad (4.126)$$

In the following we shall use the moment ratios

$$\mu_1 = \frac{M_4^{ISIS}}{(M_2{}^{IS})^2} = 3 - \frac{1}{5}\left[\frac{1}{3} I(I+1)\right]^{-1} (2I(I+1)+1) \frac{S_2}{S_1^2} \quad (4.127)$$

and

$$\mu_2 = \frac{M_4^{ISII}}{(M_2^{IS})^2} = \frac{1}{4} \left(\frac{\gamma_I}{\gamma_S}\right)^2 \frac{S_4}{S_1^2}. \quad (4.128)$$

The ratio μ_1 is usually very close to three, whereas μ_2 may become very large, depending on the ratio of γ_I/γ_S. The second moment of $K(\tau)$ may now be expressed as

$$N_2 = M_2^{IS} (\mu_2 + \mu_1 - 1). \quad (4.129)$$

The lineshape $G(\omega)$ may be calculated as outlined above if the functional form of $K(\tau)$ is known or can be represented by a sufficiently reasonable approximate expression. For simplicity and a qualitative discussion let us assume $K(\tau)$ to be Gaussian in the following.

The line width of the undecoupled spectrum can now be calculated, using Eqs. (4.122, 4.129) and

$$\delta_0 = \sqrt{\frac{\pi}{2}} \left[\frac{K^2(0)}{N_2}\right]^{1/2}$$

as derived in Appendix E. As mentioned before this line width is already reduced by the flip-flop "motion" of the I spins as expressed by the large value of N_2 [*4a, 35*]. A similar approach has been taken by R. E. Walstedt [*35*] invoking the short correlation limit, resulting in a Lorentzian line shape

(ii) $\omega_{1I} \gg \omega_{LI}$ strong decoupling $\Delta\omega \neq 0$

$$\mathbf{S}(\tau) = \mathbf{S}_1(\tau) \mathbf{S}_0(\tau)$$

where

$$\mathbf{S}_1(\tau) = e^{-i(1-P)\hat{\mathcal{H}}_{1I}\tau}$$

and

$$S_0(\tau) = T\exp[-i\int_0^{\tau} dt'(1-P)(\hat{\mathcal{H}}_{II}(t') + \hat{\mathcal{H}}_{IS}(t')]$$

with

$$(1-P)(\hat{\mathcal{H}}_{II}(t') + \hat{\mathcal{H}}_{IS}(t')) = S_1(t')^{-1}(t')\ (1-P)(\hat{\mathcal{H}}_{II} + \hat{\mathcal{H}}_{IS})\ S_1(t').$$

Due to the strong decoupling field, we neglect non-secular terms in $\tilde{\mathcal{H}}(t')$ and write

$$S_0(\tau) = \exp[-i(1-P)\hat{\mathcal{H}}^{(0)}\tau]$$

where

$$\hat{\mathcal{H}}^{(0)} = P_2(\cos\vartheta_I)\hat{\mathcal{H}}_{II} + P_1(\cos\vartheta_I)\hat{\mathcal{H}}_{IS}.$$

Here ϑ_I is the angle of the effective field in the rotating frame with respect to H_0. A tedious, but straightforward calculation evaluating $K(\tau)$ according to Eq. (4.121) leads to

$$K(\tau) = K_z(\tau) + K_x(\tau)\cos\omega_{eI}\tau \tag{4.130}$$

where

$$K_\mu(\tau) = \frac{(S_x|\hat{\mathcal{H}}_{I\mu Sz}S_0(\tau)\hat{\mathcal{H}}_{IS}|S_x)}{(S_x|S_x)} \times \begin{cases} \cos\vartheta_I;\ \mu = z \\ -\sin\vartheta_I;\ \mu = x \end{cases} \tag{4.131}$$

with

$$K_z(0) = \cos^2\vartheta_I M_2^{IS} \tag{4.132a}$$

$$N_{2z} = [P_1^2(\cos\vartheta_i)(\mu_1 - 1) + P_2^2(\cos\vartheta_I)\mu_2]M_2^{IS} \tag{4.132b}$$

where M_2^{IS}, μ_1, and μ_2 are expressed in terms of lattice sums according to Eqs. (4.126–4.128). In the case of large ω_{eI}, the case considered here, $K_x(\tau)$ does not contribute appreciably to the spectral distributions $K'(\omega)$ and $K''(\omega)$ and will therefore be neglected.

Using Eqs. (4.131) and (4.132) the line shape and the corresponding "excess" linewidth can be calculated, using the memory function approach as discussed in Appendix E. [39]

With

$$\delta = \sqrt{\frac{\pi}{2}}\left[\frac{K^2(0)}{N_2}\right]^{1/2} \tag{4.133}$$

we obtain [39]

$$\delta = \sqrt{\frac{\pi}{2}}\left[\frac{\cos^4\vartheta_I M_2^{IS}}{P_2^2(\cos\vartheta_I)\mu_2 + \cos^2\vartheta_I(\mu_1 - 1)}\right]^{1/2}. \tag{4.124}$$

A similar expression for the excess line width results if a cutoff Lorentzian line shape of the S spin signal is assumed [*4a, 6*] and the corresponding second and fourth moments are inserted. The only difference is the factor μ_1 instead of $\mu_1 - 1$ in Eq. (4.134) and a different prefactor, namely $\pi/(2\sqrt{3})$.

Equation (4.134) describes qualitatively the behavior of the S spin linewidth δ under off-resonance decoupling of the I spins. Notice, that the linewidth δ does not follow the simple $\cos\vartheta_I$ variation of the second moment, but reflects rather sensitively the scaling of the correlation time of the I spin flip-flops, represented by $P_2^2(\cos\vartheta_I) \cdot \mu_2$. Figure 4.23 demonstrates this behavior very clearly in the case of ^{13}C in adamantane under proton decoupling and in the case of ^{109}Ag in AgF under ^{19}F decoupling. No decoupling is obtained far off-resonance i.e., for $\vartheta_I = 0$. The S spin resonance line is already narrowed by the unscaled fast fluctuation of the I spins due to their flip-flop motion as mentioned before [*4a, 34*].

Approaching the "magic angle" with $P_2(\cos\vartheta_I) = 0$, the correlation time of the I spins is slowed down and is finally quenched at the magic angle, resulting in a "broadened" S spin line which is solely due to the I-S interaction. This phenomenon may be considered as a freezing of the motion in spin space, resulting in a similar linewidth broadening as is observed with the freezing of lattice motion. However, the I-S interaction is also scaled by $\cos\vartheta_I$ which is an inevitable consequence of the I spin irradiation. At the "magic angle" the S spin line shape can be calculated exactly as the Fourier transform of

$$G_{IS}(t) = \langle S_x(t) \rangle = \prod_j \cos\left(\frac{1}{2\sqrt{3}} B_j t\right) \quad \text{for } I = \frac{1}{2} \tag{4.135a}$$

where B_j is defined by Eq. (4.78c). Equation (4.135a) is exact in the sense, that no memory function approach or moment expansion is involved and the corresponding lineshape can be calculated readily as a sum of δ functions over all combinations of $P_j = \pm 1$

$$F(\omega) = \frac{1}{2^n} \sum_{\text{comb.}P_j} \delta\left(\omega - \sum_{j=1}^{n} \frac{1}{2\sqrt{3}} B_j P_j\right); \; P_j \pm 1. \tag{4.135b}$$

Notice, that this is the same spectrum as observed in the experiment of Waugh and co-workers [*33*] (preceeding section) for vanishing chemical shift.

When ϑ_I approaches $90°$, the correlation time of the I spin flip-flops decreases towards half its natural value. However, since the I-S interactions is drastically reduced by $\cos\vartheta_I$ in this region, the fluctuation rate of the I spins becomes very large with respect to the I-S interaction and the line narrowing becomes very effective. This general behavior is borne out in Fig. 4.23 and qualitatively described by Eq. (4.134). For a more thorough discussion one would have to resort to a more realistic memory function, rather than simply assuming a Gaussian as was done above (see reference 39).

Recall that $K_z(\tau)$ [Eq. (4.131)] would be sufficiently close approximated by a Lorentzian, if $\mathbf{S}_0(\tau)$ would contain only the Hamiltonian $\mathcal{H}_{II}$, as was shown in the preceeding section. On the other hand, a Gaussian approximation would be appro-

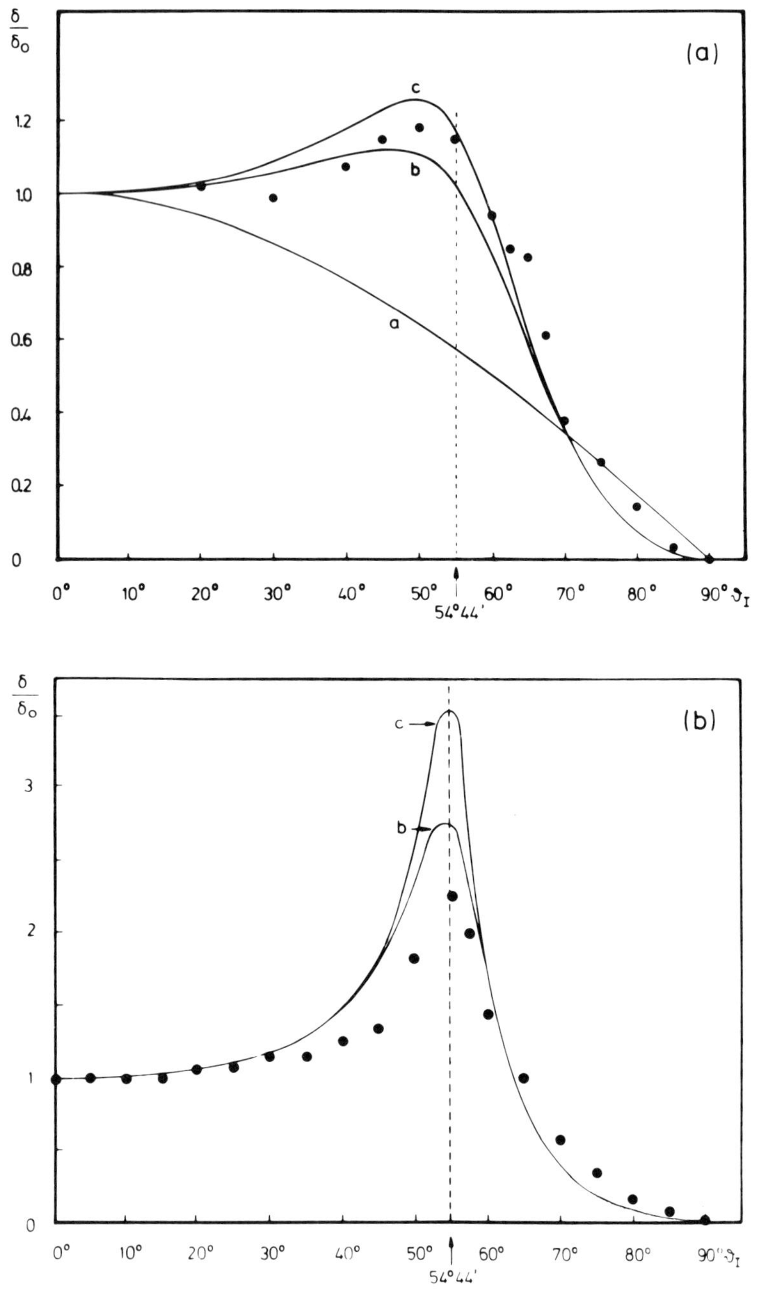

Fig. 4.23 a and b. (a) Normalized excess line width of the ^{13}C spectrum (S spins) in adamantane versus the angle ϑ_I of a strong effective field in the rotating frame of the I spins (protons) [*39*]. a: Variation of the second moment M_2^{IS} with ϑ_I. b: Calculated line width using the second and fourth moment assuming a truncated Lorentzian line. c: Linewidth calculated according to Eq. (4.134) using a memory function approach. (b) ^{109}Ag excess line width versus ϑ_I in AgF with ^{19}F being the I spin [*39*]

priate if only the term $(1 - P) \cdot \mathcal{H}_{IS}$ would be present in $\mathbf{S}_0(\tau)$. We may combine those approximations by writing:

$$K_z(\tau) = K_{II}(\tau) \cdot K_{IS}(\tau) \tag{4.136a}$$

where the second moment N_{2z} of $K_z(\tau)$ is given by Eq. (4.132b). Following the arguments above, we write

$$K_{II}(\tau) = \frac{1}{1 + (\tau/\tau_{II})^2} \tag{4.136b}$$

with

$$\frac{1}{\tau_{II}^2} = \frac{1}{2} P_2 (\cos \vartheta_I)^2 \cdot \mu_2 \cdot M_2^{IS}$$

and

$$K_{IS}(\tau) = \cos^2 \vartheta_I M_2^{IS} \exp[-\tau^2/\tau_{IS}^2] \tag{4.136c}$$

where

$$\frac{1}{\tau_{IS}^2} = \frac{1}{2} \cos^2 \vartheta_I (\mu_1 - 1) M_2^{IS}.$$

If an even higher degree of approximation is desired one has to express $K_z(\tau)$ itself by an integro-differential equation as was done by Demco, Tegenfeldt and Waugh [*11*] in the case of cross-polarization dynamics. Notice, however, that this involves higher order lattice sums and demands considerable computational effort. On the other hand, $K_{IS}(\tau)$ can be calculated numerically from

$$K_{IS}(\tau) = -\int_0^\tau dt K_{IS}(t) G'_{IS}(\tau - t) - G''_{IS}(\tau) \tag{4.136d}$$

where $G'_{IS}(t)$ and $G''_{IS}(t)$ are the first and second time derivative of $G_{IS}(t)$.

Using this memory function in the mixed memory function approach above leads to the exact lineshape at the magic angle and a decent approximation for the other values of ϑ_I. We finally remark, that the exact calculation of the linewidth $\Delta\upsilon = \delta/\pi$ at the magic angle $\vartheta_I = 54.7°$ according to Eq. (4.135) yields

^{13}C in adamantane $\Delta\upsilon = 1223$ Hz
^{109}Ag in AgF $\Delta\upsilon = 453$ Hz.

The reason that the experimental spectra do not reach these values, is the inhomogeneity of the rf field at the *I* spin resonance. This may be circumvented by using a four-pulse sequence instead. For further details consult reference [*39*].

(iii) *On resonance Decoupling* ($\Delta\omega = 0$)

In high resolution NMR experiments in solids as discussed in the preceeding sections the decoupling rf field is usually applied on resonance in order to make the spin decoupling most effective. This is highly recommended, as has been demonstrated in Fig. 4.23, since for small deviations from the resonance frequency ($\vartheta_I \neq 90^\circ$), the line broadening becomes appreciable.

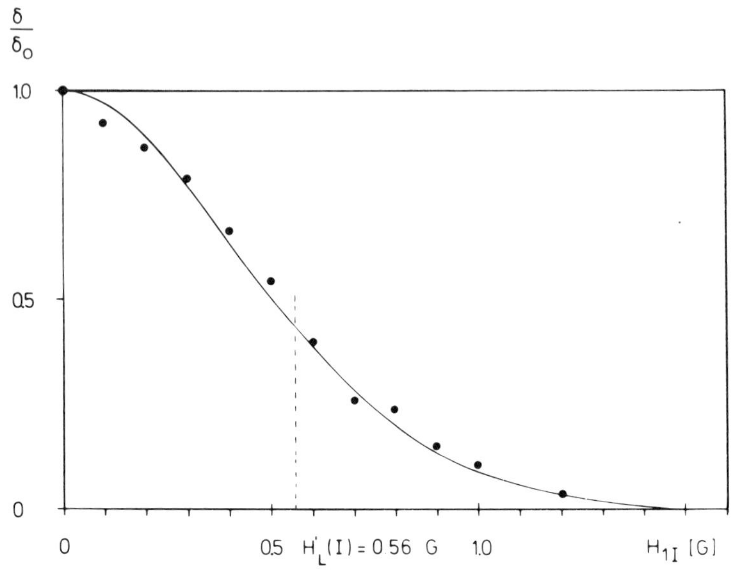

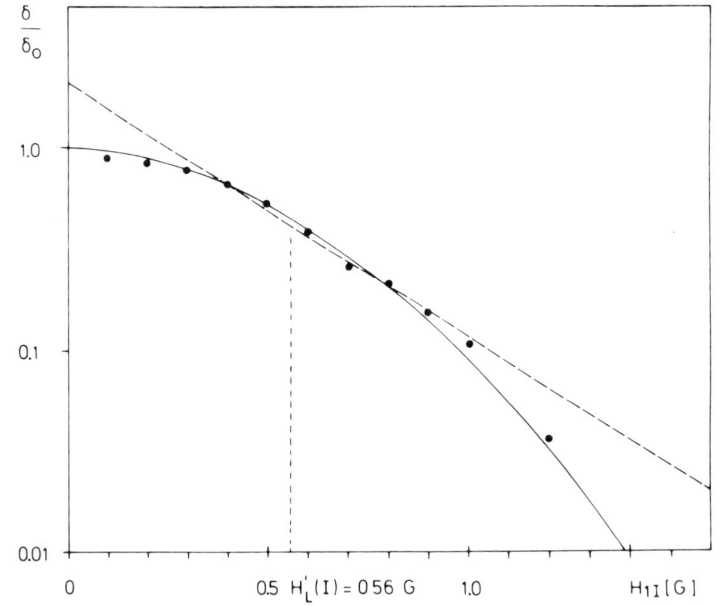

Fig. 4.24. Top: Normalized excess linewidth of the ^{13}C spectrum in adamantane applying an rf field H_{1I} at the proton resonance frequency (on-resonance decoupling [*39*]). Bottom: Logarithmic plot of top figure. The local field of the I spins (protons) is indicated. The solid line is calculated using a memory function approach as described in the text

However, as will be shown subsequently, the line narrowing due to on-resonance decoupling becomes very effective already for small rf fields which are comparable to the "local field" of the spins to be decoupled (here I spins). If we neglect the S spin dipolar coupling, we may write for the total Hamiltonian in the rotating frame

$$\mathcal{H} = \mathcal{H}_{II} + \mathcal{H}_{IS} + \mathcal{H}_{1I}$$

where

$$\mathcal{H}_{1I} = -\omega_{1I} I_x.$$

Using the memory function approach as outlined in Eqs. (4.120, 4.121), we obtain

$$K(\tau) = K_0(\tau) \cos \omega_{1I}\tau \tag{4.137a}$$

with

$$K_0(\tau) = \frac{(S_x | \mathcal{H}_{IS} \mathbf{S}_0(\tau) \mathcal{H}_{IS} | S_x)}{(S_x | S_x)} \tag{4.137b}$$

where we approximate the propagator $\mathbf{S}_0(\tau)$ for small ω_{1I} as

$$\mathbf{S}_0(\tau) = \exp[-i(1-P)(\hat{\mathcal{H}}_{II} + \hat{\mathcal{H}}_{IS})\tau] \tag{4.138}$$

using the same projector P as above. A moment expansion of $K_0(\tau)$ results in:

$$K_0(0) = M_2^{IS}$$

and

$$N_2 = M_2^{IS}(\mu_2 + \mu_1 - 1).$$

The approximation implied in Eq. (4.138) is reasonable only for $\omega_{1I} \to 0$ i.e. $\omega_{1I} \ll \| \mathcal{H}_{II} \|, \| \mathcal{H}_{IS} \|$.

The relation is therefore not expected to hold for large values of $\omega_{1I} \gg \| \mathcal{H}_{II} \|$. In this case the operator $(1-P)(\hat{\mathcal{H}}_{II} + \hat{\mathcal{H}}_{IS})$ in Eq. (4.138) has to be replaced by $-\frac{1}{2}\hat{\mathcal{H}}_{I_x I_x}$.

The corresponding second moment N_2 of the memory function $K_0(\tau)$ is than given by Eqs. (4.109) and (4.112a) with $P_2(\cos\vartheta_I) = -\frac{1}{2}$.

In order to obtain numerical values for the linewidth of the S spin resonance under on-resonance decoupling of the I spins, we assume a functional form of the memory function $K_0(\tau)$ in Eq. (4.137). The simplest functional form of $K_0(\tau)$ would be a Gaussian, which corresponds to the first term of a cumulant expansion of $K_0(\tau)$. The corresponding line width of the S spin signal for different values of ω_{1I} can be calculated as discussed before by using the lattice parameters of the system under investigation. There is no adjustable parameter in the calculation and the only approximation made concerns the assumption of the functional form of the memory function $K_0(\tau)$.

Notice, however, that the Gaussian approximation is „legitimate" only in the case $\omega_{1I} \gg \|\mathcal{H}_{II}\|$ where $K_0(\tau)$ corresponds to the cross-correlation function of the spin-locking case. For $\omega_{1I} \ll \|\mathcal{H}_{II}\|$, however, the "memory function" $K_0(\tau)$ may be better approximated by $K_z(\tau)$ of Eq. (4.136b). In order ot obtain a closed form expression for qualitative arguments we shall assume a Gaussian "memory function" $K_0(\tau)$ for the sake of simplicity.

The line shape and line width of the resonance line is readily obtained, using the corresponding equations of Appendix E.

The line width δ may be obtained by iteration from

$$\delta = K''(\delta) + [2K'(\delta = 0)K'(\delta) - [K'(\delta)]^2]^{1/2}. \tag{4.139}$$

Where because of the cosine in Eq. (4.136) $K'(\delta)$ and $K''(\delta)$ have to be replaced by

$$\begin{aligned} K'(\delta) &= \frac{1}{2}[K'(\delta + \omega_{1I}) + K'(\delta - \omega_{1I})] \\ K''(\delta) &= \frac{1}{2}[K''(\delta + \omega_{1I}) + K''(\delta - \omega_{1I})]. \end{aligned} \tag{4.140}$$

Using Gaussian approximation of $K_0(\tau)$ [Eq. (4.137)], the line width δ has been calculated in the case of ^{13}C in adamantane under proton on resonance decoupling [*39*]. This is compared with the experimental data in Fig. 4.24.

The agreement between theory and experiment is quite pleasing. Notice, that the decoupling is already very effective for fields comparable with the local field $H_L'(I) = [M_2^{II}/3]^{1/2}$ of the I spins. This was realized soon in practice [*3*] when already small decoupling fields sufficed to produce a high resolution spectrum.

In order to get a clear apprehension of the effectiveness of on-resonance decoupling we have plotted δ/δ_0 in Fig. 4.25 versus the dimensionless parameter

$$\Omega_1 = \frac{\omega_1}{\sqrt{M_2^{IS}}} \tag{4.141}$$

for different values of the moment ratio μ_2 and where μ_1 is kept sonstant, namely, $\mu_1 = 3$. These plots are generally applicable, since any specific spin system may be represented by μ_2. However, Fig. 4.25 serves only a demonstrative purpose in the light of the approximations involved in Eq. (4.138).

In a similar fashion we may treat the on-resonance pulse-decoupling. Let us apply the same approach as above, by representing the rf field Hamiltonian $\mathcal{H}_1(t)$ in Eq. (4.135) as

$$\mathcal{H}_1(t) = \pi I_x \delta(t - (2n + 1)t_c/4) \qquad n = 0, 1, 2, \ldots \tag{4.142}$$

describing the application of π pulses at the I spin resonance twice every cycle (cycle time t_c). The memory function $K(\tau)$ can now be expressed by

$$K(\tau) = K_0(\tau)\, F(t) \tag{4.143}$$

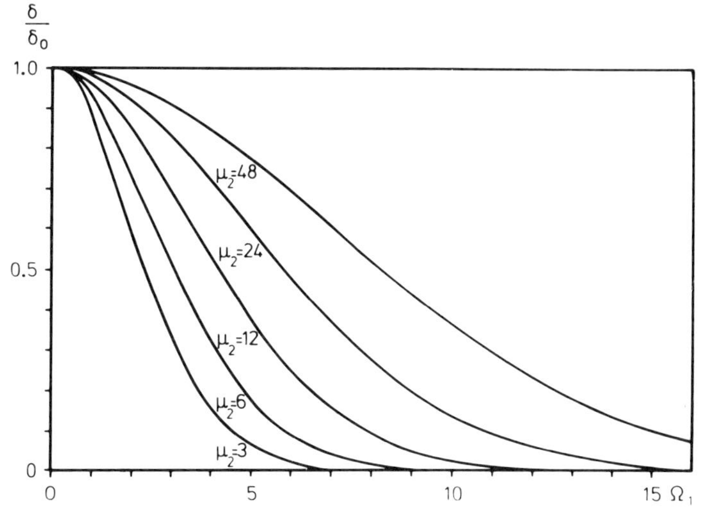

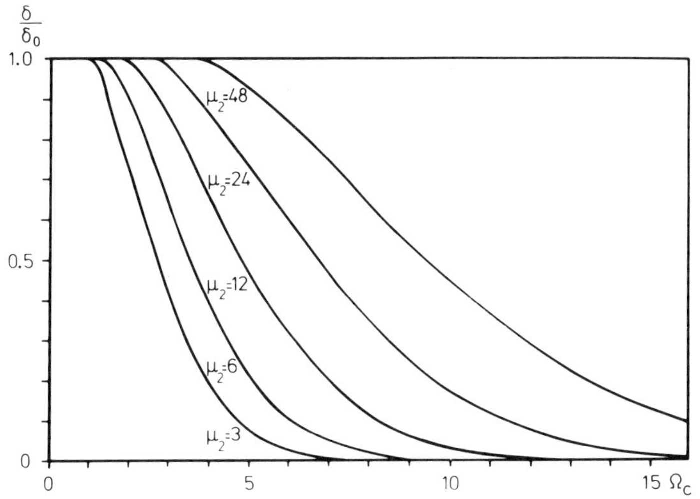

Fig. 4.25. Normalized excess linewidth of a dilute S spin system embedded in an abundant I spin system under I spin on-resonance decoupling conditions. Top: Continuos decoupling of I spins with a rate $\Omega_1 = \omega_1/(M_2^{IS})^{1/2}$. Bottom: Pulsed decoupling with a rate $\Omega_c = \omega_c/(M_2^{IS})^{1/2}$, where ω_c is the cycle frequency of the decoupling sequence (two π pulses per cycle in this case). The moment ratio μ_2 is explained in the text

where $K_0(\tau)$ is defined in Eq. (4.137) and where $F(t)$ is represented by

$$F(t) = \frac{4}{\pi}[\cos\omega_c t - \frac{1}{3}\cos 3\omega_c t + \frac{1}{5}\cos 5\omega_c t - \dots]$$
$$= \frac{4}{\pi}\sum_{k=0}^{\infty}\frac{(-1)^k}{2k+1}\cos(2k+1)\omega_c t;\ \omega_c = 2\pi/t_c. \tag{4.144}$$

Correspondingly, the function $K'(\delta)$ and $K''(\delta)$ in Eq. (4.139) have to be replaced by

$$K(\delta) = \frac{2}{\pi}\sum_{k=0}\frac{(-1)^k}{2k+1}[K(\omega + (2k+1)\omega_c) + K(\omega - (2k+1)\omega_c)] \tag{4.145}$$

where K' and K'' have to be inserted into Eq. (4.145). The iteration procedure of Eq. (4.139) using the K-functions as given by Eq. (4.145) leads to Fig. 4.25, where δ/δ_0 is plotted versus

$$\Omega_c = \frac{\omega_c}{\sqrt{M_2^{IS}}} \tag{4.146}$$

for different values of the moment ratio μ_2 and where $\mu_1 = 3$. The continuous decoupling procedure is in general more effective than the pulsed decoupling for the same cycle time ($\Omega_1 = \Omega_c$), as is demonstrated in Fig. 4.25.

Deuteron Decoupling

A further approach towards proton high resolution NMR in solids has been proposed by Pines *et al.* [*40*]. If most of the protons are exchanged by deuterons ($I = 1$) the main contribution to the linewidth of the protons is caused by the dipolar coupling of the protons to the nearby deuterons. Due to the different gyromagnetic ratios of deuterons and protons this coupling is reduced by a factor of about six, compared with the linewidth of the fully protonated compound. If strong rf irradiation is now applied to the deuteron resonance frequency, the deuterons are decoupled from the protons and a high resolution spectrum of the protons results. This is demonstrated in Fig. 4.26. If the quadrupole interaction of the deuterons vanishes, a medium size rf field with

$$\omega_1^2 \cong D^2 \tag{4.147}$$

is needed for the onset of line narrowing, if D is the size of the proton-deuteron coupling.

However, at first sight the deuteron decoupling seems to be impossible, since the deuteron spectrum is "inhomogeneously" broadened by a strong quadrupole interaction Q on the order of about 100 kHz. A first sighted argument would than imply that at least the spectral width Q of the deuteron spectrum has to be covered by the decoupling field, i.e.

$$\omega_1^2 \cong Q^2 \tag{4.148}$$

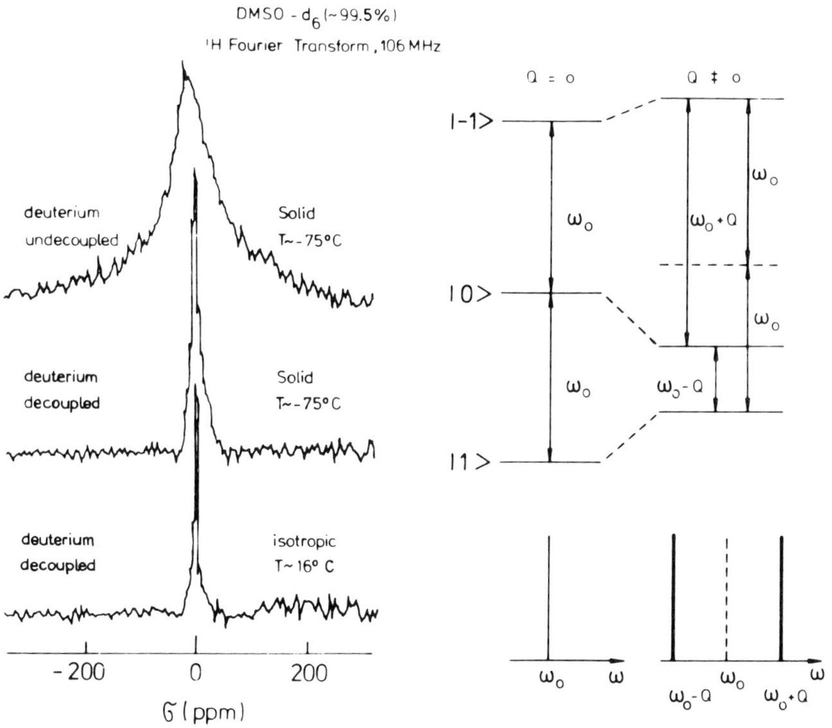

Fig. 4.26. Proton spectra of 99.5% deuterated dimethyl sulfoxide (DMSO) in the solid and liquid state respectively, with and without deuteron decoupling [*40*]. The independence of the double quantum transition on the quadrupole interaction of the deuterons is indicated in the energy level diagram of the deuterons (Courtesy of A. Pines)

in order to observe the onset of narrowing. This corresponds to an H_1 field of about 155 G and to a much higher value for obtaining a reasonable resolution. This is certainly a difficult task beyond technical practicability. Snyder and Meiboom [*41*], however, realized in the deuteron decoupling of liquid crystal spectra that a "double quantum transition" becomes very effective if the decoupling field is applied exactly at the resonance of the deuterons. Although all the three energy levels of the deuteron states in a magnetic field are shifted by the quadrupole interaction, the difference between the $|+1\rangle$ and $|-1\rangle$ states equals $2\omega_0$, independent of the quadrupole interaction in first order.

Since these levels are not coupled by rf fields, no transitions ($\Delta m = 2$) can be induced by a single photon. If, however, two photons are applied at exact resonance, rapid transitions among the $|\pm 1\rangle$ levels may be introduced. The first important point is, that the frequency is independent of the quadrupole interaction in first order and the second point, that the necessary rf field for the onset of decoupling needs to be much less than according to Eq. (4.148), namely [*40*]

$$\omega_1^2 \cong D \cdot Q.$$

This is why it is possible to decouple the deuterons with medium size rf fields, as is demonstrated in Fig. 4.26. The residual broadening of the proton line which is due to the proton-proton interaction, in a say 5% protonated compound, may be effectively reduced by applying a WHH-4 sequence in addition to the deuteron decoupling field (see Fig. 4.27). Since the dipolar coupling among the protons is already reduced considerably, no severe conditions apply to the multiple pulse sequence.

Especially the cycle time may become rather large, reducing the influence of accumulative pulse errors. A more detailed discussion may be found in Ref. [*40*].

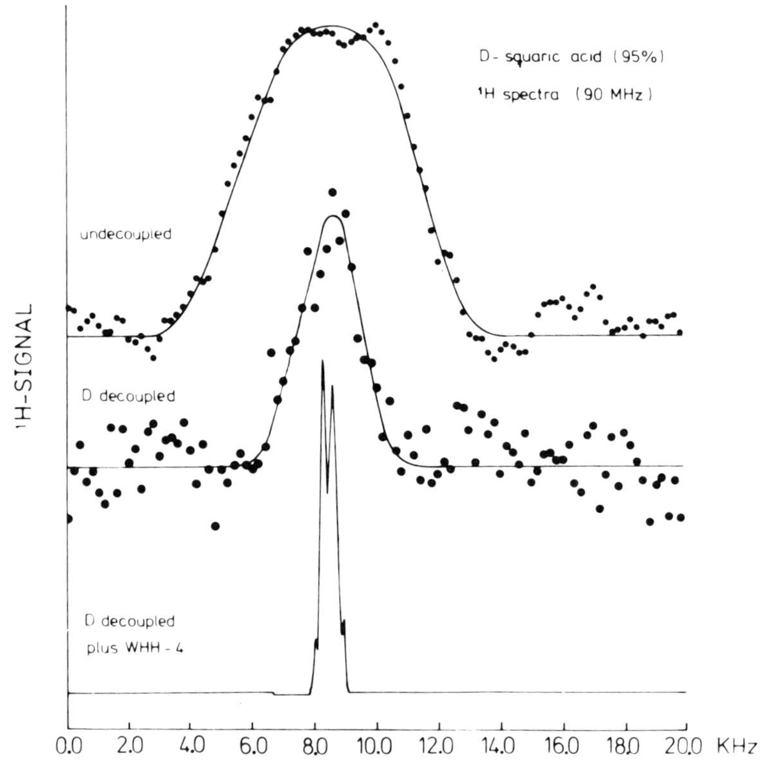

Fig. 4.27. Proton spectra of a 95% deuterated single crystal of squaric acid at 90 MHz at arbitrary orientation. A 20 *G* decoupling field is sufficient to decouple the deuterons from the protons, however, the dipolar broadening of the residual protons is still too large to resolve any fine structure (middle). If a WHH-4 sequence (t_c = 72 μs) is applied additionally, a high resolution proton spectrum results. (Courtesy of D. Suwelack)

5. Magnetic Shielding Tensor

As has been demonstrated in the preceeding chapters multiple pulse NMR and cross-polarization of dilute spins are the most prominent techniques for evaluating the chemical shielding tensor σ of nuclei in solids. Other methods for determining parts of the chemical shielding tensor like the "second moment" analysis, the NMR of partially oriented molecules dissolved in clathrates or liquid crystals will not be reviewed in this monograph because the shift tensors determined by these methods are mostly unreliable. A review of these methods including theoretical calculations has been given recently by Appleman and Dailey [*1*]. The interested reader is referred to their review article also for additional reference of chemical shielding data.

The interest in the magnetic shielding tensor can be traced back to the early days of NMR and generations of chemists have utilized the isotropic part of this quantity which is observed in a liquid, as a fingerprint of the chemical bond. In fact, the measurement of chemical shift has been developed into a tool for the investigation of molecular structure [*2, 3*]. Although the trace of this tensor is only one out of six numbers [*4*] (in case the tensor is symmetric), considerable effort has been devoted to relating the molecular electronic structure to this isotropic shift theoretically [*5, 6*]. Since now the full symmetric second rank shift tensor becomes available with the advent of the techniques described in this book, it is hoped that a better understanding of the underlying electronic structure will result and the "valid" theory will be descernable from the "invalid" theory.

Beside the actual size of the shielding tensor, its symmetry is an interesting quantity, since it is related to the molecular and crystal structure. In contrast to high resolution NMR in liquids, where the equivalence of nuclei is defined with respect to molecular symmetry alone (taking scalar coupling into account), in crystals we call those nuclei which are related by one of the symmetry elements of the space group *crystallografically equivalent.* On the other hand nuclei are called *magnetically equivalent*, if they are related by translation or/and inversion, which are elements of the space group. The tensor elements of magnetically equivalent nuclei are identical, whereas nuclei which are crystallografically equivalent are related by the point symmetry operations (rotation-reflection) of the space group, i.e. they are not necessarily magnetically equivalent [*7*].

It should be noted, that even from a single crystal study determining the complete shielding tensor including the principal axis system in general, no complete unambiguous assignment of the shielding tensor to a particular nuclear site in the unit cell can be made. This is readily visualized if two crystallografically equivalent nuclei are considered which are magnetically non-equivalent. These two nuclei result in two lines in the NMR spectrum and no symmetry argument can be applied to make an assignment. One has to rely on other arguments, based on theoretical grounds or consistency checks with related compounds. A more detailed account of symmetry can be found in Ref. [*7*].

5.1 Ramsey's Formula [4]

It is convenient to express the shift tensor as a sum of a "diamagnetic part" $\sigma^{(d)}$ and a "paramagnetic part" $\sigma^{(p)}$ [4]:

$$\sigma = \sigma^{(d)} + \sigma^{(p)}. \tag{5.1}$$

Let us consider a rigid molecule with no net electronic orbital or spin angular momentum in the absence of the magnetic field. As shown by Ramsey [4] for such as system in the state $|0\rangle$ the components $\sigma^{(d)}_{\alpha\beta}$ and $\sigma^{(p)}_{\alpha\beta}$ of the shielding tensor take the form

$$\sigma^{(d)}_{\alpha\beta} = \frac{e^2}{2mc^2} < 0 \mid \sum_i \frac{r_i^2 \delta_{\alpha\beta} - r_{i\alpha} r_{i\beta}}{r_i^3} \mid 0 > \tag{5.2}$$

and

$$\sigma^{(p)}_{\alpha\beta} = \frac{e^2}{2mc^2} \sum_k (E_0 - E_k)^{-1} \langle 0 \mid \sum_i L_{\alpha i} \mid k \rangle \langle k \mid \sum_i L_{\beta i}/r_i^3 \mid 0 \rangle$$
$$+ \langle 0 \mid \sum_i L_{\beta i}/r_i^3 \mid k \rangle \langle k \mid L_{\beta i} \mid 0 \rangle \tag{5.3}$$

where r_i is the vector from the nucleus whose shielding is to be determined to the ith electron; $L_{\alpha i}$ is the α component of the angular momentum operator of electron i about the nucleus under consideration. E_k is the energy of the state $|k\rangle$.

The "diamagnetic" term $\sigma^{(d)}_{\alpha\beta}$ contains only matrix elements involving the ground state wavefunctions, whereas the "paramagnetic" term, resulting from second order perturbation theory displays the "mixing in" of excited states into the ground state due to the vector potential of the magnetic field. The axial character of the magnetic field leads to a "dequenching" of the orbital angular momentum.

The calculation of the paramagnetic term is rather involved and in principle the knowledge of all the excited states is needed. Note, even though this term is of second order, it is by no means small compared with the diamagnetic term.

5.2 Approximate Calculations of the Shielding Tensor

A great amount of *ab initio* calculations exist, employing all kinds of approximated wave functions [1, 8, 9]. However, only simple molecules like H_2, F_2, HF etc. and molecular fragments have been treated in this fashion using Hartee-Fock functions as a suitable compromise between accuracy and simplicity [6, 9]. However, as Reid [10] states "the complexity of paramagnetic shielding is apparently sufficiently deterring that, for highly accurate calculations, molecules other than hydrogen tend to lose their appeal". Other so-called ab initio calculations using SUFO's (*SU*itably *F*udged *O*rbitals) have been applied to numerous simple molecules.

Although the isotropic shift calculations seem to theoretically reproduce the experimental values, there is usually no resemblance between the calculated anisotropies and the recently accurately measured shielding tensor elements. One approximation, usually employed, is the average energy or closure approximation [6] where $(E_0 - E_k)$ in the denominator of Eq. (5.3) is expressed by an average energy ΔE. This can be valid only in the case if one energy separation is small compared with all others. On the contrary, it has been demonstrated recently by Kempf *et al.* [11] that transitions between different states are responsible for different tensor elements. A review of these theoretical methods may be found in Refs. [1, 7].

We turn now to the approximate method of Gierke and Flygare [12, 13] (GF), which has met with some success. Especially in the case of proton shielding the GF method has been successfully applied by Haeberlen, [14–16] and Spiess and co-workers [11] to molecular fragments. The diamagnetic part of the shielding tensor is approximated in the GF-method by a multipole expansion up to the molecular quadrupole moment, whereas the paramagnetic part is mainly related to the spin rotation interaction tensor C, which is by no means easier to evaluate than the paramagnetic term itself, but may be obtained experimentally by other means. The shielding tensor in this approximation may be written as a sum of six terms with [12]

$$\sigma_{xx}^{(d)} = I + II + III + IV$$

where

$$\begin{aligned} I &= \sigma_{\text{atom}}^{(d)} \\ II &= \frac{e^2}{2mc^2} \sum_i Z_i r^{-3} (y_i^2 + z_i^2) \\ III &= \frac{e^2}{2mc^2} \sum_i \left[2 r_i^{-3} (y_i \langle y \rangle_i + z_i \langle z \rangle_i) - 3 r_i^{-5} (y_i^2 + z_i^2) \mathbf{r}_i \cdot \langle \vec{\rho} \rangle_i\right] \\ IV &= \frac{e^2}{2mc^2} \sum_i r_i^{-3} \langle \tfrac{1}{3} \rho^2 \rangle_i [2 - 3(y_i^2 + z_i^2)/r_i^2] \end{aligned} \tag{5.4}$$

and

$$\sigma_{xx}^{(p)} = V + VI$$

where

$$\begin{aligned} V &= \frac{e^2}{2mc^2} \cdot \frac{C_{xx} I_{xx} c}{e \hbar \mu_0 g_I} \\ VI &= -\frac{e^2}{2mc^2} \sum_i Z_i r_i^{-3} (y_i^2 + z_i^2) = -II. \end{aligned} \tag{5.5}$$

Here $r_i = (x_i, y_i, z_i)$ is the vector from the shielded nucleus to a neighbour nucleus i and $\langle \vec{\rho} \rangle_i$ is the expectation value of the vector from nucleus i to the electrons surrounding it

$$\langle \vec{\rho} \rangle_i = \sum_{ki}^{zi} \langle 0 | \rho_{ki} | 0 \rangle. \tag{5.6}$$

The components of $\langle\vec{\rho}\rangle_i$ are $\langle x\rangle_i$, $\langle y\rangle_i$ and $\langle z\rangle_i$.

In the same way $\langle\rho^2\rangle_i$ is defined.

$C_{\mu\upsilon}$ is the spin-rotation interaction tensor and $I_{\mu\upsilon}$ is the moment of inertia tensor. The other constants have their usual meaning. The summation is performed with the shielded nucleus omitted.

Term *I* represents the free atom contribution, whereas *II* gives the contribution of the electronic point charges centered at the other nuclei. This term *II* cancels with term *VI*. If the point charges are not centered at the *i*-th nucleus, but are displaced by a distance $\langle\vec{\rho}\rangle_i$ the third term *III* arises, which is called the dipolar term [*13*]. This term is in general quite small with respect to the others. On the contrary, term *IV*, the quadrupole term, may become rather large [*13*], because the electronic charge density at nuclues *i* is spatially extended. Note, that the quadrupole part is traceless i.e. it does not contribute to the isotropic shift tensors for simple molecules and molecular fragments. Combination with known spin-rotation tensors leads to the full shielding tensor in a semi-empiric fashion. Haeberlen, Spiess and co-workers [*11, 14–16*] have applied this method with some success to the explanation of proton shielding tensors (see also Ref. [*7*] for a detailed discussion).

The GF approach cannot really be applied to solids, since the σ values diverge logarithmically with increasing volume. Vaughan and co-workers [*17*] have therefore proposed a different approach which does not separate into "diamagnetic" and "paramagnetic" parts, which are gauge dependent, but rather separates into "geometrical" and "chemical" parts, which are separately gauge independent. Another different semiclassical approach has been proposed by Schmiedel and others [*18*].

5.3 Proton Shielding Tensors

In this most difficult domain of high resolution NMR in solids a breakthrough has been brought about by the multiple-pulse methods. Other methods such as the liquid crystal method, which claim to have obtained decent proton shift anisotropies give completely unreliable results in this field.

Multiple-pulse experiments were first applied to powder samples containing OH-groups, which show the largest proton shielding anisotropies (~ 20 ppm) observed to date [*7, 14–26*]. An example of such a spectrum was presented in Fig. 3.13, Section 3.2. As can be seen, the anisotropy range extends over the total isotropic shift observed for protons. Haeberlen and co-workers [*7, 14–16*] have pushed this field and determined a large amount of shielding tensors from powder spectra and performed very meticulous single crystal studies. A sample of such a single crystal spectrum was given in Fig. 3.11, Section 5.2. Haeberlen [*7*] and co-workers also got a clear physical apprehension of protonic shielding tensors by applying the Gierke-Flygare approach.

Since Haeberlen has dealt with these different aspects in his excellent review article extensively, we refer the interested reader to Ref. [*7*].

We shortly summarize different facts observed in protonic shielding tensors:

a) Protons in Hydrogen Bonds

(i) The shielding tensor is approximately axially symmetric, with the unique axis close to the vector joining the next nearest neighbours i.e. the "hydrogen bond direction".

(ii) The anisotropy is the largest observed among the proton shieldings and tends to be larger in ionic system than in neutral systems. Its size is about 20 ppm, with the most shielded axis being the unique axis.

(iii) In a planar structure always the least shielded axis is orthogonal to the plane.

Haeberlen [*7, 14*] and co-workers have shown, that these features can be explained by the quadrupolar term (*IV*) in Eq. (5.4). Because of its $1/r^3$ dependence only next nearest neighbours have to be considered, leading to a simple physical explanation.

A single crystal spectrum of protons in hydrogen bonds and its orientation dependence has been shown in Figs. 3.11 and 3.12 in Section 3.2.

b) Methylene Protons ($-CH_2$)

Methylene protons have been studied by Haeberlen [*7*] and co-workers [*14*]. This shielding tensor is known to be very small from powder spectra. On the other hand methylene protons have a very large dipolar coupling due to the small distance (1.7 Å), which makes the detection of shift anisotropies a formidable task. Haeberlen and co-workers solved this problem by utilizing the fact, that the dipolar interaction of the two protons vanishes when their internuclear vector is placed at the "magic angle" ($\vartheta_m = 54°44'$) with respect to the magnetic field [*7*]. The residual dipolar broadening due to other protons can be eliminated easily by multiple pulse techniques. Notice, that a conical 2π rotation of the magnetic field can be performed around the internuclear vector keeping the magic angle condition, but tracing out part of the shielding tensor. In fact, five of the six tensor elements are determined by this 2π rotation (see Section 2.3). The sixth number has been obtained by Kohlschuetter [*27*] by a rotation of the magnetic field perpendicular to the internuclear vector. It is claimed [*7*], that the shift tensor of these methylene protons is axially symmetric abount the *C-H* bond direction with $\Delta\sigma \approx$ 5–6 ppm.

c) Olefinic Protons ($-\overset{H}{C}=\overset{H}{C}-$)

These protons have been studied in three compounds so far [*7, 15, 16, 28*] (see Table 5.1). The conclusions about the shielding tensor which can be drawn tentatively are:

(i) complete non-axial symmetry,

(ii) total anisotropy $\sigma_{33} - \sigma_{11} \cong$ 6 ppm,

(iii) the least shielded direction is orthogonal to the molecular plane,

(iv) the in-plane principal axes are not always related to the bond directions.

d) Aromatic Protons

have been studied only in one case, namely ferrocene, $Fe(C_5H_5)_2$ by the Heidelberg group [*7, 28*]. The C_5H_5 rings are rapidly rotating about their 5-fold axes, leading to a motionally averaged axially symmetric shielding tensor $\Delta\sigma = -6.5 \pm 0.1$ ppm with the unique axis being the rotation axis. Because of structural phase transitions at low temperature which destroy the single crystal, it is not possible to obtain the complete shielding tensor. However, this result is remarkable, because it was demonstrated in this case that reliable shielding tensors can only be obtained when both the *shape* and the *intrinsic* anisotropy of the bulk susceptibility are properly taken into account [*28*].

Table 5.1. Proton shielding tensor σ_{11}, σ_{22}, and σ_{33} in ppm with respect to its average $\bar{\sigma}$ for different compounds. The reference for the average $\bar{\sigma}$ is σ_{TMS}. (PW, powder spectra; SC, single crystal; CW, wide line; MP, multiple pulse; DD, deuteron decoupling; RT, room temperature; $\sigma_{TMS} - \sigma_{adamantane} = 1.74$ ppm has been used)

Compound	Formula	T (° K)	σ_{11}	σ_{22}	σ_{33}	$\bar{\sigma}$	Method	Refs.
Carboxyl								
Oxalic acid	$(COOH)_2$	RT	− 6.6	− 4.5	11.1	−12.6	SC, MP	26, 19
Squaric acid	$C_4O_4H_2$	RT	− 7.3	− 7.3	14.7	−11.2	PW, MP	20
Butynedioic acid	$HOOCC_2COOH$	RT	− 6.7	− 6.7	13.3	−11.8	PW, MP	20
Trichloro acetic acid	CCl_3COOH	RT	− 6.6	− 6.6	13.3	−12	PW, MP	22
Trichloro acetic acid	CCl_3COOH	RT	− 8	− 4	12	−12	SC, CW	23
Malonic acid	$CH_2(COOH)_2$	RT	− 7.9	− 5.3	13.2	−15.6	SC, MP	14
Malonic acid	$CH_2(COOH)_2$	RT	− 8.1	− 4.1	12.2	−14.9	SC, MP	14
Maleic acid	$(CHCOOH)_2$	RT	− 8.5	− 4.2	12.7	−14.5	SC, MP	15
Maleic acid	$(CHCOOH)_2$	RT	−10	− 5.1	15.1	−16.6	SC, MP	15
Succinic acid	$(CH_2CO_2H^*)_2$	RT	− 6.3	− 6.3	12.6	−15.4	PW, MP	19
Fumaric acid	$(CHCO_2H^*)_2$	RT	− 5.6	− 5.6	11.2	−16.9	PW, MP	19
Phtalic acid	$C_6H_4(COOH^*)_2$	RT	− 7.3	− 7.3	14.6	−15	PW, MP	19
Potassium hydrogen maleate	$H^*OOCC_2H_2COOK$	RT	−11.6	− 8.7	20.2	−21	SC, MP	16
Methylene								
Malonic acid	$CH_2^*(COOH)_2$	RT	− 2	− 2	3.9	− 4.7	SC, MP	14
Malonic acid	$CH_2^*(COOH)_2$	RT	− 1.3	− 1.3	2.6	− 4.7	SC, MP	14
Succinic acid	$(CH_2^*COOH)_2$	RT	− 2.9	− 2.9	5.7	− 5.7	PW, MP	19
Succinic acid anhydride	$C_4H_4O_3$	RT	− 3.4	− 3.4	6.8	− 5.4	PW, MP	19

Table 5.1 (continued)

Compound	Formula	T (° K)	σ_{11}	σ_{22}	σ_{33}	$\bar{\sigma}$	Method	Refs.
Olefinic								
Potassium hydrogen maleate	$HOOCC_2H_2^*COOK$	RT	−2.4	−0.2	2.5	−6.6	SC, MP	16
Maleic acid	$(CH^*COOH)_2$	RT	−3	−0.3	3.3	−7.6	SC, MP	15
Fumaric acid	$(CH^*CO_2H)_2$	RT	0	0	0	−8.8	PW, MP	19
Aromatic								
Ferrocene	$(C_5H_5)_2Fe$	RT	−4.4	2.2	2.2	−6.1	SC, MP	7
Phatalic acid	$C_6H_4^*(COOH)_2$	RT	0	0	0	−9	PW, MP	19
Phtalic acid anhydride	$C_6H_4^*C_2O_3$	RT	0	0	0	−9	PW, MP	19
Others								
Ice	H^*DO	180	−11.5	−11.5	23	−5.1	PW, DD	29
KDP	KH_2PO_4	RT	−12	−12	24	−16.2	SC, MP	25
	KH^*SO_4 dimer	RT	−8.6	−7.6	16.2	−18.7	SC, MP	7, 28 d
	KH^*SO_4 chain	RT	−9.8	−8.2	18	−20.6	SC, MP	7, 28 d
	KH^*F_2	RT	−18.6	−11.4	30	−21.1	SC, MP	28 b
Calcium hydroxide	$Ca(OH)_2$	RT	−4.7	−4.7	9.3	−4.6	SC, MP	21
Magnesium sulfate	$MgSO_4H_2O$	RT	−6.5	−6.5	13	−9.3	SC, MP	24

An excellent discussion of proton shielding tensors has been given by Haeberlen and the interested reader is referred to Ref. [7] for further reading on this subject. Table 5.1 lists a number of proton shielding tensors thus far determined by multiple-pulse methods.

5.4 ^{19}F Shielding Tensors

The first application of multiple pulse techniques was performed by Waugh, Huber and Haeberlen [30] on ^{19}F in CaF_2. No shift anisotropy was to be expected, of course, since CaF_2 has cubic symmetry. However, not only the feasability of this approach to line narrowing was demonstrated by this experiment, also valuable information was obtained about the isotropic shift. Before proceeding to non degenerate shielding tensors, we take a brief look at the isotropic shift in cubic crystals thus far determined.

Vaughan and co-workers [31] have obtained a number of highly resolved spectra in group *II* difluorides such as cubic: CaF_2, SrF_2, BaF_2, CdF_2, HgF_2 and noncubic MgF_2 and ZnF_2. Figure 5.1 gives a sample of these spectra. All these compounds are closely related with nearly completely ionic bonding. Their isotropic shift, however,

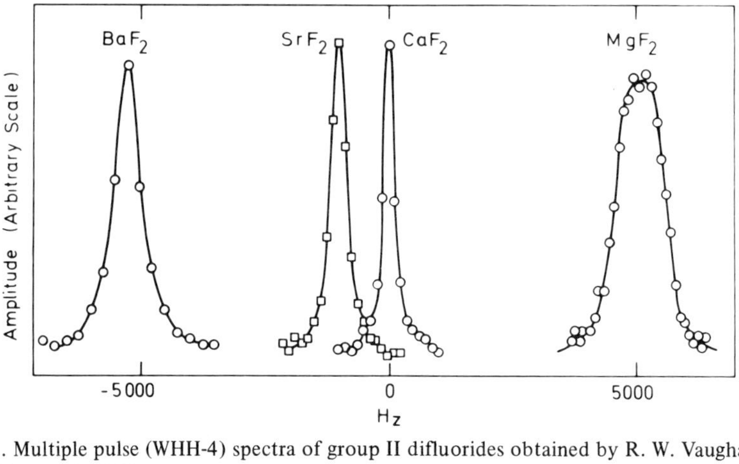

Fig. 5.1. Multiple pulse (WHH-4) spectra of group II difluorides obtained by R. W. Vaughan *et al.* [31a] at 56.4 MHz

can be correlated with electronegativities and a covalency parameter, calculated from electron spin resonance superhyperfine interaction parameters as shown in Fig. 5.2. Sears [32] has contributed to this investigation the shift of BeF_2. Since BeF_2 appears in different structures, this shift value is not expected to fit into this scheme. It is, however, a long standing discussion, which electronegativity scale should be used [33], but we are not going to discuss this point here.

Vaughan *et al.* [31] have also accounted for the shielding tensor in MgF_2 and ZnF_2 theoretically. The corresponding data may be found in Table 5.4 H. Ackermann and co-workers [34] have determined the quadrupole coupling constant of radio-

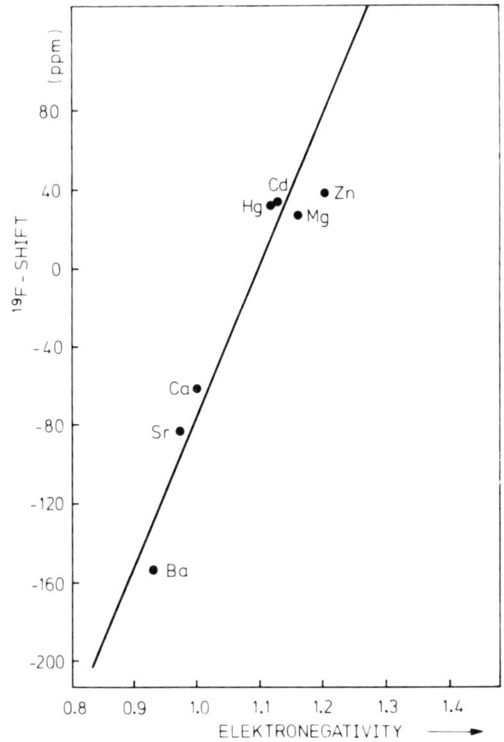

Fig. 5.2. ^{19}F shielding versus cation electronegativity on the Gordy scale [*33b*] for cubic and tetragonal group II difluorides obtained from multiple pulse spectra as demonstrated in Fig. 5.1 by R. W. Vaughan *et al.* [*31a*]

active ^{20}F in MgF_2. In order to investigate the dependence of the chemical shift on the volume, Lau and Vaughan [*35*] performed a pressure experiment on CaF_2. They observed a shift of −1.7 ppm/kbar, indicating greater overlap of the wavefunctions or "covalency" as the pressure is increased. For completeness we represent in Fig. 5.3 the isotropic shift of ^{19}F in alkalifluorides [*36*], which have been partially determined by multiple pulse techniques, employing spin decoupling. Here the same statement concerning the electronegativities applies as before. We only remark here, that the correlation with the Phillips [*32a*] ionicity scale is even worse.

Representative ^{19}F powder spectra as obtained by multiple pulse techniques have been shown already in Figs. 2.6 and 3.10. The first reliable ^{19}F shift tensor were obtained from those spectra [*37*]. Although no information about the principal axes was furnished by these experiments, still some assignments could be made by the virtue of known anisotropic molecular reorientation (see Section 2.6). Such a case is C_6F_6, in which an axially symmetric ^{19}F shielding tensor was observed at 200 °K, where the molecules are rotating about their 6-fold axis.

The immediate conclusion to be drawn, assigns the unique axis of the shielding tensor to the 6-fold axis (see Section 2.6).

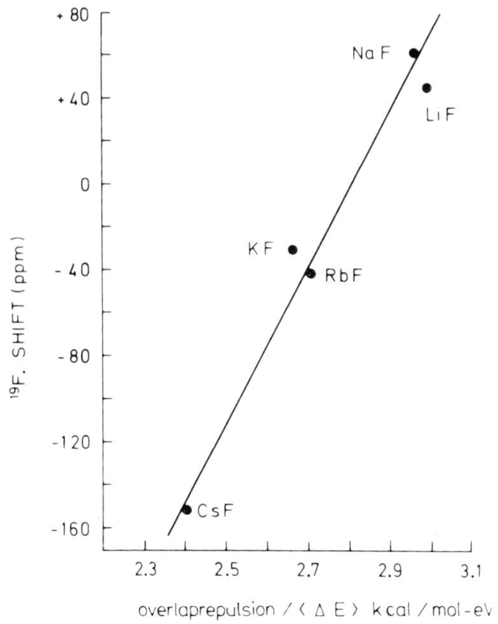

Fig. 5.3. ^{19}F shielding versus overlap repulsion energy between ions/mean excitation energy of F^-, for alkali fluorides according to R. E. J. Sears [*36*]

If the rotation is slowed down (much less than $\Delta\sigma$ at about 40 °K) the complete non averaged shielding tensor is expected to be observed. The unexpected result, however, was that the powder spectrum did not change within the limits of resolution. This leaves as the only conclusion, that the shielding tensor is axially symmetric within the limit of resolution and that the unique axis is the 6-fold axis [*37b*].

A similar conclusion was drawn from the high temperature and low temperature powder spectra of CF_3COOAg [*37b*]. An assignment of the principal axes could be made, (although not unambiguously) which turned out to be correct later as proved by a single crystal study [*39*]. The same arguments have been applied to the high and low temperature spectra of partially fluorinated benzenes [*40*], whose low temperature tensor elements are shown in Fig. 5.4. The assignment of the principle axes was supported by liquid crystal studies [*41*].

However, one should be aware, that this assignment is very preliminary, especially for the in-plane components of the shielding tensor. However, single crystal measurements on potassium tetrafluorophtaline seem to support this assignment [*42*]. Since the protons in these fluorinated benzenes lead to substantial broadening, pulsed spin decoupling had to be applied in this case (see Sections 3.2 and 4.4) The beauty of these data is reflected in the individual change of the different tensor elements due to the bonding

(i) The so-called "*ortho-effect*" is very pronounced, it shifts the σ_{33}-component (⊥ plane) by ~ +50 ppm if there is another fluorine in the ortho position,

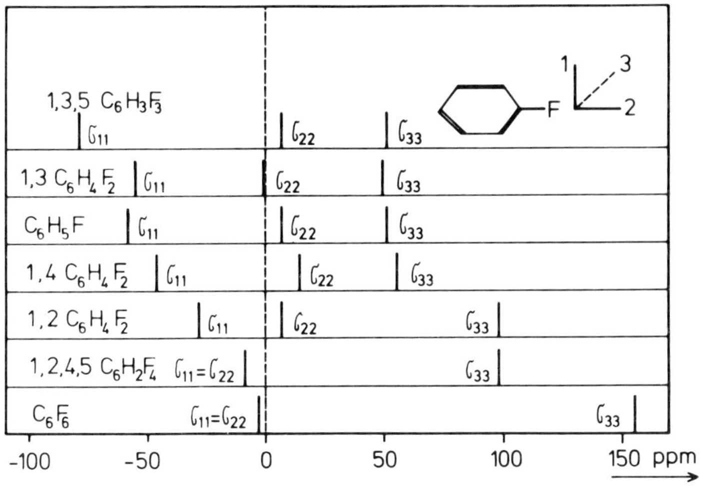

Fig. 5.4. Principal elements of the ^{19}F shielding tensor in partially fluorinated benzenes obtained from proton decoupled multiple pulse powder spectra [*40*]. The assignment of the principal axes is tentative, but in agreement with liquid crystal studies [*41*]

whereas leaving the other two components (σ_{11}, σ_{22}) more or less unchanged. *Meta-* and *para*-substitution does not lead to a similar pronounced effect.

(ii) The tensor element in the bond direction (σ_{22}) is not very sensitive at all to any kind of substitution which is also true for the σ_{33} component besides the "*ortho-effect*".

(iii) The only component which changes gradually with substitution is the σ_{11} component perpendicular to the bond direction and perpendicular to the plane.

Up to now, there is no reliable theory to account for these observations. In the early days of multiple-pulse line narrowing NMR we have tried to analyse our data along the lines of the theory of Karplus and Das [*6, 37b*]. This led to completely unacceptible conclusions about such quantities as "bond characters" and "hybridization parameters". The main outcome was to demonstrate, that this theory is completely unapplicable. But also more empiric approaches which have been successful in proton shielding, like the Gierke-Flygare method [*12*], failed completely.

This can be easily demonstrated by comparing the measured values of the shift tensor in monofluorobenzene ($\sigma_{11} = -58$ ppm, $\sigma_{22} = 8$ ppm, $\sigma_{33} = 53$ ppm) with the value calculated by Gierke and Flygare [*12*] ($\sigma_{11} = -56$ ppm, $\sigma_{22} = 123$ ppm, $\sigma_{33} = -68$ ppm). Besides the accidental agreement in the σ_{11}-component there is no resemblance between the theoretical and the experimental values. However, this can not be taken as a disproof of the GF approach, since most of the discrepancy, I believe, stems from the inaccurate values of the paramagnetic term, where one has to resort to the values of the spin rotation interaction tensor [*38*]. Also *ab initio* calculations employing SUFO's have not contributed much to the understanding of the shift tensors. Anyway, these few remarks should suffice to give pleasure to the theorist and challenge further theoretical effort towards a better understanding of this molecular quantity. Especially the paramagnetic part of the shielding tensor, which seems to be an extremely sensitive tool for studying the structure of molecular orbitals is still lacking a simple theoretical interpretation.

Leaving the complete theoretical understanding of the ^{19}F shielding tensor to the future, we now turn to another interesting example of a single crystal study, namely, the ^{19}F shielding tensor in

Silver-trifluoroacetate (CF_3COOAg) [*39*]

The first ectensive single crystal study using multiple-pulse line narrowing techniques was performed on the ^{19}F spins in the methyl group of this compound [*39*]. As part of the same investigation the complete crystal structure was determined by X-rays. The monoclinic unit cell contains four $(CF_3COOAg)_2$-dimers ($Z = 8$) as shown in Fig. 5.5. The space group determined is Cc or C_2/c, which cannot be distinguished

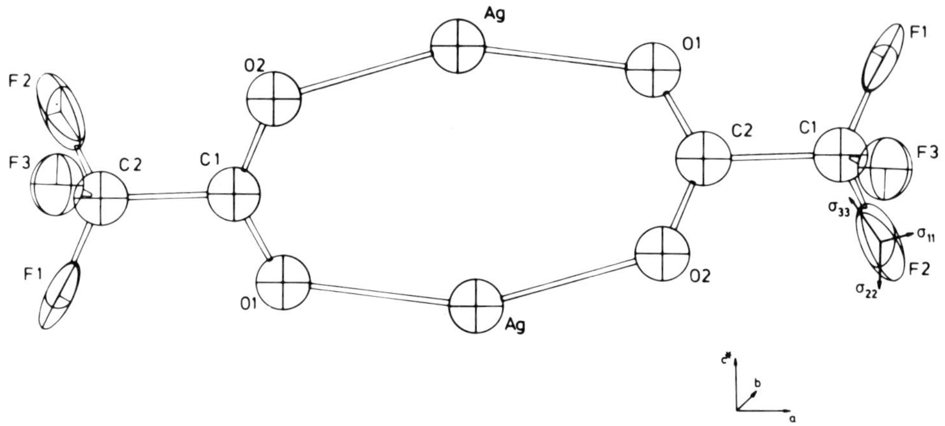

Fig. 5.5. Pictorial representation of the ^{19}F shielding tensors (aestheticized shielding tensor representation according to the prescription in Section 2.1) of the CF_3COOAg dimer as viewed along the *b*-axis of the monoclinic crystallografic unit cell. The chemical shielding tensors were determined from orientation plots of a single crystal study similar to Fig. 3.9 by Griffin *et al.* [*39*]. (Courtesy of D. Suwelack)

by the X-ray analysis nor by the NMR data. The most pronounced result is, that the apparent 3-fold axes of all CF_3-groups is parallel and approximately along the *a*-axis of the unit cell. At room temperature the CF_3-group is rapidly rotating about its 3-fold axis, leading to a single line. The shielding tensor as determined from the room temperature data, however, is not axially symmetric as expected, because it displays the average of three incongruent tensors (see Table 5.2).

This is borne out by the low temperature spectrum (see Fig. 3.8) which displays all the six lines of the six magnetically inequivalent ^{19}F nuclei. Three incongruent shielding tensors are observed for the three crystallografically inequivalent fluorine atoms in the methyl group. This is consistent with the space group C_2/c which is a special case of the space group Cc. A slight deviation of the dimer away from the two-fold axis would lead to the space group Cc with four magnetically inequivalent CF_3-groups, resulting in twelve lines at low temperature. This has not been observed,

Table 5.2. Tensor elements σ_{11}, σ_{22}, σ_{33} in ppm and direction cosines with respect to the *a, b, c** crystal axis system of ^{19}F in CF_3COOAg at room temperature from evaluation of orientation plots like Fig. 2.4 (Griffin *et al.* [*39*]). The average of the three low temperature tensors performing random jumps about the C_3-axis is included

	σ_{11}	Direction cosines	σ_{22}	Direction cosines	σ_{33}	Direction cosines	Ref. $\bar{\sigma}(C_6F_6)$
Room		−0.024		0.9985		0.0495	
temperature	−43.8	0.0203	15.7	−0.0490	29.1	0.9986	−95
		0.9995		0.0253		−0.0192	
Low temper-		0.0039		0.9999		0.0169	
ature (40 °)	−44.0	0.0027	13.0	0.0169	31.0	0.9999	−92
rapid jump		1.0000		0.0038		0.0027	
average							

but cannot be excluded because of the limited resolution. On the contrary, the ^{13}C spectra we have observed in the cross-polarization experiments (see Section 4.2) seem to support the space group Cc. However, this point is not completely settled at the time of writing.

Notice, that the symmetries of the unit cell render the fluorines, labelled F_1, F_2, and F_3 in the methyl group to be crystallografically inequivalent. However, this does not necessarily allow any conclusion to be drawn about the shielding tensor of these nuclei. In other words, the symmetries of the unit cell permit the fluorines to have incongruent shielding tensors, but symmetry is only a necessary, not a sufficient condition for the three shielding tensors to be different. Table 5.3 summarizes the low temperature single crystal results of the three incongruent shielding tensors of the CF_3-group. We may summarize the results as follows:

(i) The shielding tensor is "complete non-axially symmetric" with average values ($\sigma_{11} = -71$ ppm, $\sigma_{22} = 0.0$ ppm, $\sigma_{33} = 71$ ppm).

(ii) The most shielded direction lies approximately along the CF-bond, with a deviation of about 10°.

(iii) The least shielded axis is perpendicular to the bond and in the CCF plane.

(iv) The intermediate direction is perpendicular to the bond and the CCF plane.

A pictorial representation of these shielding tensors according to the rules stated in Section 2.2 is given in Fig. 5.5.

The difference of the three fluorine tensors is pronounced and by no means negligible. It is imposed by the crystal structure and is very likely due to the quadrupolar term of the diamagnetic shielding.

Scalar coupling (*J*-coupling) could have been observed in principle. since the three fluorine of the methyl group are magnetically inequivalent. However, the resolution was not sufficient to detect any isotropic part of the indirect spin-spin coupling. Notice, that the symmetric part of the indirect spin-spin coupling tensor is averaged by the multiple pulse experiment applied. This is so, because it has the same spin symmetry as the dipolar coupling Hamiltonian, if antisymmetric parts are neglected. We conclude this section by summarizing in Table 5.4 ^{19}F shielding tensors, which have to our knowledge thus far been determined.

Table 5.3. ^{19}F shielding tensor in solid CF_3COOAg at 40 °K according to Griffin *et al.* [*39*]. The direction cosines refer to the *a, b, c** unit cell axes. The labelling of the fluorine corresponds to Fig. 5.5, where a pictorial representation of the tensors is given. The errors are ±2 ppm and ±2° respectively

Fluorine	σ_{11}	Direction cosine	σ_{22}	Direction cosine	σ_{33}	Direction cosine	$\bar{\sigma}$ (ref. C_6F_6)	CCF bond angle	CF bond[a] direction cosines
F_1	−64.3	0.8791 −0.0638 0.4724	4.3	0.2545 0.8645 −0.4335	60.0	−0.4031 0.4986 0.7674	−89.4	114.1°	−0.3658 0.3780 0.8512
F_2	−75.0	0.9138 0.3103 −0.2622	2.0	0.0156 −0.6559 −0.7548	73.0	−0.4059 0.6883 −0.6014	−90.0	110.3°	−0.3481 0.5639 −0.7491
F_3	−73.6	0.8864 −0.4614 0.0400	−6.6	−0.0357 0.0236 0.9992	80.0	−0.4617 −0.8870 0.0152	−96.5	110.3°	−0.3875 −0.9144 −0.1182
average	−71.0		−0.1		71.0		−92.0		
Powder pattern 32 b	−67.0		−3.0		70.0		−95.5		

[a] The direction cosines of the C–C bond are: 0.9990, 0.0405, and 0.0343.

Table 5.4. Fluor (^{19}F) shielding tensor σ_{11}, σ_{22}, σ_{33} in ppm with respect to its average $\bar{\sigma}$ for different compounds. The reference for the average is $\sigma_{C_6F_6}$. (PW, powder spectra, SC, single crystal; MF, molecular frame; LC, liquid crystal; CT, clathrate; CW, wide line, FID, free induction decay; FM, firstmoment; SM, second moment; MP, multiple pulse; RT, room temperature; NP, nematic phase; $\sigma_{C_6F_6} - \sigma_{CaF_2} = 57$ ppm has been used)

Compound	Formula	T (°K)	σ_{11}	σ_{22}	σ_{33}	$\bar{\sigma}$	Method	Refs.
Aromatic								
Fluorobenzene	C_6F_6	40	− 51.7	− 51.7	103.4	7	PW, MP	37b
Fluorobenzene	C_6F_6	NP	− 53	− 53	106	0	MF, LC	41b
	1,3,5,-$C_6H_3F_3$	77	− 74	7	68	− 57	PW, MP	40
	1,3,5-$C_6H_3F_3$	NP	− 94	27	67	− 53	MF, LC	41a
	1,3-$C_6H_4F_2$	77	− 69	11	57	− 52	PW, MP	40
	1,3-$C_6H_4F_2$	NP	− 98	29	69	− 53	MF, LC	41a
	C_6H_5F	77	− 58	7	51	− 50	PW, MP	40
	C_6H_5F	NP	− 94	27	67	− 50	MF, LC	41a
	1,4-$C_6H_4F_2$	77	− 63	7	56	− 42	PW, MP	40
	1,4-$C_6H_4F_2$	NP	− 60	0	60	− 43	MF,LC	41a
	1,2-$C_6H_4F_2$	77	− 67	− 17	85	− 24	PW, MP	40
	1,2-$C_6H_4F_2$	NP	− 34	− 3	107	− 24	MF, LC	41a
	1,2,4,5-$C_6H_2F_4$	77	− 58	− 24	80	− 24	PW, MP	40
	$C_6F_3Br_3$	NP	− 75	37	37		MF, LC	47
Fluoranil	$C_6F_4O_2$	300	− 84	− 33	117	− 20	PW, MP	37
Perfluoro naphtalene	$C_{10}F_8$	300	− 53	− 53	106	− 19	PW, MP	37b
Perfluoro naphtalene	$C_{10}F_8$	83	− 50	− 50	100	− 16	PW, MP	37b
Perfluoro biphenyl		83	− 48	− 48	96	− 16.3	PW, MP	37b
Perfluoro benzophenone	$C_6F_5COC_6F_5$	83	− 66	− 31	97	− 29	PW, MP	37b
Potassium-tetrafluorophtalate	$F^*_2C_6F_2(CO_2K)_2$	77	− 55	− 43	98	− 3.5	SC, MP	42
Potassium-tetrafluorophtalate	$F_2C_6F^*_2(CO_2K)_2$	77	− 52	− 33	85	− 22	SC, MP	42

Table 5.4 (continued)

Compound	Formula	T (° K)	σ_{11}	σ_{22}	σ_{33}	$\bar{\sigma}$	Method	Refs.
Methyl								
Silver trifluoro acetate	CF_3COOAg	40	−71	0	71	−92	SC, MP	39
Silver trifluoro acetate	CF_3COOAg	RT	−44	15	29	−95	SC, MP	39
	$(CF_3CO)_2O$	182	−46	23	23	−93	PW, MP	37b
		40	−75	−1	76	−72	PW, MP	37b
	CHF_3	4	−35	−35	70	−78	CT, CW	45b, 47
	CF_2BrCF_2Br	85	−87	−87	173		SM, CW	46, 47
	$CFCl_2CFCl_2$	85	−80	−80	160		SM, CW	46, 47
Others								
Fluorine	F_2	29	−350	−350	700	−593	SM, CW, FM	43
	NF_3	4	−130	−130	260	−282	CT, CW	45b, 47
	KHF_2	RT	−36	−20	56	−23.5	SC, FID	28c
	MgF_2	RT	−15	0	15	28	SC, MP	31
	ZnF_2	RT	−22	0	22	37	SC, MP	31a
	XeF_4		−190	−190	280	−107	SM, CW	51, 52
	XeF_2		−35	−35	70	61	SM, CW	53
	$BaFPO_2$		−61	−61	122		SM, CW	54
Fluoroapatite	$Ca_5F(PO_4)_3$	298	−56	28	28	−83	SC, FID	44
	CH_3F	1.3	−44	22	22	116	CT, CW	45a, 47
	$CF_2S_2CF_2$	77	−109	54	54	−111	PW, MP	49
Teflon	$(F_2C{=}CF_2)_n$	153	−71	20	51	−41	PW, MP	37b
Teflon	$(F_2C{=}CF_2)_n$	81	−80	21	59	−40	PW, MP	37b
Teflon	$(F_2C{=}CF_2)_n$	77	−80	10	7	−35	PW, MP	37b
Teflon	$(F_2C{=}CF_2)_n$		−114	57	57		SM, CW	50

We have not included in this table, however, those values, which did not seem to be very reliable because of the method used. Only those data from liquid crystal studies are included which can be compared directly with solid state measurements. Hull and Sykes [*48*] have utilized some of the listed ^{19}F shielding tensors to investigate the spin lattice relaxation of fluor labelled protein molecules at high fields (5 Tesla).

5.5 ^{13}C Shielding Tensors

Some ^{13}C shielding tensors have been observed in solids already before the advent of line narrowing techniques, by the virtue of the small gyromagnetic ratio of ^{13}C, its low natural abundance and its relative large shielding anisotropy in noncubic surrounding [*55, 56*]. The brute force technique of determining shielding tensors directly from the NMR spectrum by applying sufficiently high magnetic fields (6–8 Tesla) and exchanging the protons by deuterons has undergone a renaissance recently [*11, 57*].

Nevertheless, the biggest supply of ^{13}C shielding tensors has been furnished by Pines, Waugh and co-workers [*58–64*] applying cross-polarization techniques or the so-called PENIS experiment (Proton Enhanced Nuclear Induction Spectroscopy) as proposed by Pines *et al.* [*58*] (see Chapter 4).

As representative examples we have shown in Fig. 4.10 of Section 4.2 some powder spectra of related compounds as determined by Waugh and co-workers [60]. ^{13}C shielding tensors as determined from these spectra are listed among others in Table 5.5.

Let us take a closer look at a representative single crystal investigation of ^{13}C shielding tensors in durene (1, 2, 4, 5-tetramethylbenzene), which was performed by Waugh and co-workers [*59, 63*]. The crystal structure of durene is monoclinic with space group C_{2h}^5 ($P2_1/a$). The asymmetric structural unit consists of one-half of a molecule. On expects therefore at most five incongruent ^{13}C shielding tensors, namely two from the methyl carbons, two from the ring carbons, which are bonded to the methylgroup and one unsubstituted ring carbon. The other tensors are related to these by the symmetry operations of C_{2h}^5.

Since the unit cell contains four asymmetric units, and at most five incongruent ^{13}C shielding tensors are allowed for each asymmetric unit, the maximum number of lines to be observed would be twenty. However, the two halves of each molecule are related by a center of inversion, leaving only ten lines, because of the invariance of the symmetric shielding tensor to inversion. That the ten lines are actually observed at some crystal orientations is demonstrated in Fig. 5.6.

Orientation plots of the shift data were obtained for three orthogonal rotation axes, from which the ^{13}C shielding tensors could be determined.

A complete pictorial representation of these tensors using the aesthetizised shielding tensor representation as proposed in Section 2.2 is given in Fig. 5.7. Numerical values may be obtained from Table 5.5. Let us summarize some conclusions which can be drawn from these investigations:

Table 5.5. Carbon (^{13}C) shielding tensor σ_{11}, σ_{22}, σ_{33} in ppm with respect to its average $\bar{\sigma}$ for different compounds. The reference for the average is C_6H_6. (PW, powder spectra; SC, single crystal; FID, free induction decay; CP, cross polarization; RT, room temperature; $\sigma_{C_6H_6} - \sigma_{CS_2} = 65$ ppm has been used)

Compound	Formula	T (° K)	σ_{11}	σ_{22}	σ_{33}	$\bar{\sigma}$	Method	Refs.
Aromatic								
Benzene	$C_6^*H_6$	223	-60	-60	120	-3	PW, CP	60
Hexafluor benzene	$C_6^*F_6$	233	-13	-13	27	-29	PW, CP	60
Hexamethyl benzene	$C_6^*(CH_3)_6$	296	-56	-56	112	-5	PW, CP	60
			-56	-56	112	-3	SC, CP	63
Hexamethyl benzene	$C_6^*(CH_3)_6$	87	-95	-17	113	-9	PW, CP	60
Toluene	$C_6^*H_5CH_3$	87	-104	-17	121	11	PW, CP	60
Durene	$C_2H_2(C^*CH_3)_4$	RT	-94	-22	116	-5	SC, CP	59
	$C_2^*H_2(CCH_3)_4$	RT	-92	8	85	-2	SC, CP	59
Hexaethyl benzene	$C_6^*(CH_2CH_3)_6$	RT	-85	-29	114	-6.5	SC, CP	63
Pentamethyl benzene	$CHC_5^*(CH_3)_5$	RT	-84	-29	112	-4	SC, CP	63
Pentamethyl benzene	$CHC_5^*(CH_3)_5$	RT	-84	1	83	-24	SC, CP	63
Methyl								
Methanol	C^*H_3OH	87	-21	-21	42	76	PW, CP	60
Ethanol	$C^*H_3CH_2OH$	103	-11	-2	13	115	PW, CP	60
Methyl formate	$HCOOC^*H_3$	87	-27	-15	41	80	PW, CP	60
Methyl acetate	$HC_3COOC^*H_3$	133	-33	-17	51	70	PW, CP	60
Dimethyl carbonate	$(C^*H_3O)_2CO$	87	-26	-20	45	76	PW, CP	60
Dimethyl oxalate	$(C^*H_3OCO)_2$	87	-27	-15	43	79	PW, CP	60
Dimethyl acetylene	$(C^*H_3C)_2$	87	-5	-5	9	118	PW, CP	60
Methyl Urea	$C^*H_3NHCONH_2$	77	-17	-5	22	-127	PW, MP	10
Dimethyl oxalate	$(C^*H_3OCO)_2$	RT	-86	15	71	43	SC, CP	62c
Hexamethyl benzene	$C_6(C^*H_3)_6$	RT	-1.1	0.1	0.9	111.8	SC, CP	63
Dimethyl sulfoxide	$(C^*H_3)_2SO$	226	-23	-1	25	86	PW, CP	60
Octomethyl cyclotetrasiloxane	$[(C^*H_3)_2SiO]_4$	87	0	0	0	128	PW, CP	60
Hexamethyl disiloxane	$[(C^*H_3)_3Si]_2O$	87	0	0	0	124	PW, CP	60
Diethyl ether	$[C^*H_3CH_2)]_2O$	133	-11	-1	13	111	PW, CP	60

Table 5.5 (continued)

Compound	Formula	T (° K)	σ_{11}	σ_{22}	σ_{33}	$\bar{\sigma}$	Method	Refs.
Acetaldehyde	C^*H_3CHO	87	− 21	− 9	31	95	PW, CP	60
Acetone	$(C^*H_3)_2CO$	87	− 17	− 17	33	98	PW, CP	60
Acetic acid	C^*H_3COOH	87	− 12	− 12	23	95	PW, CP	60
Acetic anhydride	$(C^*H_3CO)_2O$	91	− 13	− 10	25	103	PW, CP	60
Methyl acetate	$C^*H_3COOCH_3$	133	− 8	− 5	12	94	PW, CP	60
Toluene	$C_6H_5C^*H_3$	87	− 13	− 2	15	109	PW, CP	60
Durene	$C_6H_2(C^*H_3)_4$	RT	− 12	− 5	16	110	SC, CP	59
Hexamethyl benzene	$C_6(CH_2C^*H_3)_6$	RT	− 12	− 3	16	115	SC, CP	63
Hexamethyl dewar benzene	$(CC^*H_3)_4(CC^*H_3)_4$	87	0	0	0	118	PW, CP	60
Silver acetate	C^*H_3COOAg	87	− 16	− 10	26	110	PW, CP	60
	C^*F_3COOAg	296	− 6	− 6	13	6	PW, CP	60
Dimethyl dimethoxy silane	$(C^*H_3)_2(CH_3O)_2Si$	87	0	0	0	132	PW, CP	60
Thioacetic acid	C^*H_3COSH	87	− 21	− 14	35	94	PW, CP	60
Trifluor acetic anhydride	$(C^*F_3CO)_2O$	109	− 6	− 6	11	16	PW, CP	60
Dimethyl dimethoxy silane	$(CH_3)_2(C^*H_3O)_2Si$	87	− 27	− 22	50	82	PW, CP	60
Tetramethyl silane	$(C^*H_3O)_4Si$	87	− 25	− 19	45	82	PW, CP	60
Dimethyl disulfide	$(C^*H_3)_2S_2$	87	− 24	6	18	105	PW, CP	60
Methylene								
Ethanol	$CH_3C^*H_2OH$	103	− 19	− 19	38	75	PW, CP	60
Diethylether	$(CH_3C^*H_2)_2O$	133	− 30	− 18	49	61	PW, CP	60
Hexaethyl benzene	$C_6(C^*H_2CH_3)_6$	RT	− 7	− 5	13	108.5	SC, CP	63
Ammonium hydrogen malonate	methylene C	RT	− 18	− 8	26	84	SC, CP	62b
Carboxyl								
Acetic acid	CH_3C^*OOH	87	− 81	4	78	− 59	PW, CP	60
Ammonium-D tartrate	carboxyl C	RT	− 61	− 12	72	− 50	SC, CP	64, 62a
Ammonium hydrogen malonate	carboxyl C	RT	− 68.5	− 1.5	70	− 44	SC, CP	62, b
Benzoic acid	$C_6H_5C^*OOH$	RT	− 57	− 14	71	− 46	SC, FID	11, 57

Table 5.5 (continued)

Compound	Formula	T (° K)	σ_{11}	σ_{22}	σ_{33}	$\bar{\sigma}$	Method	Refs.
Carbonyl								
Acetone	$(CH_3)_2C*O$	87	−71	−57	129	−79	PW, CP	60
Acetaldehyde	CH_3C*HO	87	−77	−35	112	−70	PW, CP	60
Thioacetic acid	CH_3C*OSH	87	−72	−28	101	−73	PW, CP	60
Silver acetate	$CH_3C*OOAg$	87	−62	−25	87	−48	PW, CP	60
Silver trifluoro acetate	$CF_3C*OOAg$	296	−77	39	39	−39	PW, CP	60
Acetic anhydride	$(CH_3C*O)_2O$	91	−110	55	55	−42	PW, CP	60
Methyl formate	$HC*OOCH_3$	87	−88	29	58	−36	PW, CP	60
Methyl acetate	$CH_3C*OOCH_3$	133	−85	22	62	−53	PW, CP	60
Dimethyl carbonate	$(CH_3O)_2C*O$	87	−81	40	40	−21	PW, CP	60
Dimethyl oxalate	$(CH_3OC*O)_2$	87	−98	49	49	−19	PW, CP	60
Trifluoro acetic anhydride	$(CF_3C*O)_2O$	109	−108	46	62	−23	PW, CP	60
Benzoic acid anhydride	$C_6H_5C*O_3C*H_5C_6$	RT	−83	9	73	−24	PW, FID	11, 57
Silver benzoate	$C_6H_5C*OOAg$	RT	−75	−4	79	−43	PW, FID	11, 57
Benzophenone	$C_6H_5C*OC_6H_5$	RT	−72	−29	101	−72	SC, FID	11, 57
Others								
Nickelcarbonyl	$Ni(CO)_4$	4.2	−132	−132	264	−66	PW, FID	67
Ironcarbonyl	$Fe(CO)_5$	4.2	−142	−142	284	−85	PW, FID	67
	$K_2Pt(CN)_4Br_{0.3}3H_2O$	RT	−116	−78	193	19	SC, FID	66
Thiobenzophenone	$C_6H_5C*SC_6H_5$	RT	−148	−39	187	−107	PW, FID	11, 57
Dimethyl acetylene	$(CH_3C*)_2$	87	−67	−67	135	38	PW, CP	60
Hexamethyl dewar benzene	$(C*CH)_3)_4(CCH_3)_2$	87	−101	−6	107	−14	PW, CP	60
Hexamethyl dewar benzene	$(CCH_3)_4(C*CH_3)_2$	87	−15	0	14	72	PW, CP	60
Ammonium-D tatrate	hydroxyl C	RT	−13	−7	19	55	SC, CP	64, 62a
Acetonitrile	CH_3C*N	83	−68	−68	136		PW, CP	68
Calcite	$CaCO_3$	RT	25	25	−51	−15	PW, FID	55a, 56a
Carbondisulfide	CS_2		−144	−144	288	−65	PW, FID	56a

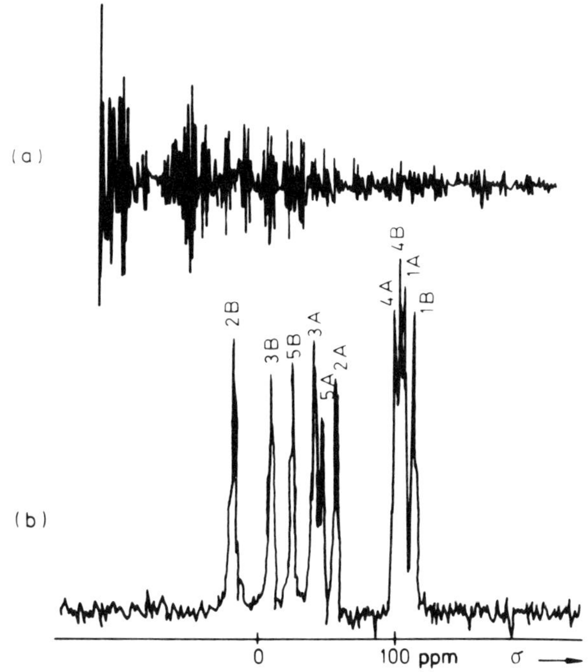

Fig. 5.6. ^{13}C spectrum in durene obtained by the cross-polarization technique by S. Pausak *et al.* [*59*]. All possible ten lines, allowed by the crystal structure are borne out at this specific crystal orientation

(i) Methyl carbons display almost axially symmetric shielding tensors, with $\Delta\sigma \cong 24$ ppm and the unique axis being the 1-axis almost parallel to the *C-C* bond. The shift tensor is most likely averaged due to tunelling motion or hindered rotation about the C_3-axis.

(ii) The ring carbons possess three distinct tensor elements with the most shielded axis perpendicular to the plane and the least shielded axis bisecting almost the *C-C-C* angle of the ring carbons.

(iii) The isotropic average of the shielding tensor is closely related to the isotropic shift in liquids, showing only little dependence on packing in the solid state.

The last symptom is also found in the eigenvalues and the principal axes of the shielding tensors, being closely related to molecular symmetry, rather than crystal symmetry. This is in contrast with ^{19}F shielding tensors which reflect subtle changes in the crystalline structure. This fact may be caused by the marginal location of the fluoratoms at the outskirts of molecular bonds, whereas the carbon atoms are buried in the interior of the molecule.

Numerous attempts have been made to relate the observed shielding tensors to the molecular orbitals involved. An interesting approach has been taken by Spiess and co-workers [*11*], relating this quantity to the different orbital transitions as known from optical absorption frequencies. Others have tried to apply *ab initio* calculations, mainly to the paramagnetic part of the shielding tensor with varying

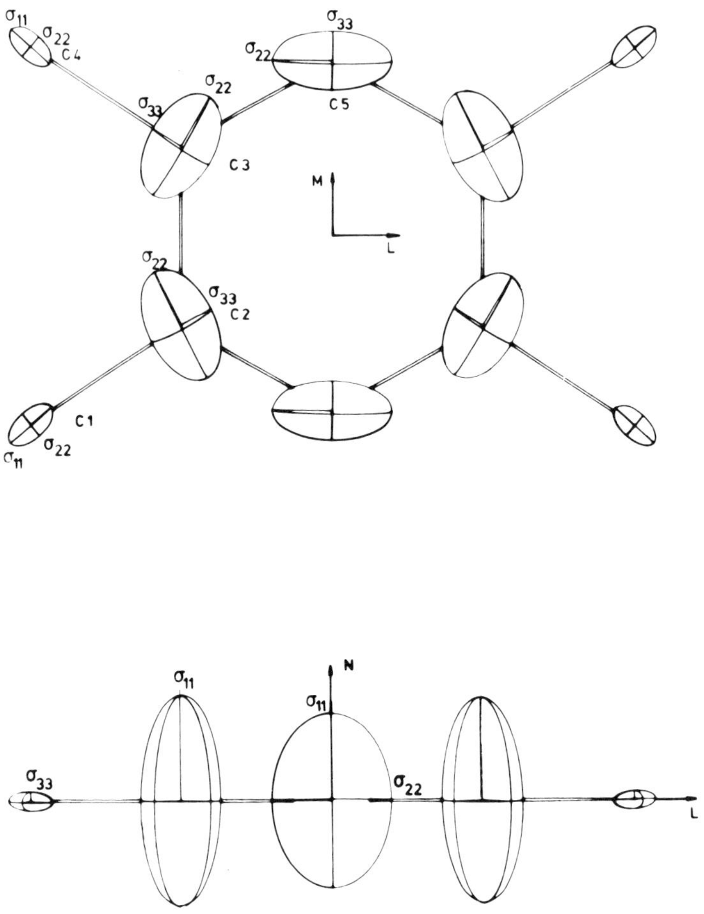

Fig. 5.7. Aestheticized chemical shielding tensor representation of ^{13}C in durene obtained by S. Pausak *et al.* [*59*]. The dimensions of the ellipsoids are directly proportional to the shielding with respect to a certain reference, according to the prescription in Section 2.1. The methyl ^{13}C ellipsoids are magnified by a factor of two. (Courtesy of D. Suwelack)

success [*65*]. Rather than going into those details, we would like to summarize some common features [*11, 57*]:

(i) The most shielded direction is:
 a) perpendicular to the ring in aromatic carbons,
 b) the C_3-axis for methyl carbons,
 c) perpendicular to the sp^2 plane for carbonyl and carboxyl carbons.

(ii) The least shielded direction is:
 a) in the plane for ring carbons, bisecting the *C-C-C* angle,
 b) for methyl carbons perpendicular to the C_3-Axis and perpendicular to a plane of symmetry to which the methyl group is connected (if not averaged),
 c) in the sp^2 plane for carboxyl and carbonyl carbons.

(iii) The intermediate shielded direction is:
   a) tangent to the ring for aromatic carbons,
   b) for non averaged methylgroups perpendicular to the C_3-axis in the plane of symmetry,
   c) in the sp^2 plane, but perpendicular to the *C-C* bond for carboxyl groups.

For further reference of ^{13}C shielding tensors and their principal axes, we list a number of data in Table 5.5.

5.6 Other Shielding Tensors

Shielding tensors of other nuclei like ^{15}N, ^{29}Si, ^{31}P, ^{77}Se, ^{125}Te etc. have also been determined [*47, 53–57*]. As demonstrated in Figs. 3.14 and 3.15 (Section 3.2) line narrowing techniques have been applied also to metals such as ^{27}Al and ^{9}Be. We are not going to discuss the details of the magnetic shielding of these compounds, but would rather like to summarize the data available in the following Tables 5.6, 5.7, and 5.8.

Table 5.6. Nitrogen (^{15}N) shielding tensor σ_{11}, σ_{22}, σ_{33} in ppm with respect to its average $\bar{\sigma}$ for different compounds. (PW, powder spectra; SC, single crystal; LC, liquid crystal; MF, molecular frame; CP, cross polarization; FID, free induction decay; NP, nematic phase; RT, room temperature)

Compound	Formula	T (°K)	σ_{11}	σ_{22}	σ_{33}	$\bar{\sigma}$	reference	Method	Refs.
Ammonium nitrate	$NH_4N^*O_3$	RT	− 83	− 57	140	352	NO_3^-	PW, CP	70
Ammonium sulfate	$(NH_4)_2SO_4$	RT	0	0	0	352	NO_3^-	PW, CW	70
Glycine		RT	0	0	0	352	NO_3^-	PW, CW	70
Nitrous oxide	NN^*O	NP	−170	−170	340	135	HNO_3	MF, LC	71
Nitrous oxide	N^*NO	NP	−123	−123	246	219	HNO_3	MF, LC	71
Nitrobenzene	$C_6H_5NO_2$	170	−265	102	164	− 8	liquid	PW, FID	69
Pyridine		105	−334	−115	448	− 21	liquid	PW, FID	72
Acetonitrile	CH_3CN	77	−326	163	163	311	NO_3^-	PW, CP	68

Table 5.7. Phosphorous (^{31}P) shielding tensor σ_{11}, σ_{22}, σ_{33} in ppm with respect to its average $\bar{\sigma}$ for different compounds. (PW, powder spectra; SC, single crystal; MF, molecular frame; FM, first moment; FID, free induction decay; MP, multiple pulse; LC, liquid crystal; NP, nematic phase; RT, room temperature)

Formula	T (°K)	σ_{11}	σ_{22}	σ_{33}	$\bar{\sigma}$	Reference	Method	Refs.
P_4	25	−270	135	135			PW, FID	67
Zn_3P_2	RT	− 80	40	40	195	Ortho phosphoric acid	PW, FID	73
Mg_3P_2	RT	− 75	− 75	150	− 95	Ortho phosphoric acid	PW, FID	73
KH_2PO_4	77	− 5.1	− 5.1	10.3	− 11.2	H_3PO_4	SC, FM	77
$P^*S_3P_3$	RT	−150	64	86	− 89	Ortho phosphoric acid	SC, FID	73
$P^*S_3P_3$	NP	−162	81	81			MF, LC	78
$PS_3P_3^*$	RT	−188	−131	319	87	Ortho phosphoric acid	SC, FID	73
$PS_3P_3^*$	NP	−177	−177	335			MF, LC	78
$P(CN)_3$		− 43	− 43	86	136	H_3PO_4	SM, CW	79
P_4O_{10}		−109	−109	218	150	H_3PO_4	SM, CW	79
P_4S_{10}		− 63	− 63	126	− 45	H_3PO_4	SM, CW	79
$BaFPO_3$		− 96	48	48			SM, CW	54

Table 5.8. Silicon (^{29}Si), Selenium (^{77}Se), and Tellurium (^{125}Te) shielding tensor σ_{11}, σ_{22}, σ_{33} in ppm with respect to its average $\bar{\sigma}$ for different compounds. The average $\bar{\sigma}$ is referenced toTMS (^{29}Se), selenic acid (^{77}Se) and $TeCl_2$ solution (^{125}Te). (PW, powder spectra; SC, single crystal; CW, wide line; CP, cross polarization; RT, room temperature)

Compound	Formula	T (°K)	σ_{11}	σ_{22}	σ_{33}	$\bar{\sigma}$	Method	Refs.
TMS	$(CH_3)_4Si$	87	0	0	0	0	PW, CP	74
Trimethyl methoxy silane	$(CH_3)_3SiOCH_3$	87	− 13.7	− 13.7	27.3	− 19	PW, CP	74
Dimethyl dimethoxy silane	$(CH_3)_2Si(OCH_3)_2$	87	− 16	− 16	32	4	PW, CP	74
Methyl trimethoxy silane	$CH_3Si(OCH_3)_3$	87	− 19	− 7	26	42	PW, CP	74
Tetramethoxy silane	$Si(OCH_3)_4$	87	0	0	0	80	PW, CP	74
	$[(CH_3)_3Si]_3CH$	87	− 13	0	13	5	PW, CP	74
	$(CH_3)_3SiC_6H_5$	87	− 12	− 6	19	10	PW, CP	74
Hexamethyl disiloxane	$[(CH_3)_3Si]_2O$	87	− 13	− 5	17	− 3	PW, CP	74
	$[(CH_3)_2SiO]_3$	87	− 26	− 18	44	18	PW, CP	74
Actamethyl cyclotetra siloxane	$[(CH_3)_2SiO]_4$	87	− 16	− 16	33	20	PW, CP	74
	^{77}Se	77	− 420	−420	840	33	SC, CW	75, 76
	^{125}Te	RT	−1190	570	620	620	SC, CW	80

6. Spin-Lattice Relaxation in Line Narrowing Experiments

In this chapter we want to account briefly for spin lattice relaxation i.e., the loss of order or the increase in entropy of the spin system due to interactions with the lattice. Rather than treating the spin lattice relaxation in general, which has been done in the fundamental books by Abragam [*1*] and Goldmann [*2*], we emphasize here the context with line narrowing experiments only. Of course, the usual spin lattice relaxation as described by the time constant T_1 is effective in these experiments too.

6.1 Spin Lattice Relaxation in Multiple-Pulse Experiments

In multiple pulse experiments, which we want to discuss first, one expects a relaxation in the rotating frame with a time constant of about $T_{1\rho}$. Let us recall, that $T_{1\rho}$ is the spin lattice relaxation in an effective field H_1 in the rotating frame. If we assume a Gaussian-Markoff process with the correlation time τ_c to cause a "random modulation" of the dipole Hamiltonian, we arrive at

$$\frac{1}{T_{1\rho}} = M_2 [\frac{\tau_c}{1 + 4\omega_1^2\tau_c^2} + \frac{5}{3}\frac{\tau_c}{1 + \omega_0^2\tau_c^2} + \frac{2}{3}\frac{\tau_c}{1 + 4\omega_0^2\tau_c^2}] \tag{6.1}$$

where

$$\omega_1 = \gamma H_1 \quad \text{and} \quad \omega_0 = \gamma H_0$$

and with

$$M_2 = \frac{3}{5}\frac{\gamma_I^4\hbar^2}{r^6} I(I+1) \tag{6.2}$$

being the powder average of the second moment of two spins I with the gyromagnetic ratio γ_I, separated by the distance r. Notice, that the maximum relaxation rate is reached for $\omega_1\tau_c = 1/2$, which is considered as being a "slow motion", since ω_1 is of the order of several kHz.

We obtain for this maximum rate

$$\frac{1}{T_{1\rho}})_{\max} \frac{1}{2} M_2 \tau_c . \tag{6.3}$$

This is of course the "worst" case in terms of resolution, i.e. the ultimate resolution in multiple-pulse experiments is limited by spin lattice relaxation [*6*]. Supposing for a while, that the effective rotating frame relaxation $T_{1\rho}$ in a multiple-pulse experiment is on the order of the spin locking $T_{1\rho}$, we want to make an estimate of the ultimate resolution which could be achieved in the presence of maximum spin lattice relaxation effect. Let us call δ_0 the "rigid line width" (for a more subtle discussion see Section 3 and Appendix E) with

$$\delta_0 \cong M_2^{1/2}$$

and

$$\delta = 1/T_{1\rho}$$

the "ultimate line width" caused by spin lattice relaxation. We then arrive under the condition of "maximum relaxation effect" according to Eq. (6.3) at

$$\delta_0/\delta = \frac{2}{\tau_c M_2^{1/2}} \tag{6.4}$$

or with $\omega_1 \tau_c = 1/2$

$$\delta/\delta_0 = 4\omega_1/M_2^{1/2} \tag{6.5}$$

Inserting $H_1 = 10\,G$ and $(M_2)^{1/2} = 1\,G$ into Eq. (6.4) results in a factor of 40 for the ultimate resolution compared with the rigid line width. This may seriously deteriorate the line narrowing efficiency in a well chosen multiple pulse experiment i.e., we do have to worry about spin lattice relaxation in multiple pulse experiments if we demand highest resolution [*7*]. On the other hand, we shall see that spin lattice relaxation can be investigated by multiple pulse experiments in a very appealing way, especially in the regime of "slow motions" [*6–8*].

Haeberlen and Waugh [*6*] have discussed extensively spin lattice relaxation in multiple-pulse experiments and here we want to review some of their results briefly. Starting from the master equation for the spin density matrix ρ [*1*]

$$\frac{d\rho}{dt} = -\int_0^\infty dt' \langle [\mathcal{H}(t), [\mathcal{H}(t-t'), \rho]] \rangle_{av} \tag{6.6}$$

where $\mathcal{H}$ is the suitably transformed dipolar Hamiltonian of a pair of spins I coupled to each other and subjected to a random motion. Haeberlen and Waugh [*6*] arrive under the following conditons to decoupled Bloch equations for the different components of the magnetization.

The conditions are

(i) A frame of reference has to be chosen in which no external field, static or oscillatory, appears explicitly.

(ii) Restriction to the secular part of the dipolar Hamiltonian.

(iii) Only spins $I = 1/2$ are considered, or else higher order tensor operator averages like $\langle T_{3m} \rangle$ are neglected.

If we denote the relaxation time of the z-component of the magnetization in the suitably transformed rotating frame $T_{1\rho}$ and the corresponding relaxation time of the components orthogonal to the z axis $T_{2\rho}$, we arrive at the following equations [6]

$$\frac{1}{T_{1\rho}} = \frac{8\pi}{3} M_2 [-\mathrm{Re}\{\mathcal{F}^1(\omega)\} + 4\,\mathrm{Re}\{\mathcal{F}^2(2\omega)\}] \tag{6.7}$$

and

$$\frac{1}{T_{1\rho}} = \frac{4\pi}{3} M_2 [3\,\mathrm{Re}\{\mathcal{F}^0(0)\} + 2\,\mathrm{Re}\{\mathcal{F}^2(2\omega)\} - 5\,\mathrm{Re}\{\mathcal{F}^1(\omega)\}] \tag{6.8}$$

where a random stationary motion has been assumed, i.e.

$$\mathcal{F}^{-m}(-\omega) = \mathcal{F}^{m*}(\omega).$$

As the motion is assumed to be due to fluctuations in the dipolar interaction, we can write

$$\mathcal{F}^m(m\omega) = (-1)^m \int_0^\infty dt' \langle Y_{2m}^*(t) Y_{2m}(t-t') \rangle e^{im\omega t} \tag{6.9}$$

where $\langle\rangle$ denotes an ensemble average to obtain the correlation function of the randomly modulated spherical harmonics. For a Gauss-Markoff process with the correlation time τ_c

$$\mathrm{Re}\{\mathcal{F}^m(m\omega)\} = (-1)^m \frac{1}{4\pi} \frac{\tau_c}{1 + (m\omega\tau_c)^2}$$

the familiar expression for T_1 and T_2 can be obtained [1]:

$$\frac{1}{T_1} = \frac{2}{3} M_2 \left[\frac{\tau_c}{1 + \omega_0^2\tau_c^2} + \frac{4\tau_c}{1 + 4\omega_0^2\tau_c^2}\right]$$

$$\frac{1}{T_2} = \frac{1}{3} M_2 \left[3\tau_c + \frac{5\tau_c}{1 + \omega_0^2\tau_c^2} + \frac{2\tau_c}{1 + 4\omega_0^2\tau_c^2}\right]$$

where T_1 represents the relaxation in the direction of the quantization axis, whereas T_2 describes the relaxation orthogonal to the quantization axis.

Let us discuss some representative experiments

a) Spinning Sample Experiment

In this case we have to replace Y_{2m} by [6]

$$d_{mm'}^{(2)}(\vartheta_r) Y_{2m'}(t) e^{im'\omega_r t}$$

where ϑ_r is the angle of the rotation axis with the magnetic field, and

$$d_{mm'}^{(2)}(\vartheta_r) = \langle 2m | e^{-i\vartheta_r I_y} | 2m' \rangle$$

is the Wigner rotation matrix. ω_r is the rotation frequency. This results for a Gauss-Markoff process in

$$\mathcal{F}^m(m\omega) = (-1)^m \frac{1}{4\pi} \sum_{m'} |d^{(2)}_{mm'}|^2 \frac{\tau_c}{1 + (m\omega_0 - m'\omega_r)^2 \tau_c^2} \tag{6.10}$$

Inserting Eq. (6.10) into Eq. (6.7) results in the spin lattice relaxation time T_1 in the spinning sample experiment. No dramatic change compared with the ordinary T_1 is expected, since $\omega_0 \gg \omega_r$ in practice. A more dramatic change is induced into the "transverse" spin lattice relaxation time T_2 as discussed by Haeberlen and Waugh [*6*] and also by Andrew and Jasinski [*9*].

If $\omega_0 \tau_c \gg 1$ is assumed and Eq. (6.10) is inserted into Eq. (6.8) we obtain under the "magic angle" condition $\vartheta_r = 54°44'$

$$\frac{1}{T_2} = \frac{1}{3} M_2 \left[\frac{2\tau_c}{1 + \omega_r^2 \tau_c^2} + \frac{\tau_c}{1 + 4\omega_r^2 \tau_c^2}\right]. \tag{6.11}$$

We notice, that the line-width in a "magic angle" spinning sample experiment may well be determined by spin lattice relaxation if $\omega_r \tau_c \cong 1$ and may cause a failure of such an experiment in terms of not achieving enough resolution. We turn now to the rotating frame relaxations, assuming throughout $\omega_0 \tau_c \gg 1$ and thus emphasizing the contribution of "slow motions".

b) Spin-Locking [10, 11] and Lee-Goldburg Experiment [12]

The relaxation takes place in an effective field $H_e = \omega_e/\gamma$ tilted by an angle ϑ with respect to the static magnetic field. With the restrictions we made so far we obtain [*6*]

$$\mathrm{Re}\{\mathcal{F}^m(m\omega_e)\} = (-1)^m |d^{(2)}_{0m}(\vartheta)|^2 \cdot \frac{1}{4\pi} \frac{\tau_c}{1 + (m\omega_e \tau_c)^2}. \tag{6.12}$$

Notice at this stage already, that replacing ϑ by $-\vartheta$ or by $\vartheta + \pi$ in Eq. (6.12) does not affect the spectral distribution function i.e., phase alternation of the effective field or the rf field does not influence the relaxation. In other words, not the "average" field is the important parameter in spin lattice relaxation, but rather the modulus of this field i.e., the rate at which it samples the stochastic motion.

In the resonance spin-locking case ($\vartheta = \pi/2$) with

$$|d^{(2)}_{0\pm2}|^2 = \frac{3}{8}; \; |d^{(0)}_{0\pm1}|^2 = 0; \; |d^{(2)}_{00}|^2 = 1/4$$

we obtain for the relaxation rate along the applied rf field

$$\frac{1}{T_{1\rho}} = M_2 \frac{\tau_c}{1 + 4\omega_e^2 \tau_c^2} \tag{6.13}$$

which is the so-called spin-lattice relaxation in the rotating frame as can be derived from Eq. (6.1) under the condition $\omega_0\tau_c \gg 1$. Similar the transverse relaxation orthogonal to the applied rf field

$$\frac{1}{T_{2\rho}} = \frac{1}{4}M_2(\tau_c + \frac{\tau_c}{1 + 4\omega_e^2\tau_c^2}) \tag{6.14}$$

is obtained, which might be observed in a rotary echo experiment.

Notice that:

(i) $T_{2\rho} = 2 \cdot T_{1\rho}$ for $\omega_e\tau_c \ll 1 \quad \omega_0\tau_c \gg 1$

(ii) $T_{1\rho} > T_{2\rho}$ for $\omega_e\tau_c \gg 1$.

For $\omega_e\tau_c \gg 1$ the correlation time τ_c approaches the flip-flop correlation time of the I spins in the rigid lattice i.e. $T_{2\rho} \to T_2$.

Lee-Goldberg-Experiment (tan $\vartheta = \sqrt{2}$) [6]

$$|d^{(2)}_{0\pm2}|^2 = \frac{1}{6};\ |d^{(2)}_{0\pm1}|^2 = \frac{1}{3};\ |d^{(2)}_{00}|^2 = 0.$$

Relaxation rate along the quantization axis (111) in the rotating frame:

$$\frac{1}{T_{1\rho}} = \frac{2}{9}M_2[\frac{\tau_c}{1 + \omega_e^2\tau_c^2} + \frac{2\tau_c}{1 + 4\omega_e^2\tau_c^2}]. \tag{6.15}$$

Relaxation rate orthogonal to the (111) direction:

$$\frac{1}{T_{2\rho}} = \frac{1}{9}M_2[\frac{5\tau_c}{1 + \omega_e^2\tau_c^2} + \frac{\tau_c}{1 + 4\omega_e^2\tau_c^2}]. \tag{6.16}$$

Haeberlen and Waugh [6] have discussed the interference which might occur if sample spinning combined with a Lee-Goldburg experiment is performed. In this case the spectral distribution function can be expressed as [6]

$$\mathrm{Re}\{\mathcal{F}^m(m\omega_0)\} = (-1)^m\,|d^{(2)}_{0m}(\vartheta)|^2 \sum_{m'} |d^{(2)}_{0m'}(\vartheta_r)|^2 \frac{1}{4\pi}\,\frac{\tau_c}{1 + (m\omega_e - m'\omega_r)^2\tau_c^2} \tag{6.17}$$

For the magic angle condition $\tan\vartheta = \tan\vartheta_r = \sqrt{2}$.

$\mathcal{F}^{(0)}$ in $1/T_{2\rho}$ drops out, because of $|d^{(2)}_{00}(\vartheta)| = |d^{(2)}_{00}(\vartheta_r)| = 0$

and terms with $(m\omega_e - m'\omega_r)$ become important, where $m, m' = -2, \ldots +2$. In the case $\omega_e \cong \omega_r$ several of those terms vanish, leading to $1/T_{2\rho} \sim \tau_c$.

If the correlation time τ_c approaches the flip-flop rate of the rigid lattice, ($T_{2\rho} = T_2$) a broad "solid" line is observed rather than a "narrowed" line. In this sense spinning of the sample counterbalances the rotation in spin space and destroys the line narrowing efficiency of each of those experiments alone [6]. It is highly recommended to make $\omega_e \neq \omega_r$, usually $\omega_e \gg \omega_r$.

c) Multiple-Pulse Experiments

Haeberlen and Waugh [6] and Mansfield [3] have discussed spin lattice relaxation in different multiple-pulse experiments, namely

MW-2 $P_y - (\tau - P_x - 2\tau - P_{\bar{x}} - \tau)_n$

WHH-4 $(\tau - P_x - 2\tau - P_{\bar{x}} - \tau - P_{\bar{y}} - 2\tau - P_y - \tau)_n$

where P_i denotes a 90° rf pulse in the i direction.

It should be noted, that MW-4 namely $P_y - (\tau - P_x - 2\tau - P_x - 2\tau - P_x - 2\tau - P_x - \tau)_n$ is governed by the same relaxation time as MW-2. Remember, that the periodic structure of the external fields is the important feature which samples the spectral distribution of the fluctuation, rather than the "average" effective field. Notice, that the "average" effective field in the MW-2 and the WHH-4 experiment performed on-resonance, is zero. Suppose off-resonance effects are disregarded and the same restrictions (i)–(iii) are applied as before with $\omega_0\tau_c \gg 1$. Haeberlen and Waugh [6] arrive after some manipulation at the following relaxation rate along the quantization axis, namely the (111) direction in the rotating frame

$$\text{WHH-4:}\ \frac{1}{T_{1\rho}} = \frac{6}{\pi^2} M_2 \sum_{n=1}^{\infty} \frac{1}{n^2} \frac{\tau_c}{1 + (n\omega_c\tau_c)^2} \tag{6.18a}$$

and for the relaxation rate orthogonal to the (111) direction at

$$\text{WHH-4:}\ \frac{1}{T_{2\rho}} = \frac{9}{2\pi^2} M_2 \left\{3 \sum_k \frac{1}{k^2} \frac{\tau_c}{1 + (k\omega_c\tau_c)^2} + \frac{1}{3} \sum_n \frac{1}{n^2} \frac{\tau_c}{1 + (n\omega_c\tau_c)^2}\right\} \tag{6.18b}$$

where $\omega_c = 2\pi/t_c$ is the "cycle frequency" as defined by the cycle time t_c. Here $k, n > 0$ and not a multiple of 3 and where k is even and n is odd i.e. $k = 2, 4, 8 \ldots$; $n = 1, 5, 7 \ldots$ Mansfield [3] has calculated $T_{1\rho}$ for the MW-2 experiment, following along those lines and obtained in the direction of the applied rf field

$$\text{MW-2:}\ \frac{1}{T_{1\rho}} = \frac{8}{\pi^2} M_2 \sum_{n=1}^{\infty} \frac{1}{n^2} \frac{\tau_c}{1 + (n\omega_c\tau_c)^2} \tag{6.19}$$

where the sum is now restricted to n odd. The factor 4 in Mansfields [3] calculation has to be replaced by 8 according to Wilsch *et al.* [26]. The main features which are displayed by Eq. (6.18) and (6.19) are:

(i) $T_{1\rho}$ in the multiple pulse experiment is always larger, that $T_{1\rho}$ in the spin locking experiment.

(ii) The maximum relaxation rate is determined by $\omega_c\tau_c \cong 1$.

For further details we refer the reader to the papers of Haeberlen and Waugh [6] and by Mansfield [3]. The approach taken can be criticized on different grounds, questioning the validity of the master-equation Eq. (6.6) in this regime ($\omega_0\tau_c \gg 1$, $\omega_c\tau_c \cong 1$) and the different approximations which are made.

Gründer [8] has attempted a more rigorous treatment by calculating directly the time development of the magnetization in a suitable direction, say the x direction for the MW-2 cycle. We shall call this decay time of the magnetization in the x-direction of the rotating frame T_2^*.

Let us start with

$$\langle I_x(t)\rangle = \frac{1}{\mathrm{Tr}\{I_x^2\}} \mathrm{Tr}\{\mathbf{U}(t) I_x \mathbf{U}^{-1}(t) I_x\} \tag{6.20}$$

where

$$\mathbf{U}(t) = \exp\{-i \sum_{n=0}^{\infty} \mathbf{F}_n(t)\}$$

is the time evolution operator due to spin lattice relaxation in a suitably defined reference frame, expressed by a Magnus expansion (see Section 3.3). Restriction to the first term in the Magnus expansion and assuming a Gauss-Markoff process leads to

$$\langle I_x(t)\rangle = \exp\Big\{\frac{1}{2\mathrm{Tr}\{I_x^2\}} \langle \mathrm{Tr}\{[\mathbf{F}_0(t), I_x]^2\}\rangle_{av}\Big\} \tag{6.21}$$

where $\langle\,\rangle_{av}$ means that an ensemble average has to be performed. Gründer [8] obtained after some algebra

$$\langle I_x(t)\rangle = \exp[-tR_1(\alpha) + \tau_c(1 - e^{-t/\tau_c})R_2(\alpha)] \tag{6.22}$$

in the case of multiple pulse experiments, where $R_1(\alpha)$ and $R_2(\alpha)$ are functions of the parameter $\alpha = \tau/\tau_c$ (2τ pulse spacing, τ_c correlation time of the Gauss-Markoff process), depending on the type of multiple pulse experiment, which is considered. Two limiting cases will be discussed, namely

(i) $t > \tau_c$, neglecting the 2. term in Eq. (6.22)

$$\langle I_x(t)\rangle = \exp(-t \cdot R_1(\alpha) \equiv \exp(-t/T_2^*) \tag{6.23}$$

(ii) $t \ll \tau_c$ series expansion of the 2. term in Eq. (6.22)

$$\begin{aligned}\langle I_x(t)\rangle &= \exp[-t(R_1(\alpha) - R_2(\alpha)) - \frac{t^2}{2\tau_c} R_2(\alpha)] \\ &\equiv \exp[-t/T_2^* + M_{2\,\mathrm{eff}} t^2/2]\end{aligned} \tag{6.24}$$

where

$$\frac{1}{T_2^*} = R_1(\alpha) - R_2(\alpha) \text{ and } M_{2\mathrm{eff}} = -\frac{R_2(\alpha)}{\tau_c}.$$

However, Eq. (6.23) is more generally applicable than it seems to be, since

in case $\tau_c \ll \tau$ and $t \gg \tau_c$ Eq. (6.23) is valid

and

in case $\tau_c \gg \tau$ and $R_2(\alpha) \ll R_1(\alpha)$ Eq. (6.23) is again valid.

The relaxation rates following this scheme are in the case of
MW-2 *and* MW-4 *experiment* [*8*]

$$\frac{1}{T_2^*} = M_2\tau_c\left(1 - \frac{\tanh\alpha}{\alpha}\right) \qquad \alpha = \tau/\tau_c \tag{6.25}$$

and for the
WHH-4 *and* HW-8 *experiment* [*8*]

$$\frac{1}{T_2^*} = \frac{2}{3}M_2\tau_c\left[1 - \frac{(5\cosh\alpha - 2)\sinh^2\alpha}{\alpha\sinh 3\alpha}\right] \tag{6.26}$$

where T_2^* is the relaxation of the x magnetization, rather than the "longitudinal" magnetization along the (111) direction. In Table 6.1 we compare the individual T_2^* values for different experiments, namely, spin locking, MW-2 and MW-4, WHH-4 and HW-8 for different values of α. Figure 6.1 gives a pictorial representation of the normalized relaxation times in the different experiments versus τ_c/τ. Notice, that $T_{1\rho}$

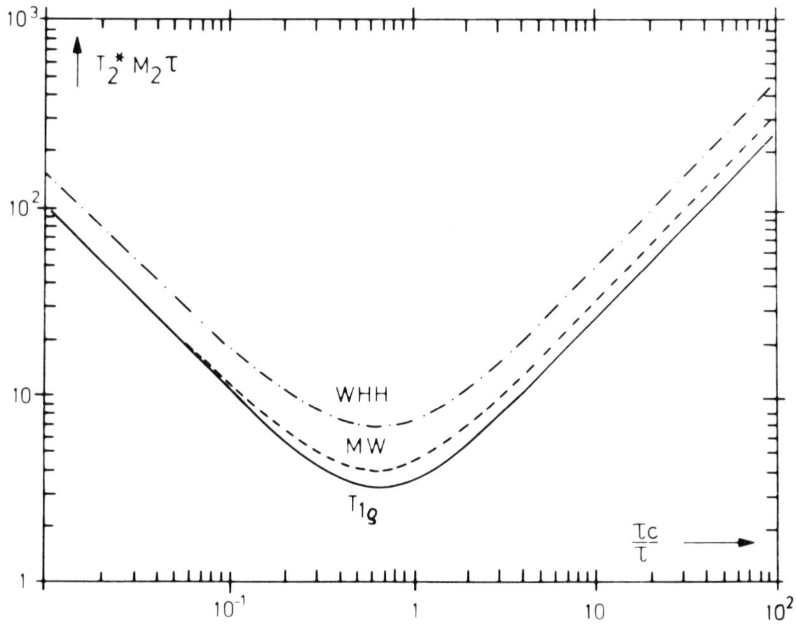

Fig. 6.1. Comparison of the normalized decay time T_2^* due to spin lattice relaxation in different multiple pulse experiments versus the normalized correlation time τ_c of the motion. (W. Gründer [*8*])

Table 6.1. Decay time constant T_2^* of magnetization in the x, y-plane of the rotating frame on resonance for different multiple-pulse experiments (δ pulses assumed) due to spin lattice relaxation with correlation time τ_c. 2τ, pulse spacing; $\alpha = \tau/\tau_c$; M_2 "rigid" second moment; t_w, pulse width; $\omega_1 = \gamma H_{1x}$; suggested preparation pulse in front of each sequence $(\frac{\pi}{2})_y$. (Gründer [8])

Multiple pulse experiment				$T_2^*(T_{1\rho})$			
	type	sequence	cycle time t_c		$\tau_c \ll \tau$	Minimum $\tau = \frac{\pi}{2}\tau_c$	$\tau_c \gg \tau$
a) Homo-nuclear dipolar interaction	MW-2	$[\tau - (\frac{\pi}{2})_x - 2\tau - (\frac{\pi}{2})_{\bar{x}} - \tau]$	4τ	$\frac{1}{T_2^*} = M_2\tau\frac{1}{\alpha}(1 - \frac{\tanh\alpha}{\alpha})$	$T_2^* = \frac{1}{M_2\tau_c}$	$T_{2\,\mathrm{min}}^* = \frac{3.8}{M_2\tau}$	$T_2^* = \frac{3\tau_c}{M_2\tau^2}$
	MW-4	$[\tau - (\frac{\pi}{2})_x - 2\tau - (\frac{\pi}{2})_x - \tau]_2$	8τ				
	WHH-4	$[\tau - (\frac{\pi}{2})_{\bar{x}} - 2\tau - (\frac{\pi}{2})_x - \tau - (\frac{\pi}{2})_y - 2\tau - (\frac{\pi}{2})_{\bar{y}}]$	6τ	$\frac{1}{T_2^*} = M_2\tau \cdot \frac{2}{3\alpha} \times [1 - \frac{(5\cosh\alpha - 2)\sinh^2\alpha}{\alpha\sinh 3\alpha}]$	$T_2^* = \frac{3}{2M_2\tau_c}$	$T_{2\,\mathrm{min}}^* = \frac{6.4}{M_2\tau}$	$T_2^* = \frac{9\tau_c}{2M_2\tau^2}$
	HW-8	WHH-4 + (–WHH4)	6τ				
b) Hetero nuclear dipolar interaction	MW-4	$[\tau - (\frac{\pi}{2})_x - 2\tau - (\frac{\pi}{2})_x - \tau]_2$	8τ	$\frac{1}{T_2^*} = M_2\tau\frac{1}{\alpha}[\frac{1 - \tanh 2\alpha}{2\alpha}]$	$T_2^* = \frac{1}{M_2\tau_c}$	$(\tau = \frac{\pi}{4}\tau_c)$ $T_{2\,\mathrm{min}}^* = \frac{1.9}{M_2\tau}$	$T_2^* = \frac{3\tau_c}{4M_2\tau^2}$
	MW-4 (π)	$[\tau - (\pi)_x - 2\tau - (\pi)_x - \tau]$	4τ	$\frac{1}{T_2^*} = M_2\tau\frac{1}{\alpha}[\frac{1 - \tanh\alpha}{\alpha}]$	$T_2^* = \frac{1}{M_2\tau_c}$	$T_2^* = \frac{3.8}{M_2\tau}$	$T_2^* = \frac{3\tau_c}{M_2\tau^2}$

Similar relations have been obtained by Wilsch *et al.* [26]

in the spin locking experiment is always shorter as compared with T_2^* in the MW or WHH experiments. The T_2^*-minimum occurs always at $\alpha = \tau/\tau_c = \pi/2$, from which the correlation time can be determined very accurately. The range of correlation times, which can be determined with the multiple-pulse method, corresponds to that of the spin-locking $T_{1\rho}$ experiment. However, the advantage of the multiple pulse method is, that the full relaxation curve can be observed in one shot. Other advantages of the multiple pulse method, connected with the quenching of spin diffusion will be discussed later. The MW-4 sequence was used to test the applicability of multiple-pulse experiments to the investigation of slow motions [8].

6.2 Application of Multiple-Pulse Experiments to the Investigation of Spin-Lattice Relaxation

Roeder and Douglass [13] have measured $T_{1\rho}$ in cyclohexane (C_6H_{12}) at different temperatures using spin locking. The MW-4 sequence was applied to this substance by Gründer [8] and Fig. 6.2 represents the different T_2^* data versus inverse temperature for different pulse spacing 2τ.

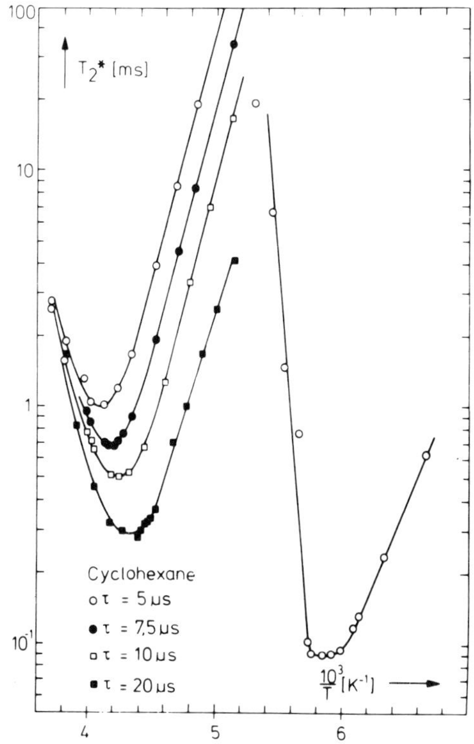

Fig. 6.2. Measured T_2^* values in a MW-4 experiment on protons in cyclohexane versus inverse temperature for different pulse spacing τ, according to W. Gründer [8]

The T_2^*-minima are expected according to Eq. (6.25) and Table 6.1 at

$$T_{2\min}^* = 3.8/M_2\tau \tag{6.27}$$

or with $\tau/\tau_c = \pi/2$

$$T_{2\min}^* = 7.6/(\pi M_2 \tau_c). \tag{6.28}$$

Equation (6.27) allows an independent check of the theory, using no adjustable parameter. Using the value of $M_2 = 7.8\ 10^8 \text{sec}^{-2}$ as calculated by Roeder and Douglass [*13*] the calculated values of $T_{2\min}^*$ according to Eq. (6.27) are listed in the second column of Table 6.2 and are to be compared with the measured values as listed in the third column of the same table.

Table 6.2. Comparison of theoretical (Eq. (6.27), $M_2 = 7.8 \cdot 10^8\ \text{sec}^{-2}$) and experimental $T_{2\min}^*$ values of cyclhexane for different values of the pulse spacing τ in a MW-4 experiment (see Fig. 6.2) according to Gründer [*8*]

τ (μs)	$T_{2\min}^*$ (theor.)	$T_{2\min}^*$ (exptl.)
5	0.98	0.98
7.5	0.65	0.67
10	0.49	0.51
20	0.24	0.28

The agreement is quite convincing. The correlation time τ_c as determined by this method for varying temperature T is plotted in Fig. 6.3 on a logarithmic scale. The data can be fitted by an Arrhenius equation [*8*] to

$$\tau_c = 5.5 \cdot 10^{-15} \exp(9.6 \cdot 10^3/\text{RT})$$

where the activation energy $\Delta E = 9.6$ kcal/Mol agrees well with the value of $\Delta E =$ 9.9 kcal/Mol as determined by Roeder and Douglass [*13*] by means of a spin locking experiment. Further aspects, expecially the influence of pulse width are discussed by Gründer [*8*].

We are now going to discuss another interesting application of multiple-pulse experiments for the investigation of the orientation dependence of T_1. Suppose, the shielding tensor of a nuclear site is known, this knowledge may be exploited to monitor the orientation dependence of some nuclear quantity with respect to the static field in case of a powder sample. We have applied this method to ^{19}F in CF_3COOAg where the complete ^{19}F shielding tensor is known [*14, 15*]. At a high enough temperature this tensor is axially symmetric with the symmetry axis being the C_3-axis, which is also the "rotation" axis [*15*]. If we apply a π pulse to the spins a time τ prior to the application of a high resolution multiple-pulse burst, the spectrum observed after Fourier transformation of the multiple-pulse decay will be

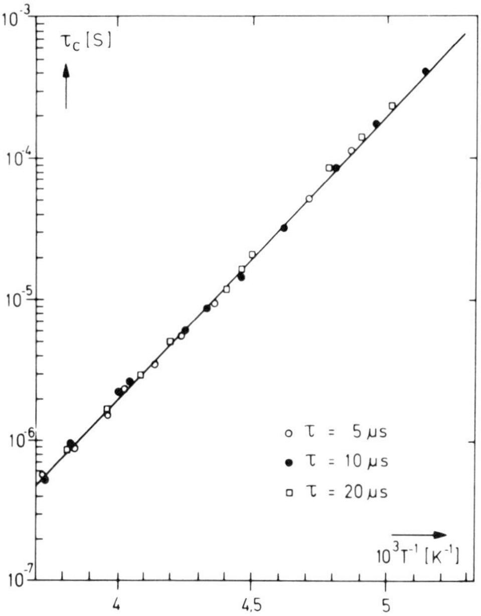

Fig. 6.3. Correlation time τ_c versus inverse temperature obtained from Fig. 6.2 by W. Gründer [*8*] for different pulse spacings in a MW-4 experiment

partially relaxed i.e., it will be inverted if $\tau \ll T_1$, it will be an ordinary spectrum if $\tau > T_1$ and it will be intermediate in between. This is demonstrated in Fig. 6.4, where the partially relaxed powder spectra of ^{19}F in a CF_3COOAg powder sample are shown for different values of τ.

By plotting the amplitude of the signal versus τ the relaxation function $R(\tau)$ can be obtained. This recovery signal will be in the case of exponential relaxation proportional to $1-\exp(-\tau/T_1)$.

As clearly seen from Fig. 6.4 the relaxation is anisotropic, since for $\tau = 1.1$ sec the unique axis component i.e., $C_3 \parallel H_0$ relaxes faster than the component with $C_3 \perp H_0$. The complete evaluation of the relaxation function $R(\tau)$ renders a non-exponential process which is most pronounced for $C_3 \parallel H_0$ i.e., $\beta = 0$. This has been demonstrated earlier in the case of a single crystal relaxation study. The interested reader is referred to Ref. [*16*].

Evaluating only the initial slope of the relaxation function $R(\tau)$ leads to the relaxation rate $1/T_1$, which is plotted versus the angle β of the C_3 axis with respect to the static field H_0 in Fig. 6.5 for two different temperatures. The temperatures are chosen so as to fulfill the conditions $\omega_0\tau_c \ll 1$ ($T = 22°$ C) in one case and $\omega_0\tau_c \gg 1$, in the other case but with $\omega_c\tau_c \ll 1$ ($T = -85°$ C), where ω_c is the "cycle frequency". The relaxation of a 3-spin group has been calculated previously [*17*]

$$\frac{1}{T_1^{(3)}} = \frac{9}{16}\frac{\gamma^4\hbar^2}{r^6}\left[(1-\cos^4\beta)\frac{\tau_c}{1+\omega_0^2\tau_c^2} + (1+6\cos^2\beta+\cos^4\beta)\frac{\tau_c}{1+(2\omega_0\tau_c)^2}\right]. \tag{6.29}$$

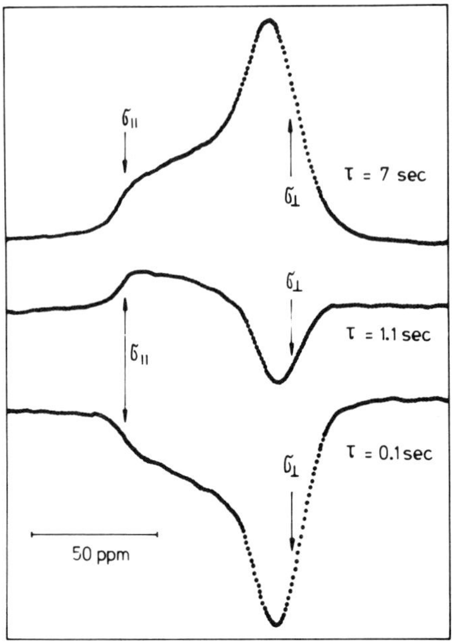

Fig. 6.4. Partially relaxed powder spectra of ^{19}F in CF_3COOAg at room temperature [*14*]. Notice, that the $\sigma_{\|}$ component relaxes faster than the $\sigma_{\perp}$ component, i.e., the spin lattice relaxation is much faster when the methyl C_3 axis is parallel to the static magnetic field H_0

If we call the total relaxation rate $1/T_1 = 1/T_1^{(3)} + A$ to account for intermolecular relaxation and other types of relaxation mechanisms, such as spin-rotation interaction and shift anisotropies (those contributions can be shown to be less than 3% [*18*], we obtain

$$\frac{1}{T_1} = \frac{9}{16} T_0^{-1}(5 + 3\cos 2\beta) + A_1 \quad \text{for } \omega_0\tau_c \ll 1 \tag{6.30}$$

and

$$\frac{1}{T_1} = \frac{9}{4} T_0^{-1}(\omega_0\tau_c)^{-2}(5 + 6\cos^2\beta - 3\cos^4\beta) + A_2 \quad \text{for } \omega_0\tau_c \gg 1 \tag{6.31}$$

where

$$T_0^{-1} = \gamma^4\hbar^2\tau_c/r^{-6}.$$

The solid lines in Fig. 6.5 are obtained by fitting Eqs. (6.30) and (6.31) to the experimental data. This results in $A_1 = 0.29$ sec^{-1} and $\tau_c = 4.10^{-11}$ sec at $T = 22°$ C and $A_2 = 0.181$ sec^{-1} and $\tau_c = 3.1 \cdot 10^{-8}$ sec at $T = -85°$ C. The agreement between the experimental data and the solid lines in Fig. 6.5 is quite convincing.

The example chosen is, however, a very fortitious one, because all the CF_3-groups in the sample are magnetically equivalent (see Fig. 6.6) i.e., the C_3 axes of all CF_3-

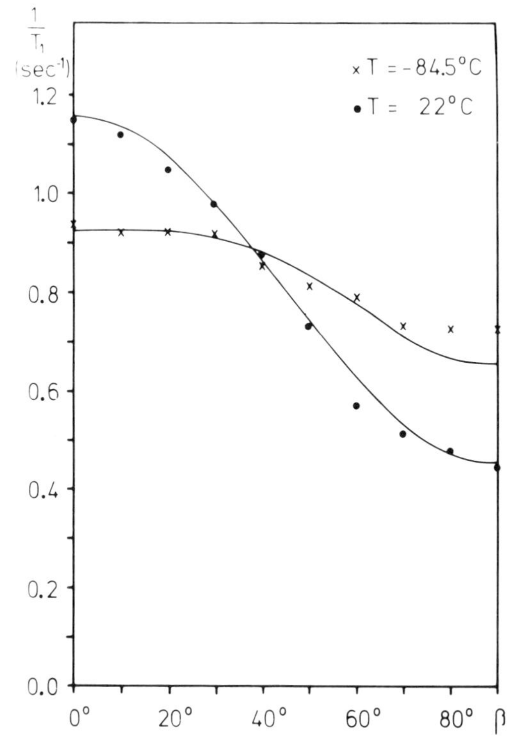

Fig. 6.5. Orientation dependence of the spin lattice relaxation rate $1/T_1$ of ^{19}F in CF_3COOAg for two different temperatures (above and below the T_1 minimum), obtained from powder samples by evaluating spectra as shown in Fig. 6.4 [*14*]. β is the angle between the C_3-axis of the methyl group and the magnetic field H_0. The solid lines correspond to theoretical expressions discussed in the text [Eqs. (6.30, 6.31)]

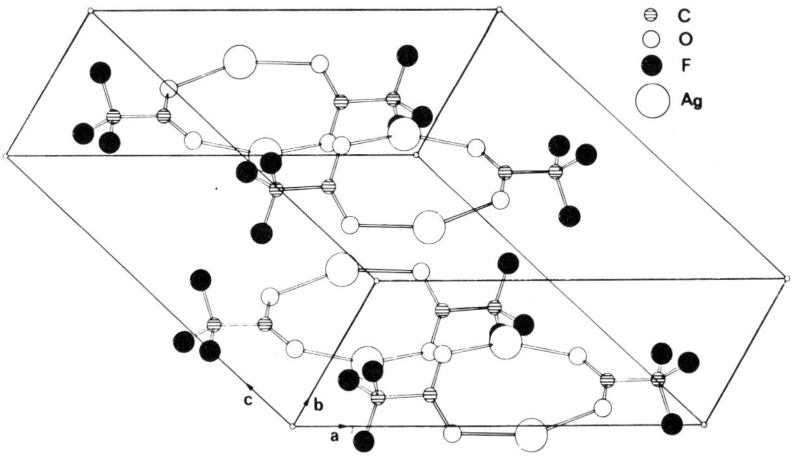

Fig. 6.6. Unit cell of CF_3COOAg according to the space group Cc. Internuclear distances and angles may be found in Ref. [*15b*]. (Courtesy of D. Suwelack)

groups are parallel [*15*]. If differently oriented CF_3-groups would be present in the unit cell, the whole beauty of this experiment would have been spoiled by spin diffusion. This is why there is so little orientation dependence of T_1 observed in methyl naphtalene [*19*]. There is, of course, a way around this, by applying a locking field at the magic angle and thus quenching the spin diffusion process.

A multiple-pulse experiment, which serves the same goal, namely, the quenching of spin diffusion has been applied to the relaxation of ^{19}F spins by paramagnetic centers. Hartmann *et al.* [*20*] have applied a spin locking burst at the magic angle to serve the same goal. The relaxation of nuclear spins during the application of multiple-pulse experiments was studied earlier in this chapter, neglecting completely any contribution due to spin diffusion among different magnetically inequivalent nuclei. In the relaxation of nuclear spins via electronic paramagnetic centers, the process of spin diffusion becomes very important for remote nuclei. The direct relaxation rate of these nuclei is usually very small since $1/T_1 \sim r^{-6}$ where r is the distance to the paramagnetic center. General aspects of relaxation by paramagnetic centers may be found in Refs. [*21, 22*].

We only remark here, that "remote" nuclei usually communicate via spin diffusion (by the nuclear dipole reservoir or the electronic spin dipole reservoir [*22*]) with "closer" nuclei, resulting in a more or less exponential relaxation of the nuclei. It was realized, however, that in the spin diffusion vanishing case, the relaxation should follow $\exp[-(t/\tau_1)^{1/2}]$ [*20, 21*].

Suppose we have quenched spin diffusion by some sort of experiment, the relaxation rate of an individual nuclear spin i will be

$$\frac{1}{T_1^{(i)}} = C \cdot r_i^{-6}$$

The total decay of the magnetization of the sample will follow

$$R(t) = \frac{1}{N} \sum_i \exp(-t/T_1^{(i)} \tag{6.33}$$

excluding all the nuclei which are in a critical volume V_c around the paramagnetic center i.e., nuclei which are shifted by more than the resonance line width are excluded. The sum in Eq. (6.33) can be casted into an integral as

$$R(t) = \frac{1}{V} \int_{V_c}^{V} dV \exp(-t/T_1(r)). \tag{6.34}$$

Hartmann and co-workers [*20*] have evaluated this integral for different cases. All these cases may be expressed by [*20–23*]

$$R(t) = \exp[-(t/\tau_1)^{1/2} \{\psi_1(t) + \psi_2(t)\}] \tag{6.35a}$$

with

$$\omega_0 \tau_c \ll 1: \frac{1}{\tau_1^{1/2}} = 1.3 \frac{2}{3} \pi^{3/2} N_p (\gamma_p \gamma_n \hbar) [S(S+1)\tau_c]^{1/2} \tag{6.35b}$$

and

$$\omega_0\tau_c \gg 1: \frac{1}{\tau_1^{1/2}} = \frac{16}{9\sqrt{3}}\pi^{3/2}N_p C_e^{1/2} \tag{6.35c}$$

where

$$C_e = \frac{2}{9}S(S+1)(\gamma_p\gamma_n\hbar)^2 \frac{\tau_c}{1+\omega_e^2\tau_c^2}$$

Here N_p is the number of electronic spins, S their quantum number and γ_p their gyromagnetic ratio, whereas γ_n is the gyromagnetic ratio of the nuclei. τ_c is the correlation time of the electronic spins.

$\psi_1(t)$ and $\psi_2(t)$ are "modifying functions" which depend on nuclear spin and electronic spin parameters. We remark, that $\psi_1(t) \to 0$ and $\psi_2(t) \to 1$ for $t \to \infty$. Plotting $\ln R(t)$ versus $t^{1/2}$ should therefore result in a straight line for large t. To demonstrate this behavior an experiment, which quenches spin diffusion was performed on ^{19}F in CaF_2 doped with 0.05 mol% Eu^{2+} (see Fig. 6.7) [*14*]. In order to eliminate the oscillations due to the resonance offset in an ordinary WHH-4 experiment, in every second WHH-4 cycle all phases were inverted, leading to an 8-pulse cycle. This eight pulse cycle has the property of quenching spin diffusion on one hand and eliminating resonance offset oscillations on the other hand. By shifting the spectrometer frequency by an amount of $\Delta\omega$ away from resonance, however, an effective field $H_{\text{eff}} = \frac{1}{\sqrt{3}}\,\Delta\omega/\gamma$ can be applied to the spins. Since the local field is virtually zero, extremely slow motions which are not accessible by ordinary $T_{1\rho}$ and T_{1D} measure-

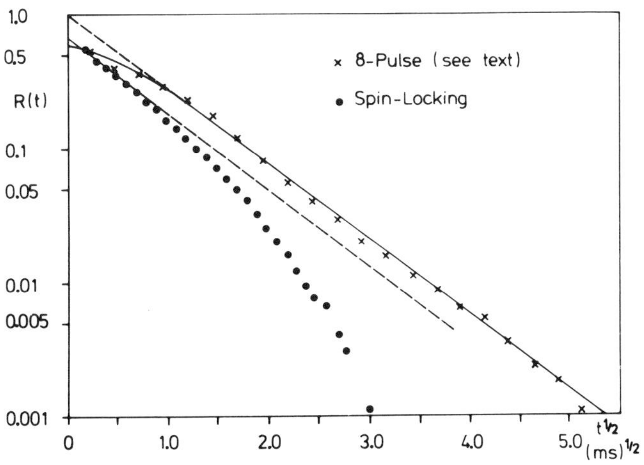

Fig. 6.7. Decay of magnetization of ^{19}F in CaF_2 (0.05 mol.-% Eu^{2+}) versus $t^{1/2}$. In the case of spin locking exponential relaxation is observed, whereas, in a multiple-pulse experiment (8-pulse), which quenches spin diffusion among the ^{19}F spins, the magnetization decays for large times as $\exp[-(\frac{t}{\tau_1})^{1/2}]$. The solid line corresponds to Eq. (6.35) (see also Ref. [*14*])

ments can be studied by this method. An investigation of this kind is currently in progress in our laboratory.

Coming back to the relaxation by paramagnetic impurities (Fig. 6.7) we notice a linearity between $\ln R(t)$ versus $t^{1/2}$ over several decades for large t. The behavior of the relaxation function is completely different for the ordinary $T_{1\rho}$ as measured by spin locking. The interested reader is referred, for a further discussion, to Refs. [*20–23*].

6.3 Spin-Lattice Relaxation in Dilute Spin Systems

Finally, we want to draw attention to the study of spin lattice relaxation anisotropy of dilute spins, as proposed by Gibby, Pines and Waugh [*24*].
A scheme for studying the spin lattice relaxation of dilute spins is drawn schematically in Fig. 6.8.

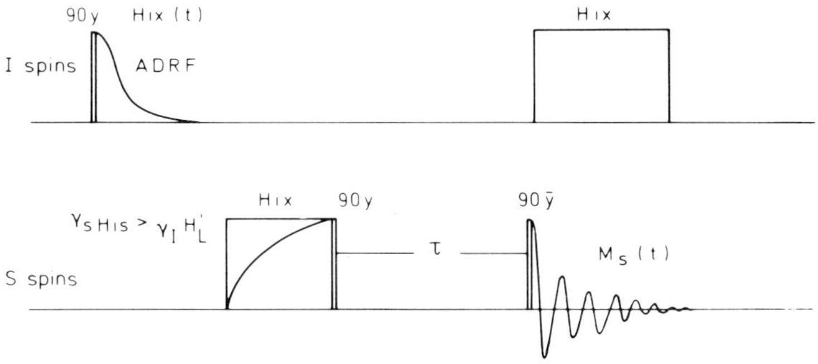

Fig. 6.8. Scheme for measuring spin lattice relaxation of dilute spins as proposed by Pines *et al.* [*25*]

Suppose we are dealing with a powder sample of benzene (C_6H_6) and are observing the spectrum of the dilute ^{13}C in the sample applying the PENIS experiment [*25*]. The observed shielding tensor is axially symmetric at the applied temperatures with the unique axis being the 6-fold rotation axis. We can therefore use the high resolution ^{13}C powder spectrum to monitor any anisotropic relaxation process such as spin lattice relaxation and the corresponding Overhauser effect [*24*].

Following along the lines of Gibby *et al.* [*24*] we assume that the dominant contribution to the relaxation of the ^{13}C spins (S spins) comes from the proton (I spins) directly bonded to it. The molecular motion is governed by random jumps about the 6-fold axis with a correlation time τ_c (Gauss-Markoff process). Coupled rate equations can be obtained for the I and S magnetization. Averaging over the six equivalent orientations of the benzene ring results in the following relaxation rates [*24*]

$$\frac{1}{T_1^{SS}} = \frac{3}{16}\frac{\gamma_S^2\gamma_I^2\hbar^2 I(I+1)}{r^6}\left[y_0(\beta)\frac{\tau_c}{1+(\omega_S-\omega_I)^2\tau_c^2} + y_1(\beta)\frac{\tau_c}{1+\omega_S^2\tau_c^2} + y_2\frac{\tau_c}{1+(\omega_S+\omega_I)^2\tau_c^2}\right] \tag{6.36}$$

and

$$\frac{1}{T_1^{IS}} = \frac{3}{16}\frac{\gamma_S^2\gamma_I^2\hbar^2 S(S+1)}{r^6}\left[y_2(\beta)\frac{\tau_c}{1+(\omega_S+\omega_I)^2\tau_c^2} - y_0(\beta)\frac{\tau_c}{1+(\omega_S-\omega_I)^2\tau_c^2}\right] \tag{6.37}$$

where

$$y_0(\beta) = \sin^4\beta$$

$$y_1(\beta) = 2\sin^2\beta(1+\cos^2\beta)$$

$$y_2(\beta) = 1 + 6\cos^2\beta + \cos^4\beta$$

and β is the angle between the C_6 axis and the static magnetic field H_0.

Two interesting special cases will be discussed

(i) The S spins (^{13}C) will be disturbed e.g. by a π pulse, which does not affect the $\langle I_z\rangle$ polarization. A relaxation of the S spins with T_1^{SS} will be observed,

(ii) The I spins (^1H) are saturated, $\langle I_z\rangle = 0$ and the S spin magnetization will reach a steady state with the time constant T_1^{SS}, leading to an Overhauser enhancement $1+\eta$ in the S spin polarization, with

$$\eta = \frac{\gamma_I}{\gamma_S}\frac{T_1^{SS}}{T_1^{IS}}\,. \tag{6.38}$$

Since both T_1^{SS} and T_1^{IS} show strong angular dependence, we also expect η to vary strongly with the angle β. This angular dependence can be studied in a powder sample as demonstrated in Figs. 6.10 and 6.11. Instead of the angle β one may find it convenient to use the parameter

$$\cos^2\beta = (\sigma - \sigma_\perp)/(\sigma_\parallel - \sigma_\perp) \tag{6.39}$$

to describe the angular dependence of T_1^{SS} and η. From Eq. (6.36) it is expected, that the anisotropy of T_1^{SS} changes sign by going from one side of the T_1 minimum to the other i.e., at high temperature the right hand side of the powder spectrum relaxes more rapidly than the left hand side, whereas at low temperatures the reverse is true. Evaluating partially relaxed powder spectra as in Fig. 6.9 leads to the T_1^{SS} anisotropy as shown in Fig. 6.10. The temperature dependence of the correlation time τ_c has been determined previously by deuterons relaxation as

$$\tau_c = 9.2\cdot 10^{-15}\exp(-4.2\ 10^3/\mathrm{RT}).$$

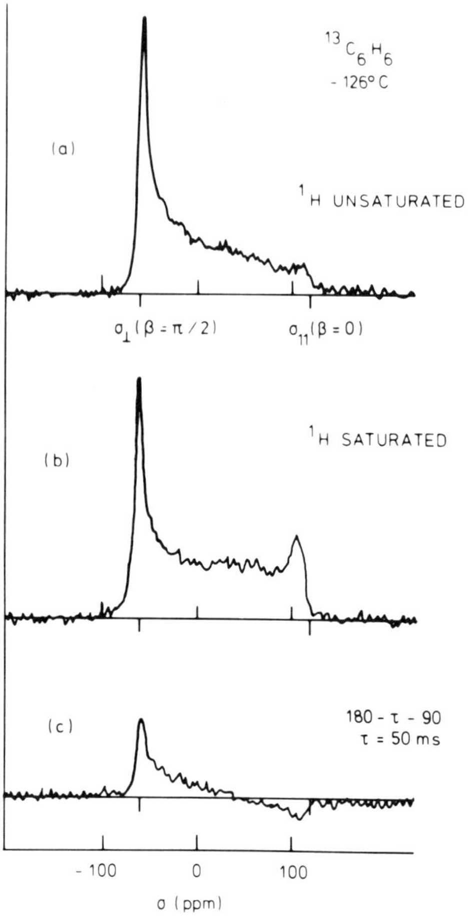

Fig. 6.9 a–c. Relaxation anisotropy of ^{13}C spins due to anisotropic molecular motion in solid benzene according to Gibby, Pines and Waugh [*24*]. (a) Proton decoupled ^{13}C spectrum of benzene with 4% abundance of ^{13}C. (b) Same as (a) but with a 1 *G* saturation field applied at the protons, demonstrating the anisotropic nuclear Overhauser enhancement. The powder pattern shown is expected from theory and depends strongly on temperature (see also Fig. 6.11). (c) Anisotropic T_1 relaxation after a recovery time of 50 ms. Molecules with $\beta = \frac{\pi}{2}$ $(\sigma = \sigma_\perp)$ show a seven times faster relaxation than those with $\beta = 0$ $(\sigma = \sigma_{||})$ (see also Fig. 6.10)

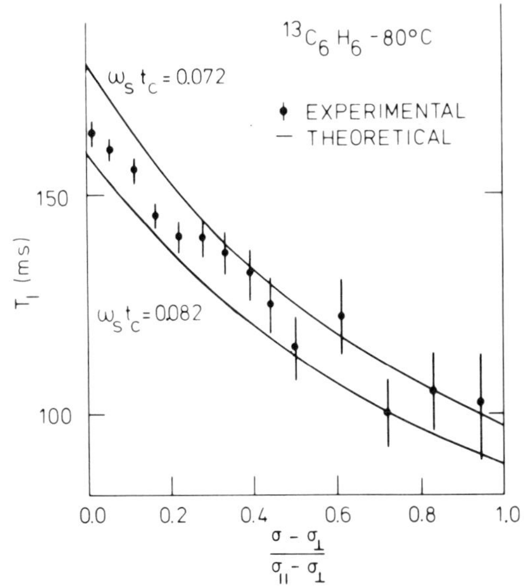

Fig. 6.10. Orientation dependence of the spin lattice relaxation time T_1 of ^{13}C in benzene according to Gibby, Pines and Waugh [*24*]. The experimental points were obtained by the analysis of spectra like in Fig. 6.9c. The theoretical curves correspond to Eq. (6.36) using r = 1.085 Å

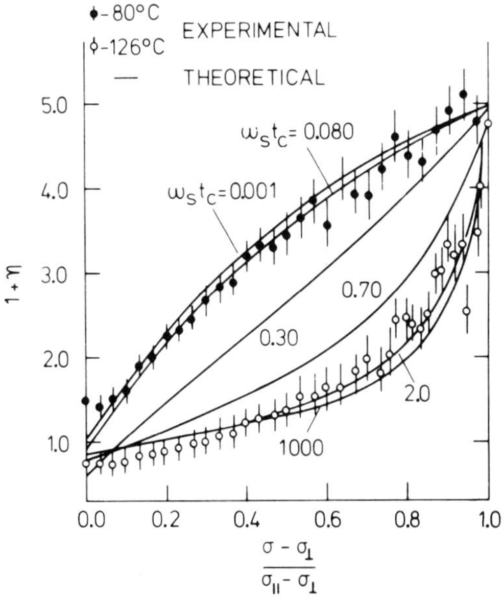

Fig. 6.11. Orientation dependence of the ^{13}C nuclear Overhauser enhancement $1 + \eta$ in benzene according to Gibby, Pines and Waugh [*24*] for two different temperatures. The experimental points were obtained from spectra like Fig. 6.9b, whereas the theoretical curves are derived from Eqs. (6.36–6.39)

At $T = -80°$ C, which is above the T_1 minimum, a value of $\omega_S \tau_c = 0.075$ is predicted in excellent agreement with Fig. 6.10. A similar agreement can be obtained for the nuclear Overhauser effect for temperatures above and below the T_1 minimum as demonstrated in Fig. 6.11.

Thus the study of direct spin lattice relaxation anisotropies of dilute spins can give valuable information on the molecular processes involved. In a similar investigation it has been shown by GPW [*24*] in adamantane, that those ^{13}C nuclei which have two rather than one directly bound proton relax more rapidly than the others.

These relaxation anisotropies should not be confused, however, with anisotropies of the cross relaxation rates $1/T_{IS}$ which have been discussed in Chapter 4 and which can of course result in distorted line shapes too. Another distinction has to be made concerning line shape changes due to anisotropic molecular reorientation, which show up when $\Delta\omega\tau_c \cong 1$, where $\Delta\omega$ is the size of the anisotropy (see Section 2.7).

7. Appendix

A. Irreducible Tensor Representation of Spin Interactions

Although Cartesian tensors lead to a direct physical apprehension in terms of direction cosines it might be more convenient to express tensors in a spherical basis, when rotations are involved. We therefore introduce a spherical representation of second rank tensors here.

The nine components T_{ij} of a Cartesian tensor of second rank can be decomposed into a [*1*]

scalar

$$\mathbf{T}_0 = \frac{1}{3} \mathrm{Tr}\,\{T_{ij}\} = \frac{1}{3} \sum_i T_{ii} \tag{A.1}$$

an *antisymmetric* tensor of first rank

$$\mathbf{T}_1 : T'_{ij} = \frac{1}{2}(T_{ij} - T_{ji}) \tag{A.2}$$

with three components and zero trace, and into a tracless second rank tensor, which is

symmetric

$$\mathbf{T}_2 : T''_{ij} = \frac{1}{2}(T_{ij} + T_{ji}) - \frac{1}{3} \mathrm{Tr}\,\{T_{ij}\} \tag{A.3}$$

with five components. This leads to a sum of three irreducible tensors as

$$T_{ij} = \frac{1}{3} \mathrm{Tr}\,\{T_{ij}\} + T'_{ij} + T''_{ij}. \tag{A.4}$$

The components of the three quantities $\mathbf{T}_0$, $\mathbf{T}_1$, and T_2 transform in the same way as the spherical harmonics of order zero, one and two, respectively.

When tensors are to be expressed in a new coordinate system (primed axes) obtained by a rotation of the old (unprimed) coordinate system about the Euler angles (α, β, γ) it is found convenient to use a spherical representation. In this spherical representation the irreducible spherical tensor $\mathbf{T}_k$ of rank k with $2k + 1$ components $\mathbf{T}_{kq}$ transforms according to the irreducible representation D_k of the rotation group

$$T'_{kq} = \mathbf{R}(\alpha\beta\gamma) T_{kq} \mathbf{R}^{-1}(\alpha\beta\gamma) = \sum_{p=-k}^{+k} T_{kp} D^k_{pq}(\alpha\beta\gamma). \tag{A.5}$$

With the sperical unit vectors

$$\mathbf{e}_{10} = \mathbf{e}_z; \ \mathbf{e}_{1\pm1} = \mp(1/\sqrt{2})(\mathbf{e}_x \pm i\mathbf{e}_y) \tag{A.6}$$

the components of a first rank irreducible tensor T_{1q} in terms of the Cartesian components (T_x, T_y, T_z) are given by [2]

$$T_{1q} = \mathbf{e}_{1q} \cdot \mathbf{T} = \mathbf{e}_{iq}(\mathbf{e}_x T_x + \mathbf{e}_y T_y + \mathbf{e}_z T_z) \tag{A.7}$$

or

$$T_{10} = T_{1z}; \ T_{1\pm1} = \mp(1/\sqrt{2})(T_x \pm iT_y). \tag{A.8}$$

A second rank Cartesian tensor is decomposed into the three irreducible spherical tensors in a similar fashion. First we have to find the corresponding spherical unit vectors e_{kq}. These may be contructed from the rule for the product of two irreducible tensor operators [1, 2]:

$$e_{kq} = [\mathbf{e}_1 \times \mathbf{e}_1]_{kq} = (2k+1)^{1/2} \sum_{q_1 q_2} (-1)^q \begin{pmatrix} 1 & 1 & k \\ q_1 & q_2 & -q \end{pmatrix} \mathbf{e}_{1q_1} \mathbf{e}_{1q_2}. \quad (k = 0, 1, 2) \tag{A.9}$$

The irreducible spherical tensors in terms of the Cartesian tensors are obtained by

$$T_{kq} = \mathbf{e}_{kq} \cdot \mathbf{T} \tag{A.10}$$

where

$$\mathbf{T} = \sum_{ij} \mathbf{e}_i \mathbf{e}_j T_{ij} \quad (i, j = x, y, z) \tag{A.11}$$

as [2]

$$\begin{aligned}
T_{00} &= -(1/\sqrt{3})[T_{xx} + T_{yy} + T_{zz}] \\
T_{10} &= -(i/\sqrt{2})[T_{xy} - T_{yx}] \\
T_{1\pm1} &= -\frac{1}{2}[T_{zx} - T_{xz} \pm i(T_{zy} - T_{yz})] \\
T_{20} &= (1/\sqrt{6})[3T_{zz} - (T_{xx} + T_{yy} + T_{zz})] \\
T_{2\pm1} &= \mp\frac{1}{2}[T_{xz} + T_{zx} \pm i(T_{yz} + T_{zy})] \\
T_{2\pm2} &= \frac{1}{2}[T_{xx} - T_{yy} \pm i(T_{xy} + T_{yx})].
\end{aligned} \tag{A.12}$$

In cases where the Cartesian tensor is symmetric and traceless only the second rank irreducible tensor is non vanishing, e.g. in the case of dipole-dipole interaction and quadrupole interaction.

Hamiltonians are usually expressible as scalar products of tensors. The scalar product of two irreducible tensors $\mathbf{A}_k$ and $\mathbf{T}_k$ with the components A_{kq} and T_{kp}, respectively, is defined as [*1, 2*]

$$\mathbf{A}_k \cdot \mathbf{T}_k = \sum_{q=-k}^{+k} (-1)^q A_{kq} T_{k-q} = \sum_{q=-k}^{+k} (-1)^q A_{k-q} T_{kq}. \tag{A.13}$$

In Section 2.1 we had expressed spin interactions by second rank tensors as

$$\mathcal{H} = \mathbf{X} \cdot \mathbf{A} \cdot \mathbf{Y} = \sum_{i,\,j} A_{ij} X_i Y_j \tag{A.14}$$

or

$$\mathcal{H} = \sum_{i,j} A_{ij} T_{ij} \tag{A.15}$$

where

$$T_{ij} = X_i Y_j.$$

In terms of irreducible spherical tensors, we may express the spin interaction Hamiltonian as

$$\mathcal{H} = \sum_{k=0}^{2} \sum_{q=-k}^{+k} (-1)^q A_{kq} T_{k-q} \tag{A.16}$$

where the spherical tensor components A_{kq} and T_{kq} can be expressed by the Cartesian tensor components A_{ij} and T_{ij} according to Eq. (A.12). Applying this to the shift interaction yields:

$$\mathcal{H}_S = \mathbf{I} \cdot \mathbf{S} \cdot \mathbf{H}_0 = \sum_{i,j} S_{ij} I_i H_{0j} \qquad (i, j = x, y, z)$$

or

$$\mathcal{H}_S = \sum_{k=0}^{2} \sum_{q=-k}^{+k} (-1)^q A_{kq} T_{k-q}$$

where

$$\begin{aligned} A_{00} &= -(1/\sqrt{3})[S_{xx} + S_{yy} + S_{zz}] = -(1/\sqrt{3})\,\mathrm{Tr}\{S_{ij}\} \\ A_{10} &= -(i/\sqrt{2})[S_{xy} - S_{yx}] \\ A_{1\pm 1} &= -\tfrac{1}{2}[S_{zx} - S_{xz} \pm i(S_{zy} - S_{yz})] \end{aligned} \tag{A.17}$$

$$A_{20} = (1/\sqrt{6})[3S_{zz} - \mathrm{Tr}\{S_{ij}\}]$$

$$A_{2\pm1} = \mp\frac{1}{2}[S_{xz} + S_{zx} \pm i(S_{yz} + S_{zy})]$$

$$A_{2\pm2} = \frac{1}{2}[S_{xx} - S_{yy} \pm i(S_{xy} + S_{yx})]$$

and

$$\begin{aligned}
T_{00} &= -(1/\sqrt{3})[I_x H_{0x} + I_y H_{0y} + I_z H_{0z}] \\
T_{10} &= -(i/\sqrt{2})[I_x H_{0y} - I_y H_{0x}] \\
T_{1\pm1} &= -\frac{1}{2}[I_z H_{0x} - I_x H_{0z} \pm i(I_z H_{0y} - I_y H_{0z})] \\
T_{20} &= (1/\sqrt{6})[3 I_z H_{0z} - (I_x H_{0x} + I_y H_{0y} + I_z H_{0z})] \\
T_{2\pm1} &= \mp\frac{1}{2}[I_x H_{0z} + I_z H_{0x} \pm i(I_y H_{0z} + I_z H_{0y})] \\
T_{2\pm2} &= \frac{1}{2}[I_x H_{0x} - I_y H_{0y} \pm i(I_x H_{0y} + I_y H_{0x})].
\end{aligned} \tag{A.18}$$

This is readily simplified with $\mathbf{H}_0 = (0, 0, H_0)$ as

$$\begin{aligned}
T_{10} &= -(1/\sqrt{3}) I_z H_0 \\
T_{10} &= 0;\ T_{1\pm1} = \frac{1}{2}(I_x \pm i I_y) H_0 \\
T_{20} &= (2/\sqrt{6}) I_z H_0 \\
T_{2\pm1} &= \mp\frac{1}{2}(I_x \pm i I_y) H_0 \\
T_{2\pm2} &= 0
\end{aligned} \tag{A.19}$$

from which follows

$$\mathcal{H}_S = A_{00} T_0 - (A_{11} T_{1-1} + A_{1-1} T_{11}) + A_{20} T_{20} - (A_{2-1} T_{21} + A_{21} T_{2-1}). \tag{A.20}$$

In case S_{ij} is symmetric i.e., $S_{ij} = S_{ji}$, in addition $A_{1\pm1}$ vanishes and we are left with

$$\mathcal{H}_S = A_{00} T_0 + A_{20} T_{20} - (A_{2-1} T_{21} + A_{21} T_{2-1}) \tag{A.21}$$

where A_{00} and A_{20} are given by Eq. (A.17) and $A_{2\pm1}$ is further simplified

$$A_{2\pm1} = \mp(S_{xz} \pm i S_{yz}). \tag{A.22}$$

When rotations are applied to the symmetric part of $\mathcal{H}_S$ in real space, only the components A_{2q} have to be transformed. The separation of the shielding Hamiltonian $\mathcal{H}_S$

into the second rank spherical tensor products is not as convenient as it appears to be, since rotations are usually performed, whether in real space or in spin space. For rotations in real space the former representation is fine, but rotations in spin space cannot be applied to the T_{kq}, since they also contain H_0.

We find it therefore more convenient to separate strictly geometrical and spin variables

$$\mathcal{H}_S = \mathbf{I} \cdot \mathbf{S} \cdot \mathbf{H}_0 = \mathbf{I} \cdot \mathbf{H}_S \tag{A.23}$$

where

$$\mathbf{H}_S = \mathbf{S} \cdot \mathbf{H}_0 .$$

It follows

$$\mathcal{H}_S = \sum_{q=-1}^{1} (-1)^q A_{1q} T_{1-q} \tag{A.24}$$

where

$$A_{10} = H_{Sz};\, A_{1\pm 1} = \mp(1/\sqrt{2})\,(H_{Sx} + i\,H_{Sy}) \tag{A.25}$$

and

$$T_{10} = I_z;\, T_{1\pm 1} = \mp(1/\sqrt{2})(I_x \pm i I_y). \tag{A.26}$$

With $\mathbf{H}_0 = (0, 0, H_0)$ we obtain further

$$H_{Sx} = S_{xz}H_0;\, H_{Sy} = S_{yz}H_0;\, H_{Sz} = S_{zz}H_0. \tag{A.27}$$

Spin-spin interactions which are bilinear in the spin variable are conveniently expressed by irreducible spherical tensor products as

$$\mathcal{H}_{IS} = \mathbf{I} \cdot \mathbf{D} \cdot \mathbf{S} = \sum_{ij} D_{ij} I_i S_j$$

$$\mathcal{H}_{IS} = \sum_{k=0}^{2} \sum_{q=-k}^{k} (-1)^q A_{kq} T_{k-q}$$

where **D** represents the direct as well as the indirect spin-spin interaction. If we assume $\mathbf{I} \equiv \mathbf{S}$, even the quadrupole interaction with $\mathbf{D} = \mathbf{Q}$ can be represented in this fashion. The sperical tensor components A_{kq} and T_{kq} can be readily obtained from Eq. (A.12) in terms of their Cartesian counterparts. Now spin and geometrical variable are well separated, since the A_{kq} represent only geometrical variables, whereas the T_{kq} contain only spin variables.

If only symmetric components have to be considered we obtain

$$
\begin{aligned}
A_{00} &= -\frac{1}{\sqrt{3}}\,\mathrm{Tr}\{D_{ij}\} \\
A_{10} &= A_{1\pm 1} = 0 \\
A_{20} &= (3/\sqrt{6})\,D_{zz} \\
A_{2\pm 1} &= \mp(D_{xz} \pm i D_{yz}) \\
A_{2\pm 2} &= \frac{1}{2}(D_{xx} - D_{yy} \pm 2i D_{xy})
\end{aligned}
\tag{A.28}
$$

and

$$
\begin{aligned}
T_{20} &= (1/\sqrt{6})(3 I_z S_z - \mathbf{I} \cdot \mathbf{S}) \\
T_{2\pm 1} &= \mp\frac{1}{2}(I_z S_\pm + I_\pm S_z) \\
T_{2\pm 2} &= \frac{1}{2} I_\pm S_\pm
\end{aligned}
\tag{A.29}
$$

As a specific example let us consider the dipolar interaction. With $\mathrm{Tr}\{D_{ij}\} = 0$ and

$$D_{ij} = \frac{\gamma_I \gamma_S \hbar}{r^3}(\delta_{ij} - 3\mathbf{e}_i \cdot \mathbf{e}_j)\ (i, j = x, y, z) \tag{A.30}$$

with $e_i (i = x, y, z)$ are the x, y and z components of a unit vector pointing from one spin to the other we obtain

$$A_{2q} = -\sqrt{6}\,\frac{\gamma_I \gamma_S \hbar}{r^3}\sqrt{\frac{4\pi}{5}}\,Y_{2q} \tag{A.31}$$

where Y_{kq} are spherical harmonics. Sometimes it might be convenient to use the modified spherical harmonics.

$$C_{kq} = \left[\frac{4\pi}{2k+1}\right]^{1/2} Y_{kq}. \tag{A.32}$$

B. Rotations

We want to summarize some general rotation properties for the convenience of the reader. A more rigorous treatment can be found in standard texts. We adopt here the sign conventions and the Euler angle definition of Rose [*1a*]. If a positive rotation is to be applied to a frame (x, y, z) about the Euler angles (α, β, γ) as

$$\mathbf{R}(\alpha\beta\gamma) = \mathbf{R}_{z''}(\gamma)\,\mathbf{R}_{y'}(\beta)\,\mathbf{R}_z(\alpha) \tag{B.1}$$

where α is a rotation about the original z axis, β is about the new y axis, and γ is about the final z axis, the product of the three rotation matrices in Eq. (B.1) leads to

$$\mathbf{R}(\alpha\beta\gamma) = \begin{Bmatrix} \cos\alpha\cos\beta\cos\gamma - \sin\alpha\sin\gamma & \sin\alpha\cos\beta\cos\gamma + \cos\alpha\sin\gamma & -\sin\beta\cos\gamma \\ -\cos\alpha\cos\beta\sin\gamma - \sin\alpha\cos\gamma & -\sin\alpha\cos\beta\sin\gamma + \cos\alpha\cos\gamma & \sin\beta\sin\gamma \\ \cos\alpha\sin\beta & \sin\alpha\sin\beta & \cos\beta \end{Bmatrix} \tag{B.2}$$

The product of rotations in Eq. (B.1) may be readily expressed in terms of rotations about the original axes (x, y, z) as:

$$\mathbf{R}(\alpha\beta\gamma) = \mathbf{R}_z(\alpha)\mathbf{R}_y(\beta)\mathbf{R}_z(\gamma). \tag{B.3}$$

These rotations in Cartesian coordinates can be transformed to spherical coordinates as discussed similarily in Appendix A. We come to this later.

Let us now summarize some positive rotations (counter clockwise) in spin space:

$$e^{-i\alpha I_x}\begin{Bmatrix} I_x \\ I_y \\ I_z \end{Bmatrix}e^{i\alpha I_x} = \begin{Bmatrix} I_x \\ I_y\cos\alpha + I_z\sin\alpha \\ I_z\cos\alpha - I_y\sin\alpha \end{Bmatrix}$$

$$e^{-i\alpha I_y}\begin{Bmatrix} I_x \\ I_y \\ I_z \end{Bmatrix}e^{i\alpha I_y} = \begin{Bmatrix} I_x\cos\alpha - I_z\sin\alpha \\ I_y \\ I_z\cos\alpha + I_x\sin\alpha \end{Bmatrix} \tag{B.4}$$

$$e^{-i\alpha I_z}\begin{Bmatrix} I_x \\ I_y \\ I_z \end{Bmatrix}e^{i\alpha I_z} = \begin{Bmatrix} I_x\cos\alpha + I_y\sin\alpha \\ I_y\cos\alpha - I_x\sin\alpha \\ I_z \end{Bmatrix}$$

From these relations, combined rotations can be constructed e.g.

$$e^{-i\alpha I_x} = e^{i\frac{\pi}{2}I_z}\, e^{-i\alpha I_y}\, e^{-i\frac{\pi}{2}I_z} \tag{B.5}$$

or

$$e^{-i\alpha I_x} = e^{-i\frac{\pi}{2}I_y}\, e^{-i\alpha I_z}\, e^{i\frac{\pi}{2}I_y} \tag{B.6}$$

and in a similar way

$$e^{-i\alpha(I_x + I_y + I_z)/\sqrt{3}} = e^{-i\frac{\pi}{4}I_z}\, e^{-i\beta_m I_y}\, e^{-i\alpha I_z}\, e^{i\beta_m I_y}\, e^{i\frac{\pi}{4}I_z} \tag{B.7}$$

where

$$\tan\beta_m = \sqrt{2}.$$

Since in the definition of the Euler angles [Eq. (B.3)] only rotations about the z and y axis are allowed, we may write any x-rotation according to Eqs. (B.5, B.6) as

$$e^{-i\alpha I_x} = \mathbf{R}(-\tfrac{\pi}{2}\ \alpha\ \tfrac{\pi}{2}) \tag{B.8}$$

or

$$e^{-i\alpha I_x} = \mathbf{R}(0\ \tfrac{\pi}{2}\ 0)\ \mathbf{R}_z(\beta)\ \mathbf{R}(0\ -\tfrac{\pi}{2}\ 0) \tag{B.9}$$

and in a similar way

$$e^{-i\alpha(I_x+I_y+I_z)/\sqrt{3}} = \mathbf{R}(\tfrac{\pi}{4}\ \beta_m\ 0)\ \mathbf{R}_z(\beta)\ \mathbf{R}(0\ -\beta_m\ -\tfrac{\pi}{4}). \tag{B.10}$$

We turn now to rotations of irreducible tensor operators, by writing

$$T'_{kq} = \mathbf{R}(\alpha\beta\gamma) T_{kq} \mathbf{R}^{-1}(\alpha\beta\gamma) = \sum_{p=-k}^{+k} T_{kp} D^k_{pq}(\alpha\beta\gamma) \tag{B.11}$$

with the Wigner rotation matrices

$$D^k_{pq}(\alpha\beta\gamma) = \langle kp |\, e^{-i\alpha I_z}\ e^{-i\beta I_y}\ e^{-i\gamma I_z} \,| kq \rangle \tag{B.12}$$

or

$$D^k_{pq}(\alpha\beta\gamma) = e^{-i\alpha p}\ d^k_{pq}(\beta)\ e^{-i\gamma q}. \tag{B.13}$$

The reduced rotation matrices d^k_{pq} (β) are real and are expressed explicitly for the values $k = 1/2, 1, 3/2, 2$ in Table B.1.

As a representative example let us consider

$$I'_z = e^{-i\alpha I_x}\ I_z\ e^{+i\alpha I_x}$$

or

$$T'_{10} = \mathbf{R}(-\tfrac{\pi}{2}\ \beta\ \tfrac{\pi}{2})\ T_{10}\ \mathbf{R}^{-1}(-\tfrac{\pi}{2}\ \beta\ \tfrac{\pi}{2})$$

$$T'_{10} = \sum_q T_{1q} D^1_{q0}(-\tfrac{\pi}{2}\ \beta\ \tfrac{\pi}{2})$$

$$T'_{10} = \sum_q T_{1q}\, e^{iq\frac{\pi}{2}} d^1_{q0}(\beta).$$

With the help of table B.1 and $T_{11} + T_{1-1} = -i\sqrt{2}\, I_y$ this leads to

$$T'_{10} = I_z \cos\alpha - I_y \sin\alpha$$

as in Eq. (B.4).

Table B. 1. Expressions of $d^j_{mm'}(\beta)$ for $j = \frac{1}{2}, 1, \frac{3}{2}$, and 2 (Brink and Satchler [1])

$d^{1/2}_{1/2\,1/2} = d^{1/2}_{-1/2\,-1/2} = \cos(\frac{\beta}{2})$	$d^2_{22} = d^2_{-22}\cos^4(\frac{\beta}{2})$
$d^{1/2}_{-1/2\,1/2} = -d^{1/2}_{1/2\,-1/2} = \sin(\frac{\beta}{2})$	$d^2_{21} = -d^2_{12} = -d^2_{-2-1}$
$d^1_{11} = d^1_{-1-1} = \cos^2(\frac{\beta}{2})$	$= d^2_{-1-2} = -\frac{1}{2}\sin\beta\,(1+\cos\beta)$
$d^1_{1-1} = d^1_{-11} = \sin^2(\frac{\beta}{2})$	$d^2_{20} = d^2_{02} = d^2_{-20} = d^2_{0-2} = \sqrt{\frac{3}{8}}\sin^2\beta$
$d^1_{01} = d^1_{-10} = -d^1_{0-1} = -d^1_{10} = \sin\beta/\sqrt{2}$	$d^2_{2-1} = d^2_{1-2} = -d^2_{-21} = -d^2_{-12} = \frac{1}{2}\sin\beta(\cos\beta - 1)$
$d^1_{00} = \cos\beta$	$d^2_{2-2} = d^2_{-22} = \sin^4(\frac{\beta}{2})$
$d^{3/2}_{3/2\,3/2} = d^{3/2}_{-3/2-3/2} = \cos^3(\frac{\beta}{2})$	$d^2_{11} = d^2_{-1-1} = \frac{1}{2}(2\cos\beta - 1)(\cos\beta + 1)$
$d^{3/2}_{3/2\,1/2} = d^{3/2}_{-1/2\,-3/2} = -d^{3/2}_{1/2\,3/2}$	$d^2_{1-1} = d^2_{-11} = \frac{1}{2}(2\cos\beta + 1)(1 - \cos\beta)$
$= -d^{3/2}_{-3/2\,-1/2} = -\sqrt{3}\cos^2(\frac{\beta}{2})\sin(\frac{\beta}{2})$	$d^2_{10} = d^2_{0-1} = -d^2_{01} = -d^2_{-10} = -\sqrt{\frac{3}{2}}\sin\beta\cos\beta$
$d^{3/2}_{3/2\,-1/2} = d^{3/2}_{-1/2\,3/2} = d^{3/2}_{1/2\,-3/2}$	$d^2_{00} = \frac{1}{2}(3\cos^2\beta - 1)$
$= d^{3/2}_{-3/2\,1/2} = \sqrt{3}\cos(\frac{\beta}{2})\sin^2(\frac{\beta}{2})$	
$d^{3/2}_{3/2-3/2} = -d^{3/2}_{-3/2\,3/2} = -\sin^3(\frac{\beta}{2})$	
$d^{3/2}_{1/2\,1/2} = d^{3/2}_{-1/2\,1/2} = \cos(\frac{\beta}{2})[3\cos^2(\frac{\beta}{2}) - 2]$	
$d^{3/2}_{1/2\,-1/2} = -d^{3/2}_{-1/2\,1/2} = \sin(\frac{\beta}{2})[3\sin^2(\frac{\beta}{2}) - 2]$	

References for additional tables for $d^j_{mm'}(\beta)$ are as follows:
$j = 2, 4, 6$: Buckmaster, H. A., Can. J. Phys. **42**, 386 (1964)
$j = 1, 3, 5$: ibid. **44**, 2525 (1966)
$j = 3$ Ying-Nan Chiu, J. Chem. Phys. **45**, 2969 (1966)

C. Contribution of Non-Secular Shielding Tensor Elements to the Resonance Shift

The shift of the nuclear resonance line is given by the "size" of the shift Hamiltonian $\mathcal{H}_S$, namely

$$\|\mathcal{H}_S\| = \left[\frac{(\mathcal{H}_S\,|\,\mathcal{H}_S)}{(I_z\,|\,I_z)}\right]^{1/2} \tag{C.1}$$

In the laboratory frame we have

$$\mathcal{H}_S = \omega_0[\sigma_{xz}I_x + \sigma_{yz}I_y + \sigma_{zz}I_z]. \tag{C.1}$$

In the rotating frame, where we perform our measurements we can write

$$\tilde{\mathcal{H}}_S(t) = e^{-i\omega_0 t I_z}\, \mathcal{H}_S\, e^{i\omega_0 t I_z} \tag{C.3}$$

where only the term $\omega_0 \sigma_{zz} I_z$ is stationary.

We are going to express the shielding Hamiltonian in the rotating frame by an average Hamiltonian $\bar{\mathcal{H}}_S$ according to the Magnus expansion, with

$$\bar{\mathcal{H}}_S = \bar{\mathcal{H}}^{(0)} + \bar{\mathcal{H}}^{(1)} + \ldots \tag{C.4}$$

where we use the corresponding expressions of the Magnus expansion.

$$|\bar{\mathcal{H}}_S^{(0)}) = \omega_0 \sigma_{zz} | I_z) \tag{C.5}$$

$$|\bar{\mathcal{H}}_S^{(1)}) = -\frac{i}{2t} \int_0^t dt_2 \int_0^{t_2} dt_1\, e^{-i\omega_0 t_2 \hat{I}_z} \hat{\mathcal{H}}_S\, e^{i(t_2 - t_1)\omega_0 \hat{I}_z} | \mathcal{H}_S). \tag{C.6}$$

Using Eqs. (C.1) and (C.4) we write

$$\| \mathcal{H}_S \|^2 = \frac{(\bar{\mathcal{H}}_S^{(0)} | \bar{\mathcal{H}}_S^{(0)}) + 2\,(\bar{\mathcal{H}}_S^{(0)} | \bar{\mathcal{H}}_S^{(1)}) + (\bar{\mathcal{H}}_S^{(1)} | \bar{\mathcal{H}}_S^{(1)})}{(I_z | I_z)}. \tag{C.7}$$

We readily obtain

$$\frac{(\bar{\mathcal{H}}_S^{(0)} | \bar{\mathcal{H}}_S^{(0)})}{(I_z | I_z)} = \omega_0^2 \sigma_{zz}^2 \tag{C.8}$$

$$\frac{(\bar{\mathcal{H}}_S^{(0)} | \bar{\mathcal{H}}_S^{(1)})}{(I_z | I_z)} = -\frac{i}{2t} \int_0^t dt_2 \int_0^{t_2} dt_1\, \omega_0 \sigma_{zz} \left\{ \frac{(I_z | e^{-i\omega_0 t_2 \hat{I}_z} \hat{\mathcal{H}}_S\, e^{i(t_2 - t_1)\omega_0 \hat{I}_z} | \mathcal{H}_S)}{(I_z | I_z)} \right\}$$

$$= \frac{\omega_0 \sigma_{zz}}{2t} \int_0^t dt_2 \int_0^{t_2} dt_1 [-i\{\}] \tag{C.9}$$

with

$$-i\left\{\right\} = \omega_0 \Big[\sigma_{xz} \frac{(I_y | e^{i(t_2 - t_1)\omega_0 \hat{I}_z} | \mathcal{H}_S)}{(I_z | I_z)}$$

$$- \sigma_{yz} \frac{(I_x | e^{i(t_2 - t_1)\omega_0 \hat{I}_z} | \mathcal{H}_S)}{(I_z | I_z)} \Big] \tag{C.10}$$

or

$$i\left\{\right\} = \omega_0^2 (\sigma_{xz}^2 + \sigma_{yz}^2) \sin \omega_0 (t_2 - t_1). \tag{C.11}$$

Integration over a "cycle" $\omega_0 t = 2\pi$ leads to

$$-2 \cdot \frac{(\bar{\mathcal{H}}_S^{(0)} | \bar{\mathcal{H}}_S^{(1)})}{(I_z | I_z)} = \omega_0^2 \sigma_{zz} (\sigma_{xz}^2 + \sigma_{yz}^2). \tag{C.12}$$

The contribution of $(\overline{\mathcal{H}}_S^{(1)} | \overline{\mathcal{H}}_S^{(1)})$ is next higher order and will be neglected here. It can be of course included straightforwardly. Inserting Eqs. (C.8, C.12) into Eq. (C.7) leads to

$$\|\mathcal{H}_S\| = \omega_0 \sigma_{zz} [1 - \frac{\sigma_{xz}^2 + \sigma_{yz}^2}{\sigma_{zz}}]^{1/2}. \tag{C.13}$$

Since σ is a small number on the order of $\leqslant 10^{-4}$, the relative contribution of non-secular terms to the resonance shift $\omega_0 \sigma_{zz}$ is of the same order of magnitude. The same argument applies to the antisymmetric tensor elements, of course, since the σ_{xz}, σ_{yz} contain the symmetric and the antisymmetric constituents of the shielding tensor. Eq. (C.13) justifies that we have neglected antisymmetric tensor contributions before. Eq. (C 13) may be further approximated by [*4*]

$$\|\mathcal{H}_S\| = \omega_0 \sigma_{zz} (1 - \frac{\sigma_{xz}^2 + \sigma_{yz}^2}{2\sigma_{zz}}) = \omega_0 (\sigma_{zz} - \frac{1}{2}(\sigma_{xz}^2 + \sigma_{yz}^2)) \tag{C.14}$$

let us separate the shielding tensor σ_{ij} into a symmetric part $\sigma_{ij}^{(s)}$ and an antisymmetric part $\sigma_{ij}^{(a)}$

$$\sigma_{ij} = \sigma_{ij}^{(s)} + \sigma_{ij}^{(a)}. \tag{C.15}$$

The symmetric part may be diagonalized and is given in its principal axis frame (1, 2, 3) by

$$\sigma_{ii}^{(s)} = \begin{pmatrix} \sigma_{11}^{(s)} & 0 & 0 \\ 0 & \sigma_{22}^{(s)} & 0 \\ 0 & 0 & \sigma_{33}^{(s)} \end{pmatrix} \tag{C.16}$$

whereas the antisymmetric part $\sigma_{ij}^{(a)}$ is given in the same frame as

$$\sigma_{ij}^{(a)} = \begin{pmatrix} 0 & \sigma_{12}^{(a)} & \sigma_{13}^{(a)} \\ -\sigma_{12}^{(a)} & 0 & \sigma_{23}^{(a)} \\ -\sigma_{13}^{(a)} & -\sigma_{23}^{(a)} & 0 \end{pmatrix}. \tag{C.17}$$

With the transformation $\mathbf{R}(\alpha, \beta, \gamma)$ we transform the shielding tensor from the principal axis frame into the magnetic field frame (lab.-frame) as

$$\sigma_{ij} = \mathbf{R}(\alpha\beta\gamma)\, \sigma_{ij}^{(1,2,3)}\, \mathbf{R}^{-1}(\alpha\beta\gamma),$$

$$\sigma_{ij} = \sum_{k,l} r_{ik} \sigma_{kl}^{(1,2,3)} r_{jl}. \tag{C.19}$$

We obtain

$$\sigma_{zz} = \sum_k r_{3k}^2 \sigma_{kk}^{(s)} + \sum_{k,e} r_{3k} r_{3e} \sigma_{ke}^{(a)}. \tag{C.20}$$

Since $\sigma_{kl}^{(a)} = -\sigma_{lk}^{(a)}$ the last term in Eq. (C.20) vanishes, leaving

$$\sigma_{zz} = \sum_k r_{3k}^2 \sigma_{kk}^{(s)}. \tag{C.21}$$

The non-secular components are correspondingly

$$\sigma_{xz} = \sum_{k,e} r_{1k} r_{3e} \sigma_{ke}, \tag{C.22}$$

$$\sigma_{yz} = \sum_{k,e} r_{2k} r_{3e} \sigma_{ke}. \tag{C.23}$$

This can be combined to obtain the non-secular contribution to the shift as

$$\frac{1}{2}(\sigma_{xz}^2 + \sigma_{yz}^2) = \sum_{i,j,k,e} r_{3j} r_{3e} \frac{1}{2}(r_{1i} r_{1k} + r_{2i} r_{2k}) \sigma_{ij} \sigma_{ke}. \tag{C.24}$$

Eq. (C.24) is quite a general formula, from which special cases can be obtained readily.
Let us now discuss some specific examples:

(i) The shielding tensor σ is purely symmetric

$$\sigma_{ij}^{(s)} = \sigma_{ii}^{(s)} \delta_{ij} \tag{C.25}$$

$$\frac{1}{2}(\sigma_{xz}^{(s)^2} + \sigma_{yz}^{(s)^2}) = \sum_{i,k} r_{3i} r_{3k} \frac{1}{2}(r_{1i} r_{1k} + r_{2i} r_{2k}) \sigma_{ii}^{(s)} \sigma_{kk}^{(s)}.$$

(ii) The tensor σ is purely antisymmetric

$$\sigma_{ij}^{(a)} = -\sigma_{ji}^{(a)}(1 - \delta_{ij}) \tag{C.26}$$

$$\frac{1}{2}(\sigma_{xz}^{(a)^2} + \sigma_{yz}^{(a)^2}) = \sum_{i<j;k<e} [r_{3j} r_{3e} \frac{1}{2}(r_{1i} r_{1k} + r_{2i} r_{2k}) - r_{3i} r_{3k} \frac{1}{2}(r_{1j} r_{1e} + r_{2j} r_{2e})] \sigma_{ij}^{(a)} \sigma_{ke}^{(a)}.$$

(iii) Special orientation dependence: $\alpha = \gamma = 0$

$$R = \begin{pmatrix} \cos\beta & 0 & -\sin\beta \\ 0 & 1 & 0 \\ \sin\beta & 0 & \cos\beta \end{pmatrix}$$

$$\sigma_{zz} = \sin^2\beta \sigma_{11}^{(s)} + \cos^2\beta \sigma_{33}^{(s)}$$

$$\frac{1}{2}(\sigma_{xz}^{(s)^2} + \sigma_{yz}^{(s)^2}) = \frac{1}{2}\sin^2\beta \cos^2\beta (\sigma_{33}^{(s)} - \sigma_{11}^{(s)})^2$$

$$\frac{1}{2}(\sigma_{xz}^{(a)^2} + \sigma_{yz}^{(a)^2}) = \frac{1}{2}\cos^2\beta \sigma_{23}^{(a)^2} - \frac{1}{2}\sin^2\beta \sigma_{12}^{(a)^2} + \frac{1}{2}(\cos^4\beta - \sin^4\beta) \sigma_{13}^{(a)^2}.$$

Mixed terms have to be considered in the general case, if both symmetric and anti-symmetric constituents are present. This can be done straightforwardly by starting from Eq. (C.24).

D. Bloch-Siegert Shift [5]

Irradiation of a spin system by an rf field

$$\mathcal{H}_1 = -2\omega_1 I_x \cos \omega t \tag{D.1}$$

causes a rotation of the spins in a suitable reference frame. Suppose for example, the irradiation is performed

(i) on-resonance $\omega = \omega_0$ Larmor frequency

$\mathcal{H}_1$ is transformed into the rotating frame as

$$\tilde{\mathcal{H}}_1(t) = -[\omega_1 I_x + \omega_1 I_x \cos 2\omega_0 t + \omega_1 I_y \sin 2\omega_0 t]. \tag{D.2}$$

The secular part $-\omega_1 I_x$ causes a rotation abount the x axis with the frequency ω_1. The non-secular term is due to the counter rotating part of the rf field and contributes to the rotation only in higher order. This non-secular contribution is called the "Bloch-Siegert shift" and can be usually neglected as follows.

Expressing $\mathcal{H}_1$ by an average Hamiltonian as

$$\bar{\mathcal{H}}_1 = \bar{\mathcal{H}}_1^{(0)} + \bar{\mathcal{H}}_1^{(1)} + \ldots \tag{D.3}$$

we arrive at

$$\bar{\mathcal{H}}_1^{(0)} = -\omega_1 I_x \tag{D.4}$$

$$\bar{\mathcal{H}}_1^{(1)} = \frac{-i}{2\omega_0^2 t} \int_0^{\omega_0 t} d\alpha_2 \int_0^{\alpha_2} d\alpha_1 [\tilde{\mathcal{H}}_1(\alpha_2), \tilde{\mathcal{H}}(\alpha_1)] \tag{D.5}$$

where $\alpha_i = \omega_0 t_i \quad i = 1, 2$.

If we define a cycle by $\omega_0 t = 2\pi$ ("stroboscopic observation") [4], we arrive at

$$\bar{\mathcal{H}}_1^{(1)} = \omega_1^{(1)} I_z \tag{D.6}$$

where

$$\omega_1^{(1)} = -\omega_0 \left(\frac{\omega_1}{2\omega_0}\right)^2. \tag{D.7}$$

The relative Bloch-Siegert shift [4, 5]

$$\epsilon = \omega_1^{(1)}/\omega_0 = -[\frac{\omega_1}{2\omega_0}]^2 \tag{D.8}$$

is usually negligible.

However, in decoupling experiments, where a strong rf field is applied at a frequency usually much closer to the Larmor frequency of the spins to be observed, than the counter rotating component, remarkable resonance shifts can be expected.

(ii) *off-resonance* $|\omega - \omega_0| \gg \omega_1$

In the "resonance frame" rotating with the frequency ω_0, we can express $\widetilde{\mathcal{H}}_1(t)$ as:

$$\widetilde{\mathcal{H}}_1(t) = -\omega_1 I_x[\cos(\omega + \omega_0)t + \cos(\omega - \omega_0)t]$$

$$-\omega_1 I_y[\sin(\omega + \omega_0)t - \sin(\omega - \omega_0)t]. \tag{D.9}$$

We express now the propagator of $\widetilde{\mathcal{H}}_1(t)$ by an average Hamiltonian

$$\bar{\widetilde{\mathcal{H}}}_1 = \bar{\mathcal{H}}_1^{(0)} + \bar{\mathcal{H}}_1^{(1)} + \ldots$$

where we have to choose convenient cycles.

It can be shown, that a cycle time can be always defined with $\omega t_c = m2\pi$ and $\omega_0 t_c = n2\pi$ ("stroboscopic observation") [4].

Using this cyclic property of $\widetilde{\mathcal{H}}_1(t)$ we obtain

$$\bar{\mathcal{H}}_1^{(0)} = 0$$

$$\bar{\mathcal{H}}_1^{(1)} = \frac{-i}{2t_c}\int_0^{t_c} dt_2 \int_0^{t_2} dt_1 [\widetilde{\mathcal{H}}_1(t_2), \widetilde{\mathcal{H}}_1(t_1)]$$

and by choosing "proper" cycles

$$\widetilde{\mathcal{H}}_1^{(1)} = \omega_1^{(1)} I_z$$

where

$$\omega_1^{(1)} = \frac{1}{2}\omega_1^2[\frac{1}{\omega_0 + \omega} + \frac{1}{\omega_0 - \omega}] \tag{D.10}$$

or

$$\omega_1^{(1)} = \omega_0[\frac{\omega_1^2}{\omega_0^2 - \omega^2}] \tag{D.11}$$

with the relative shift

$$\epsilon = \omega_1^{(1)}/\omega_0 = \frac{\omega_1^2}{\omega_0^2 - \omega^2} \tag{D.12}$$

Examples: a) ^{1}H decoupled, ^{13}C observed

$$\upsilon_0(^{13}C) = 23 \text{ MHz}; \upsilon(^1H) = 90 \text{ MHz}; \upsilon_1(^1H) = 100 \text{ kHz } (23.5\, G)$$

results in

$$\epsilon = -1.32 \text{ ppm}$$

b) ^{1}H decoupled, ^{19}F observed

$$\upsilon_0(^{19}F) = 84.6 \text{ MHz}; \upsilon(^1H) = 90 \text{ MHz}; \upsilon_1(^1H) = 100\text{kHz } (23.5\, G)$$

results in

$$\epsilon = -10.6 \text{ ppm}.$$

Notice, that the sign of the Bloch-Siegert shift changes, according to Eq. (D.11) depending, whether irradiation is performed above (+) or below (–) resonance of the observed spins. In decoupling experiments this shift has to be taken into account. This is usually done by measuring the resonance frequency of a reference sample with the decoupling field on, i.e. the reference signal is affected by the same shift and the frequency difference to the spectral line is about the same. Notice, however, that different lines in a spectrum are affected differently. A simple calculation shows, however, that this distinction is usually negligible. Suppose two lines in the spectrum are separated by $\Delta\omega$.

The change of this separation by the Bloch-Siegert shift is according to Eq. (D.11)

$$\Delta\omega_1^{(1)} \cong \Delta\omega \cdot \epsilon$$

i.e. this absolute shift is of the order of some 10^{-12}. This is even in "Highest Resolution NMR" undetectable.

E. General Line Shape Theory

In this context we consider the "line-shape" in a general sense, covering the static resonance line shape of the NMR spectrum as well as spectral distribution functions which are involved in relaxation, cross-relaxation, cross-polarization as well as spin decoupling processes, thus providing the basic formalism for Sections 3, 3.8, 4.3, 4.4. Only the "non-trivial" case of many body interactions, such as dipolar interaction is considered [*6*].

Let us first define a Liouville space by considering the ordinary operators of quantum mechanics to be state vectors in this Liouville space i.e., if the Hilbert space of state vectors ψ has the dimension n, the corresponding Liouville space will have the dimension $n \times n$.

The Liouville-v. Neumann equation of motion for the density matrix ρ may now be expressed in Liouville space as [7]

$$\frac{d}{dt} | \rho) = -i\hat{\mathcal{H}} | \rho) \tag{E.1}$$

with the formal solution $|\rho(t)) = T \exp[-i \int_0^t dt' \hat{\mathcal{H}}(t)] | \rho(0))$ where $\hat{\mathcal{H}}$ is a superoperator or Liouville operator (Liouvillian) acting on the state vector $|\rho)$ in Liouville space and T the usual Dyson time ordering operator.

$\hat{\mathcal{H}}$ is defined by

$$\hat{\mathcal{H}} | \mathbf{A}) = |[\mathcal{H}, \mathbf{A}]) \tag{E.2}$$

correspondingly

$$\hat{\mathcal{H}}^n | \mathbf{A}) = |[\mathcal{H}, [\mathcal{H}, ..., [\mathcal{H}, \mathbf{A}]]...]_n) \tag{E.3}$$

and

$$e^{\alpha \hat{\mathbf{A}}} | \mathbf{B}) = | e^{\alpha \mathbf{A}} \mathbf{B} e^{-\alpha \mathbf{A}}). \tag{E.4}$$

A scalar product between two Liouville state vectors is defined as [7]

$$(\mathbf{A} | \mathbf{B}) \underset{\text{def}}{=} \operatorname{Tr}\{\mathbf{A}^+ \mathbf{B}\} \tag{E.5}$$

where $\mathbf{A}^+$ is the Hermitian adjoint of $\mathbf{A}$. This definition [Eq. (E.5)] is appropriate for finite dimensional Liouville spaces.

Let us summarize some more relations

$$\begin{aligned} &(\mathbf{A} | \mathbf{B})^* = (\mathbf{B} | \mathbf{A}) \\ &(\mathbf{A} | \beta \mathbf{B}) = \beta (\mathbf{A} | \mathbf{B}) \quad \beta \text{ complex number} \\ &(\mathbf{A} | \mathbf{B} + \mathbf{C}) = (\mathbf{A} | \mathbf{B}) + (\mathbf{A} | \mathbf{C}) \\ &(\mathbf{A} | \mathbf{A}) \geqslant 0 \quad \text{equals zero only if } \mathbf{A} = 0. \end{aligned} \tag{E.6}$$

We define the expectation value of an operator Q as

$$\langle \mathbf{Q} \rangle = (\rho | \mathbf{Q}) = (\mathbf{Q} | \rho) \tag{E.7}$$

or

$$\langle \mathbf{Q}(t) \rangle = (\mathbf{Q} | T \exp\{-i \int_0^t dt' \hat{\mathcal{H}}(t')\} | \rho(0)). \tag{E.8}$$

Let us always consider the case where an appropriate interaction representation has been chosen to give the time evolution operator a simple form. Moreover we shall always represent it by an average Hamiltonian by means of a Magnus expansion

$$T \exp[-i \int_0^t dt' \mathcal{H}(t')] = \exp[-it (\overline{\mathcal{H}}^{(0)} + \overline{\mathcal{H}}^{(1)} + ...)]. \tag{E.9}$$

We arrive at a simple form of Eq. (E.8) as

$$\langle \mathbf{Q}(t)\rangle = (\mathbf{Q}\ |\exp[-it\hat{\overline{\mathcal{H}}}]|\ \rho(0)) \tag{E.10}$$

with

$$\overline{\mathcal{H}} = \overline{\mathcal{H}}^{(0)} + \overline{\mathcal{H}}^{(1)} + \ldots$$

according to a Magnus expansion which keeps $\overline{\mathcal{H}}$ Hermetian at every stage of approximation.

The corresponding line shape is formally defined as

$$J(\omega) = \int_{-\infty}^{+\infty} dt\, \langle \mathbf{Q}(t)\rangle\, e^{i\omega t} = (\mathbf{Q}\ |\ \delta(\omega - \hat{\overline{\mathcal{H}}})|\ \rho(0)) \tag{E.11}$$

where $\delta(\omega - \hat{\overline{\mathcal{H}}})$ is a δ function operator.

As an example we treat the Bloch decay in a solid with only one spin species I coupled to its neighbors via the dipolar interaction. The zeroth order Hamiltonian $\overline{\mathcal{H}}^{(0)}$ is in this case the truncated or secular dipolar Hamiltonian $\mathcal{H}'_D$, which produces a Bloch decay of the e.g. x-component of the magnetization as

$$G(t) = \frac{(I_x\ |\exp[-it\hat{\mathcal{H}}'_D]|\ \rho(0))}{(I_x\ |\ \rho(0))} \tag{E.12}$$

where the exponential operator may be expanded to yield

$$G(t) = \sum_{n=0}^{\infty} \frac{(-it)^n}{n!} M_n \tag{E.13}$$

with the n-th Moment M_n defined as

$$M_n = \frac{(I_x\ |\ \hat{\mathcal{H}}'^{n}_D|\ \rho(0))}{(I_x\ |\ \rho(0))}. \tag{E.14}$$

The moment expansion, however, is not rapidly converging and the calculation of higher order moments (higher than the fourth moment) becomes a formidable task. Different approximations have therefore been devised to calculate the Bloch decay due to the many body dipolar interaction more accurately.

We are now going to discuss a very convenient way of approximation which is based on the Mori formalism. First the equation of motion for an operator is rewritten in a different form, which is still exact, but more susceptible to approximations i.e., the hierarchy of approximations is always pertained.

Let us expand the density matrix into a set of m "orthogonal" operators $\mathbf{Q}_k$ $k = 1, \ldots, m$ as

$$|\rho) = Z^{-1}[1 + \beta_1\mathbf{Q}_1 + \beta_2\mathbf{Q}_2 + \ldots \beta_m\mathbf{Q}_m + \mathbf{O}_{\text{residual}}] \tag{E.15}$$

where "orthogonal" means

$$(\mathbf{Q}_i | \mathbf{Q}_j) = 0 \quad \text{for } i \neq j \tag{E.16}$$

and

$$(\mathbf{Q}_i | \mathbf{O}_{\text{residual}}) = 0.$$

We do not consider the system to be at equilibrium explicitly, thus avoiding the concept of a spin temperature. We may, however, define thermodynamic coordinates (which are not temperatures necessarily) by

$$\beta_k = \frac{(\rho | \mathbf{Q}_k)}{(\mathbf{Q}_k | \mathbf{Q}_k)} = \frac{\langle \mathbf{Q}_k \rangle}{\mathrm{Tr}\{\mathbf{Q}_k^{+} \mathbf{Q}_k\}}. \tag{E.17}$$

Nevertheless β_k is assumed to be small.

A suitable projection operator is defined which projects the "relevant" part of the density matrix onto the individual operators $\mathbf{Q}_k$ by

$$P \underset{\text{def}}{=} \sum_{k=1}^{m} \frac{|\mathbf{Q}_k)(\mathbf{Q}_k|}{(\mathbf{Q}_k | \mathbf{Q}_k)} \tag{E.18}$$

with $P^2 = P$ (idempotent).

The density matrix is now properly separated into the "relevant" part and the "irrelevant" part by

$$|\rho) = P|\rho) + (1 - P)|\rho) \tag{E.19}$$

which results in a Liouville-v. Neumann equation as

$$\frac{d}{dt}|\rho) = -i\hat{\mathcal{H}}P|\rho) - i\,\hat{\mathcal{H}}(1 - P)|\rho) \tag{E.20}$$

and a corresponding equation of motion for the expectation value $\langle \mathbf{Q}_k \rangle$ of the operator $\mathbf{Q}_k$ as

$$\frac{d}{dt}(\mathbf{Q}_k | \rho) = -i(\mathbf{Q}_k^t | \hat{\mathcal{H}}P|\rho) - i(\mathbf{Q}_k | \hat{\mathcal{H}}(1 - P)|\rho). \tag{E.21}$$

The first term represents the direct or trivial contribution to the motion of $\langle \mathbf{Q}_k(t) \rangle$, whereas the second term contains "memory" effects as will be seen later. In order to calculate the second term we write

$$\frac{d}{dt}(1 - P)|\rho) = -i(1 - P)\,\hat{\mathcal{H}}P|\rho) - i(1 - P)\hat{\mathcal{H}}(1 - P)|\rho) \tag{E.22}$$

which has the formal solution

$$(1 - P)|\rho(t)) = \mathbf{S}(t, 0)(1 - P)|\rho(0)) - i\int_0^t dt' \mathbf{S}(t, t')(1 - P)\hat{\mathcal{H}}(t')P|\rho(t)) \tag{E.23}$$

where

$$\mathbf{S}(t, t') = T \exp[-i \int_{t'}^{t} d\tau (1 - P) \hat{\mathcal{H}}(\tau)]. \tag{E.24}$$

We insert Eq. (E.23) into Eq. (E.21) to obtain

$$\frac{d}{dt} \langle \mathbf{Q}_k(t) \rangle = \mathbf{K} + \mathbf{L} + \mathbf{M} \tag{E.25}$$

where

$$\mathbf{K} = -i(\mathbf{Q}_k \mid \hat{\mathcal{H}} P \mid \rho) = -i \sum_{j=1}^{m} \frac{(\mathbf{Q}_k^{+} \mid \hat{\mathcal{H}}(t) \mid \mathbf{Q}_j)}{(\mathbf{Q}_j \mid \mathbf{Q}_j)} \langle \mathbf{Q}_j(t) \rangle. \tag{E.26}$$

$$\mathbf{L} = -i(\mathbf{Q}_k \mid \hat{\mathcal{H}} \mathbf{S}(t, 0)(1 - P) \mid \rho(0)). \tag{E.27}$$

$$\mathbf{M} = -\sum_{j=1}^{m} \int_0^t dt' \frac{(\mathbf{Q}_k \mid \hat{\mathcal{H}}(t) \mathbf{S}(t, t')(1 - P) \hat{\mathcal{H}}(t') \mid \mathbf{Q}_j)}{(\mathbf{Q}_j \mid \mathbf{Q}_j)} \langle \mathbf{Q}_j(t) \rangle. \tag{E.28}$$

If the complete Liouville space is exhausted by the m operators $\mathbf{Q}_k$ i.e. dim $\mathcal{L} = m$, **L** and **M** vanish and Eq. (E.25) reduces to the Liouville equation again. In other words there is no irreversibility if dim$\mathcal{L} = m$. Irreversibility, therefore, can occur inly if $m <$ dim$\mathcal{L}$ i.e., the m orthogonal operators $\mathbf{Q}_k$ do not span the complete Liouville space. The term L represents the initial condition and this term is, in general, not zero. However, in most experiments the initial condition can be prepared so as to let **L** vanish [*9*]. The first term **K** as defined in Eq. (E.26) can be readily shown to vanish if $[\mathbf{Q}_k, \mathbf{Q}_j] = 0$, which leads to $P\hat{\mathcal{H}}P = 0$.

We always assume in the following that the conditions for $\mathbf{K} = \mathbf{L} = 0$ are fulfilled and summarize [*11*]

$$\frac{d}{dt} \langle \mathbf{Q}_k(t) \rangle = -\int_0^t dt' \sum_{j=1}^{m} K_{kj}(t, t') \langle \mathbf{Q}_j(t) \rangle \tag{E.29}$$

with

$$K_{kj}(t, t') = \frac{(\mathbf{Q}_k \mid \hat{\mathcal{H}}(t) \mathbf{S}(t, t')(1 - P) \hat{\mathcal{H}}(t') \mid \mathbf{Q}_j)}{(\mathbf{Q}_j \mid \mathbf{Q}_j)} \tag{E.30}$$

The Kernel $K_{kj}(t, t')$ represents the "memory" effects and takes the history of earlier times into account. $K_{kj}(t, t')$ is therefore called the "memory function". We note that Eq. (E.29) is still exact, but is of course by no means easier to evalute than the formal solution Eq. (E.8). However, approximations may be applied to the "memory function" $K_{kj}(t, t')$ which amounts to moving a step upwards in the hierarchiy of approximations.

Let us introduce further simplifications. We assume, that a convenient interaction representation has been chosen, such that the interaction Hamiltonian can be considered time independent i.e., to a good approximation the Hamiltonian is represented by an average Hamiltonian.

Following Eq. (E.29), where we suppose $[\mathbf{Q}_k, \mathbf{Q}_j] = 0$ i.e. $P\hat{\mathcal{H}}P = 0$, we obtain

$$K_{kj}(\tau) = \frac{(\mathbf{Q}_k \mid \hat{\mathcal{H}} \exp[-i\tau(1-P)\hat{\mathcal{H}}]\, \hat{\mathcal{H}} \mid \mathbf{Q}_j)}{(\mathbf{Q}_j \mid \mathbf{Q}_j)} \tag{E.31}$$

with

$$K_{kj}(0) = \frac{(\mathbf{Q}_k \mid \hat{\mathcal{H}}^2 \mid \mathbf{Q}_j)}{(\mathbf{Q}_j \mid \mathbf{Q}_j)} = M_{2kj}. \tag{E.32}$$

We may expand $K_{kj}(\tau)$ as

$$K_{kj}(\tau) = \sum_{n=0}^{\infty} \frac{(-i\tau)^n}{n!} K_n$$

with

$$K_n = \frac{(\mathbf{Q}_k \mid \hat{\mathcal{H}}\{(1-P)\, \hat{\mathcal{H}}\}^n \hat{\mathcal{H}} \mid \mathbf{Q}_j)}{(\mathbf{Q}_j \mid \mathbf{Q}_j)}. \tag{E.33}$$

We specialize now in the case, where the corresponding spectral distribution function is an even function with frequency i.e., $J(\omega) = J(-\omega)$ and consequently all odd order moments in $K_{kj}(\tau)$ vanish

$$K_{kj}(\tau) = K_{kj}(0)\left[1 - \frac{N_2}{2!}\tau^2 + \frac{N_4}{4!}\tau^4 - + \ldots\right]$$

with $K_{kj}(0)$ as before, [Eq. (E.32)] and $[\mathbf{Q}_k, \mathbf{Q}_j] = 0$ (implying $P\hat{\mathcal{H}}P = 0$), we obtain:

$$N_2 \cdot K_{kj}(0) = \frac{(\mathbf{Q}_k \mid \hat{\mathcal{H}}^4 \mid \mathbf{Q}_j)}{(\mathbf{Q}_j \mid \mathbf{Q}_j)} - \frac{(\mathbf{Q}_k \mid \hat{\mathcal{H}}^2 P \hat{\mathcal{H}}^2 \mid \mathbf{Q}_j)}{(\mathbf{Q}_j \mid \mathbf{Q}_j)}. \tag{E.34}$$

$$N_4 K_{kj}(0) = \frac{(\mathbf{Q}_k \mid \hat{\mathcal{H}}^6 - \hat{\mathcal{H}}^4 P \hat{\mathcal{H}}^2 - \hat{\mathcal{H}}^2 P \hat{\mathcal{H}}^4 + \hat{\mathcal{H}}^2 P \hat{\mathcal{H}}^2 P \hat{\mathcal{H}}^2 \mid \mathbf{Q}_j)}{(\mathbf{Q}_j \mid \mathbf{Q}_j)} \tag{E.35}$$

If we restrict ourselves to only one observable, we finally obtain

$$\frac{d}{dt} \langle \mathbf{Q} \mid (t) \rangle = -\int_0^t dt' K(t-t') \langle \mathbf{Q}(t') \rangle \tag{E.36}$$

with

$$\begin{aligned} K(0) &= M_2 \\ N_2 &= M_2(\mu - 1) \\ N_4 &= M_2^2(\nu - 2\mu + 1) \end{aligned} \tag{E.37}$$

where

$$M_n = \frac{(\mathbf{Q} \mid \hat{\mathcal{H}}^n \mid \mathbf{Q})}{(\mathbf{Q} \mid \mathbf{Q})} \tag{E.38}$$

and

$$\mu = M_4/M_2^2\,;\ \nu = M_6/M_2^3. \tag{E.39}$$

The corresponding line shape $I(\omega)$ can be obtained from Eq. (E.36) with

$$I(\omega) = \int_0^\infty dt\,\langle \mathbf{Q}(t)\rangle \cos\omega t \tag{E.40}$$

by formal integration as [11]

$$I(\omega) = \langle \mathbf{Q}(0)\rangle \cdot \frac{K'(\omega)}{[\omega - K''(\omega)]^2 + [K'(\omega)]^2} \tag{E.41}$$

where

$$K'(\omega) = \int_0^\infty d\tau K(\tau)\cos\omega\tau \tag{E.42a}$$

and

$$K''(\omega) = \int_0^\infty d\tau K(\tau)\sin\omega\tau. \tag{E.42b}$$

The half width at half height of the line shape $I(\omega)$ according to Eq. (E.41) can be obtained by iteration from

$$\delta = K''(\delta) + [2K'(0)K'(\delta) - K'^2(\delta)]^{1/2} \tag{E.43}$$

So far no approximation other than the average Hamiltonian approximation in a suitably chosen interaction representation has been used. In order to calculate actual line shapes, we will have to approximate the memory function $K(\tau)$. A very convenient approximation to $K(\tau)$ would be a Gaussian

$$K(\tau) = K(0)\exp[-\frac{N_2}{2}\tau^2] \tag{E.44}$$

since only the second moment of $K(\tau)$ is involved.

In fact, it has been shown by several authors, that this approximation is indeed widely applicable and that it covers a wealth of different physical situations (see also Sections 3, 3.8, 4.3, 4.4).

The lineshape $I(\omega)$ according to Eq. (E.41) is readily obtained by insertion of the sine- and cosine-transform of $K(\tau)$ according to Eq. (E.44) as [11]

$$K'(\omega) = \sqrt{\frac{\pi}{2}}\frac{K(0)}{\sqrt{N_2}}\exp(-\omega^2/2N_2). \tag{E.45a}$$

$$K''(\omega) = \frac{K(0)\omega}{N_2}\exp(-\omega^2/2N_2)\cdot F(\frac{1}{2};\frac{3}{2};\omega^2/2N_2) \tag{E.45b}$$

where $K(0)$ and N_2 are given by Eq. (E.37). $F(\frac{1}{2};\frac{3}{2};\omega^2/2N_2)$ is the confluent hypergeometric function, which is tabulated and can be calculated numerically. Notice that no adjustable parameter is involved at all and the only approximation being the functional form of the memory function.

The linewidth δ which is a suitable experimental parameter can now be expressed according to Eqs. (E.43) and (E.45) as

$$\delta = \sqrt{\frac{\pi}{2}}\left(\frac{K^2(0)}{N_2}\right)^{1/2} \cdot f(x) \tag{E.46}$$

with

$$f(x) = [2 - e^{-x^2}]^{1/2} \cdot e^{-x^2/2} + \frac{2x}{\sqrt{\pi}} \cdot e^{-x^2} \cdot F(\frac{1}{2};\frac{3}{2};x^2) \tag{E.47}$$

where $f(x)$ is a function close to one with

$$x = \delta/(2N_2)^{1/2}.$$

Another relation between $\mu = M_4/M_2^2$ and $f(x)$ is readily found with

$$\mu - 1 = \frac{\sqrt{\pi}}{2}\frac{f(x)}{x}. \tag{E.48}$$

This allows $f(\mu)$ to be expressed by μ as is shown in Fig. E 1. For large values of μ the function $f(\mu)$ approaches the value 1 and the linewidth δ may be expressed according to Eqs. (E.43, E.45, E.46) as

$$\delta = \sqrt{\frac{\pi}{2}}\left[\frac{M_2}{\mu - 1}\right]^{1/2}. \tag{E.49}$$

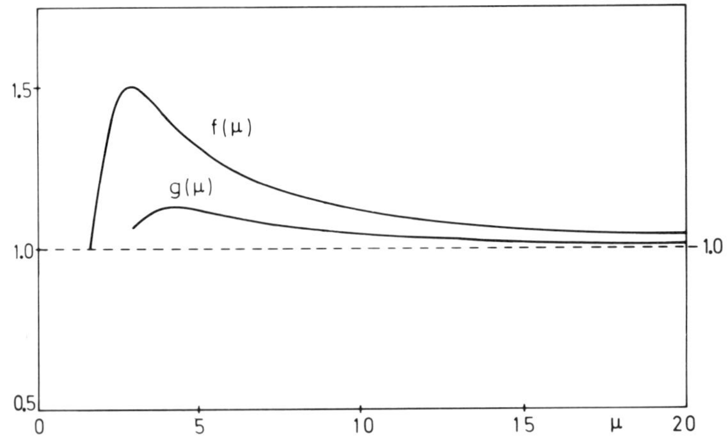

Fig. E. 1. Line width correction functions $g(\mu)$ and $f(\mu)$ versus moment ratio $\mu = M_4/M_2^2$ (see text)

The lineshape $I(\omega)$ is shown to be Lorentzian in this case. This linewidth may be contrasted with the cut-off Lorentzian linewidth. A better approximation to $f(\mu)$ for smaller values of μ is expressed as

$$f(\mu) = \left(\frac{\mu - 1}{\mu - 2}\right)^{1/2} g(\mu)$$

where now $g(\mu)$ is always close to one. Along these lines we propose the linewidth δ_G with a Gaussian memory function approximation to be

$$\delta_G = \sqrt{\frac{\pi}{2}} \left[\frac{M_2}{\mu - 2}\right]^{1/2}. \tag{E.50a}$$

In order to obtain the correct linewidth in the case of a Gaussian ($\mu = 3$) and a close enough approximation for Lorentzians ($\mu > 3$) we suggest to use as a universal linewidth

$$\delta = \sqrt{\frac{\pi}{2}} \left[\frac{M_2}{\mu - 1.87}\right]^{1/2}. \tag{E.50b}$$

We finally remark that higher order approximations of the memory function $K(\tau)$ are readily obtained by writing down an integro-differential Eq. (E.36) for $K(\tau)$ itself, followed by an approximation of the corresponding memory function of $K(\tau)$. This extension is straightforward and will not be discussed any further here.

Another approximation of $K(\tau)$, does not assume a particular functional form of $K_{kj}(\tau)$, but rather assumes the so-called "short correlation limit" to hold, where the correlation time of $K_{kj}(\tau)$ is assumed to be very short compared with any change of $\langle \mathbf{Q}_j(t) \rangle$. This is usually the case in relaxation and also in line narrowing multiple-pulse experiments (see Section 3.8). In this limit $\langle \mathbf{Q}_j(t) \rangle$ can be separated from the integral in Eq. (E.29) and the integration limit is extended to infinity

$$\frac{d}{dt} \langle \mathbf{Q}_k(t) \rangle = -\int_0^{\infty} d\tau \sum_{j=1}^{m} K_{kj}(\tau) \cdot \langle \mathbf{Q}_j(t) \rangle \tag{E.51}$$

which results in a sum of exponentials or corresponding Lorentzian line shape functions. The individual relaxation rates are defined as

$$\delta = \frac{1}{T_{kj}} = \int_0^{\infty} d\tau \, K_{kj}(\tau). \tag{E.52}$$

Other functional forms of $K(\tau)$ may be appropriate for special interactions, as was employed in Section 4.4.

References

Chapter 1

1. Abragam, A.: The Principles of Nuclear Magnetism. London: Oxford Univ. Press 1961.
2. Friebolin, H.: NMR-Spektroskopie, p. 74. Weinheim: Physik-Verlag 1974
3. Andrew, E. R., Hinshaw, W. S., Jasinski, A.: Chem. Phys. Letters **24**, 399 (1974).
4. a) Andrew, E. R., Eades, R. G.: Proc. Roy. Soc. London **A216**, 398 (1953).
   b) Andrew, E. R., Clough, S., Farnell, L. F., Gledhill, T. D., Roberts, I.: Phys. Letters **19**, 6 (1966).
5. Lowe, I. J.: Phys. Rev. Letters **2**, 285 (1959).
6. a) Andrew, E. R.: Prog. NMR Spectroscopy **8**, 1 (1971).
   b) Andrew, E. R.: Biennial Rev. Magn. Res. (ed. by C. A. Mc Dowell, Vancouver) (1974).
7. Waugh, J. S., Huber, L. M., Haeberlein, U.: Phys. Rev. Letters **20**, 180 (1968).
8. a) Ellett, J. D., Haeberlen, U., Waugh, J. S.: J. Am. Chem. Soc. **92**, 411 (1970).
   b) Ellett, J. D., Gibby, M. G., Haeberlen, U., Huber, L. M., Mehring, M., Pines, A.: Advan. Magn. Res. **5**, 117 (1971).
9. a) Mehring, M., Griffin, R. G., Waugh, J. S.: J. Am. Chem. Soc. **92**, 7222 (1970).
   b) Mehring, M., Griffin, R. G., Waugh, J. S.: Chem. Phys. **55**, 746 (1971).
10. Stacey, L. M., Vaughan, R. W., Elleman, D. D.: Phys. Rev. Letters **26**, 1153 (1971).
11. Griffin, R. G., Ellett, J. D., Mehring, M., Bullitt, J. G., Waugh, J. S.: J. Chem. Phys. **57**, 2147 (1972).
12. Haeberlen, U., Kohlschuetter, U., Kempf, J. , Spiess, H. W., Zimmermann, H.: Chem. Phys. **3**, 248 (1974).
13. Silberzyc, W.: Acta Phys. Satyr. **11**, 1111 (1970).
14. a) Pines, A., Gibby, M. G., Waugh, J. S.: J. Chem. Phys. **56**, 1776 (1972).
    b) Pines, A., Waugh, J. S.: J. Magn. Res. **8**, 354 (1972).
    c) Pines, A., Gibby, M. G., Waugh, J. S.: J. Chem. Phys. **59**, 569 (1973).
    d) Pines, A., Gibby, M. G., Waugh, J. S.: Chem. Phys. Letters **15**, 373 (1972).
15. Pines, A., Schattuck, T. W.: J. Chem. Phys. **61**, 1255 (1974).
16. Mansfield, P.: Progr. NMR Spectroscopy **8**, 41 (1971).
17. Vaughan, R. W.: Annual Reviews in Materials Science, Vol. 4, (ed. by Huggins, R. A., Bube, H. R., Roberts, R. W.) (Annual Reviews, Palo Alto 1974).
18. Haeberlen, U.: Advan. Magn. Res. supplement (1976)

Chapter 2

1. Abragam, A.: The Principles of Nuclear Magnetism. London: Oxford Univ. Press 1961.
2. Poole, C. P., Farach, H. A.: The Theory of Magnetic Resonance, New York: Wiley Interscience 1972.
3. Rose, M. E.: Elementary Theory of Angular Momentum. New York: John Wiley 1967.
4. Brink, D. M., Satchler, G. R.: Angular Momentum. Oxford: Clarendon Press 1968.
5. Haeberlen, U., Waugh, J. S.: Phys. Rev. **185**, 420 (1969).
6. Haeberlen, U.: Advan. Magn. Res. (ed. by Waugh, J. S.) supplement (1976)
7. Griffin, R. G., Ellett, J. D., Mehring, M., Bullitt, J. G., Waugh, J. S.: J. Chem. Phys. **57**, 2147 (1972).
8. Bloembergen, N., Rowland, J. A.: Acta Met. **1**, 731 (1953).
9. Mehring, M., Griffin, R. G., Waugh, J. S.: J. Chem. Phys. **55**, 746 (1971).

10. Andrew, E. R., Hinshaw, W. S.: Phys. Letters **43A**, 113 (1973).
11. Andrew, E. R., Eades, R. G.: Proc. Roy. Soc. (London) **A216**, 398 (1953).
12. Lowe, I. J.: Phys. Rev. Letters **2**, 285 (1959)
13. Andrew, E. R.: Prog. NMR Spectroscopy **8**, 1 (1971).
14. Andrew, E. R.: Biennial Rev. Magn. Res. (ed. by C. A. McDowell, Vancouver) (1974).
15. Andrew, E. R., Farnell, L. F., Gledhill, T. D.: Phys. Rev. Letters **19**, 6 (1967).
16. Andrew, E. R., Hinshaw, W. S., Riffen, R. S.: Phys. Letters **46A**, 57 (1973).
17. Mehring, M., Raber, H.: Solid. State Commun. **13**, 1637 (1973).
18. a) Schnabel, B.: Wiss. Z. Univ. Jena **22**, p. 335 (1973).
b) Schnabel, B., Taplick, T.: Phys. Letters **27a**, 310 (1968).
19. a) Cohn, M., Kowalsky, A., Leigh, H., Maricić, S.: Magnetic Resonance in Biological System, p. 45. New York: Pergamon Press Inc. (1967)
b) Babka, J., Doskocilova, D., Pivcova, H., Ruzieka, Z., Schneider, B.: Proc. 16th Congress Ampere Bucharest, Rumania 1970, 785 (Publ. House Academy R. S. Rumania).
20. Freed, J. H., Bruno, G. V., Polnaszek, C.: J. Phys. Chem. **75**, 3386 (1971).
21. Freed, J. H., Fraenke, G. K.: J. Chem. Phys. **39**, 326 (1963).
22. Sillescu, H., Kivelson, D.: J. Chem. Phys. **48**, 3493 (1968).
23. Sillescu, H.: J. Chem. Phys. **54**, 2111 (1971).
24. Sillescu, H.: Ber. Bunsenges. Phys. Chem. **75**, 283 (1971).
25. Hensen, K., Riede, W. O., Sillescu, H., Wittgenstein, A. v.: J. Chem. Phys. **61**, 4365 (1974).
26. Goldman, S. A., Bruno, G. V., Polnaszek, C. F., Freed, J. H.: J. Chem. Phys. **56**, 716 (1972).
27. Spiess, H. W.: Chem. Phys. **6**, 217 (1974).
28. Spiess, H. W., Grosescu, R., Haeberlen, U.: Chem. Phys. **6**, 226 (1974).
29. Gordon, R. G., McGinnis, R. P.: J. Chem. Phys. **49**, 2455 (1968).
30. Wilkinson, J. H.: The Algebraic Eigenvalue Problem, Chap. 8. London: Oxford Univ. Press 1965.
31. Becker, H. J.: Diploma work, Dortmund 1975.
32. a) Alexander, S., Baram, A, Luz, Z.: Mol. Phys. **27**, 441 (1974).
b) Baram, A., Luz, Z., Alexander, S.: J. Chem. Phys. (in press).
33. Pines, A., Wemmer, D. E., Ruben, D. J.: J. Chem. Phys. (in press).

Chapter 3

1. Hahn, E. L.: Phys. Rev. **80**, 580 (1950).
2. Carr, H. Y., Purcell, E. M.: Phys. Rev. **94**, 630 (1954).
3. Gill, D., Meiboom, S.: Rev. Sci. Instr. **29**, 688 (1958).
4. Ostroff, E. D., Waugh, J. S.: Phys. Rev. Letters **16**, 1097, (1966).
5. Mansfield, P., Ware, D.: Phys. Letters **22**, 133 (1966).
6. Waugh, J. S., Wang, C. H.: Phys. Rev. **162**, 209 (1967).
7. Mansfield, P., Ware, D.: Phys. Rev. **168**, 318 (1968).
8. Mansfield, P., Richards, K. H. B., Ware, D.: Phys. Rev. **B1**, 2048 (1970).
9. Waugh, J. S., Huber, L. M., Haeberlen, U.: Phys. Rev. Letters **20**, 180 (1968).
10. Abragam, A.: The Pinciples of Nuclear Magnetism. London: Oxford Univ. Press 1961.
11. Goldman, M.: Spin Temperature and Nuclear Magnetic Resonance in Solids. London: Oxfod Univ. Press 1970.
12. Haeberlen, U., Waugh, J. S.: Phys. Rev. **175**, 453 (1968).
13. Van Vleck, J. H.: Phys. Rev. **74**, 1168 (1948).
14. a) Zwanzig, R., In: Lectures in Theoretical Physics. New York: Interscience 1961.
b) Mori, H.: Progr. Theor. Phys. Jap. **34**, 399 (1965).
c) Bosse, J.: (private communication).
15. a) Mori, H., Kawasaki, K.: Progr. Theoret. Phys. (Kyoto) **27**, 529 (1962).
b) Mori, H.: Progr. Theoret. Phys. (Kyoto) **33**, 423 (1965).
16. Lado, F., Memory, J. D., Parker, G. W.: Phys. Rev. **B4**, 1406 (1971).
17. Tjon, J. A.: Phys. Rev. **143**, 259 (1966).
18. a) Robertson, B.: Phys. Rev. **144**, 151 (1966).

b) Robertson, B.: Phys. Rev. **153**, 391 (1967).
c) Andersson, P. W., Weiss, P. R.: Rev. Mod. Phys. **25**, 269 (1953).
19. Argyres, P. N., Kelley, P. L.: Phys. Rev. **134**, A93 (1964).
20. Kivelson, D., Ogan, K.: Advan. Mag. Res. **7**, 71 (1974).
21. Parker, G. W., Lado, F.: Phys. Rev. **B8**, 3081 (1973).
22. Parker, G. W., Lado, F.: Phys. Rev. **B9**, 22 (1974).
23. Demco, D., Tegenfeld, J., Waugh, J. S.: Phys. Rev. **B11**, 4133 (1975).
24. Ellett, J. D., Jr., Waugh, J. S.: J. Chem. Phys. **51**, 2581 (1969).
25. Haeberlen, U.: Advan. Magn. Res. (in press).
26. a) Mansfield, P.: J. Phys. C: Solid State Phys. **4**, 1444 (1971).
b) Mansfield, P., Orchard, M. J., Stalker, D. C., Richards, K. H. B.: Phys. Rev. **B7**, 90 (1973).
27. Stacey, L. M., Vaughan, R. W., Elleman, D. D.: Phys. Rev. Letters **26**, 1153 (1971).
28. Griffin, R. G., Ellett, J. D., Mehring, M., Bullitt, J. G., Waugh, J. S.: J. Chem. Phys. **57**, 2147 (1972).
29. Mehring, M., Griffin, R. G., Waugh, J. S.: J. Am. Chem. Soc. **92**, 7222 (1970).
30. Mehring, M., Griffin, R. G., Waugh, J. S.: J. Chem. Phys. **55**, 746 (1971).
31. Mansfield, P., Orchard, M. J., Stalker, D. C., Richards, K. H. B.: Phys. Rev. **B7**, 90 (1973).
32. Griffin, R. G., Yeung, H. N., LaPrade, M. D., Waugh, J. S.: J. Chem. Phys. **59**, 777 (1973).
33. Rhim, W.-K., Elleman, D. D., Vaughan, R. W.: J. Chem. Phys. **59**, 3740 (1973).
34. a) Vaughan, R. W., Elleman, D. D., Rhim, W.-K., Stacey, L. M.: J. Chem. Phys. **57**, 5383 (1972).
34. b) Vaughan, R. W.: Annual Review in Materials Science, Vol. 4, (ed. by Huggins, Bube, R. A., Roberts, R. W.) Annual Review, Palo Alto, 1974.
35. Haeberlen, U., Kohlschütter, U.: Chem. Phys. **2**, 76 (1973).
36. Haeberlen, U., Kohlschütter, U., Kempf, J., Spiess, H. W., Zimmermann, H.: Chem. Phys. **3**, 248 (1974).
37. Grosescu, R., Achlama, A. M., Haeberlen, U., Spiess, H. W.: Chem. Phys. **5**, 119 (1974).
38. Achlama, A. M., Kohlschütter, U., Haeberlen, U.: Chem. Phys. **7**, 287 (1975).
39. Raber, H., Brünger, G., Mehring, M.: Chem. Phys. Letters **23**, 400 (1973).
40. a) Haeberlen, U., Schnabel, B.: 18th Ampere Congress, p. 545. Nottingham 1974.
b) Döhler, H., Neubauer, R., Schnabel, B.: Phys. Stat. Sol. **b65**, K141 (1974).
41. Schreiber, L. B., Vaughan, R. W.: Chem. Phys. Letters **28**, 586 (1974).
42. a) Van Hecke, P., Weaver, J. C., Neff, B. L., Waugh, J. S.: J. Chem Phys. **60**, 1668 (1974).
b) Van Hecke, P., Weaver, J. C., Neff, B. L., Waugh, J. S.: Proc. First Specialised Colloque Ampere, Krakow, Poland 143 (1973).
43. a) Haubenreisser, U., Schnabel, B.: Proc. First Specialised Colloque Ampere, Krakow, Poland 140 (1973).
b) Haubenreisser, U., Schnabel, B., Scheler, G., Burghoff, U., Müller, R., Wilsch, R.: Phys. Stat. Sol. **a20**, K45 (1973).
c) Wilsch, R., Burghoff, U., Müller, R., Rosenberger, H., Scheler, G., Pettig, M., Schnabel, B.: 18. Ampere Congress, Vol. 2, p. 553, Nottingham 1974.
44. Burghoff, U., Scheler, G., Müller, R.: Phys. Stat. Sol. **a25**, k31 (1974).
45. Mehring, M., Raber, H.: Solid State Commun. **13**, 1637 (1973).
46. Kanert, O., Mehring, M.: NMR: Basic Principles and Progress, Vol. 3. Berlin-Heidelberg-New York: Springer.
47. a) Mehring, M., Kotzur, D., Kanert, O.: Phys. Stat. Sol. **b35**, K25 (1972).
b) Kotzur, D., Mehring, M., Kanert, O.: Z. Naturforsch. **28a**, 1607 (1973).
48. Mehring, M., Pines, A., Rhim, W.-K., Waugh, J. S.: J. Chem. Phys. **54**, 3239 (1971).
49. a) Mehring, M., Raber, H., Sinning, G.: Proceedings of the 18th. Congress Ampere, p. 35. Nottingham (1974).
b) Ackermann, H., Fujara, F.: spring meeting of the DPG, Freudenstadt, Germany (1976).
c) Fujara, F.: Diploma work, Heidelberg 1975.
50. Bloch, F., Siegert, A.: Phys. Rev. **57**, 522 (1940).
51. Wang, C. H., Ramshaw, J. D.: Phys. Rev. **B6**, 3253, 1972, also Wang, C. H.: J. Chem. Phys. **59**, 225 (1973).
52. Magnus, W.: Com. Pure Apll. Math. **7**, 649 (1954).

53. Pechukas, P., Light, F. C.: J. Chem. Phys. **44**, 3897 (1966).
54. Evans, W. A. B.: Ann. Phys. (N. Y.) **48**, 72 (1968).
55. Bialnicky-Birula, Mielnik, B.: Ann. Phys. (N. Y.) **51**, 187 (1969).
56. Mehring, M., Waugh, J. S.: Phys. Rev. **B5**, 3459 (1972).
57 a) Rhim, W.-K., Pines, A., Waugh, J. S.: Phys. Rev. Letters **25**, 218 (1970).
b) Rhim, W.-K., Pines, A., Waugh, J. S.: Phys. Rev. **B3**, 684 (1971).
58. Garroway, A. N., Mansfield, P., Stalker, D. C.: Phys. Rev. **B11**, 121 (1975).
59. Silberszyc, W.: Acta Phys. Satyr. **11**, 1111 (1970).
60. Rhim, W.-K., Ellemann, D. D., Vaughan, R. W.: J. Chem. Phys. **58**, 1772 (1973).
61. Mansfield, P.: Proceedings of the 17th. Congress Ampere, Turku 1972.
62. Rose, M. E.: Elementary Theory of Angular Momentum, p. 77. New York: J. Wiley 1967.
63. Brink, D. M., Satchler, G. R.: Angular Momentum, p. 48. London: Oxford Univ. Press 1968.
64. a) Mehring, M.: Z. Naturforsch. **27a**, 1634 (1972), **28a**, 804 (1973).
b) Mehring, M.: Rev. Sci. Instr. **44**, 64 (1973).
65. a) Ernst, H., Fenzke, D., Heink, W.: Colloq. Ampere, Krakow, Poland 1973.
b) Ernst, H., Fenzke, D., Heink, W.: W. Wiss. Z. Karl-Marx-Univ. Math.-Naturw. **R23**, 545 (1974).
c) Haubenreisser, U., Schnabel, B.: 18th. Ampere Congress, p. 551. Nottingham 1974.
66. Haeberlen, U., Ellett, J. D.,Jr., Waugh, J. S.: J. Chem. Phys. **55**, 53 (1971).
67. Pines, A., Waugh, J. S.: J. Magn. Res. **8**, 354 (1972).
68. Rhim, W.-K., Elleman, D. D., Schreiber, L. B., Vaughan, R. W.: J. Chem. Phys. **60**, 4595 (1974).
69. Lan, N. Q., Pfeifer, H., Schmiedel, H.: Wiss. Z. Karl-Marx- Univ. Leipzig, Math.-Naturw. **R23**, 498 (1974).
70. Garroway, A. N., Mansfield, P., Stalker, D. C.: Phys. Rev. **B11**, 121 (1975).
71. Fenzke, D., Schmiedel, H.: Wiss. Z. Karl-Marx-Univ. Leipzig, Math.-Naturw. **R23**, 519 (1974).
72. Schmiedel, H.: Wiss. Z. Karl-Marx-Univ. Leipzig, Math.-Naturw. **R23**, 506 (1974).
73. Ernst, H., Fenzke, D., Heink, W.: Wiss. Z. Karl-Marx-Univ. Leipzig, Math.-Naturw. **23**, 530. (1974).
74. Lee, M., Goldberg, W. I.: Phys. Rev. **A140**, 1261 (1965).

Chapter 4

1. a) Hartmann, S. R., Hahn, E. L.: Phys. Rev. **128**, 2042 (1962).
b) Jones, E. P., Hartmann, S. R.: Phys. Rev. **36**, 757 (1972).
2. a) Lurie, F. M., Slichter, C. P.: Phys. Rev. **133**, A1108 (1964).
b) Slichter, C. P., Holton, W. C.: Phys. Rev. **122**, 1701 (1961).
3. a) Pines, A., Gibby, M. G., Waugh, J. S.: J. Chem. Phys. **56**, 1776 (1972).
b) Pines, A., Gibby, M. G., Waugh, J. S.: J. Chem. Phys. **59**, 569 (1973).
c) Pines, A., Gibby, M. G., Waugh, J. S.: Chem. Phys. Letters **15**, 373 (1972).
4. a) Abragam, A.: The Principles of Nuclear Magnetism. London: Oxford Univ. Press 1961.
b) Abragam, A., Proctor, W. G.: Phys. Rev. **109**, 1441 (1958).
5. Redfield, A. G.: Phys. Rev. 98, 1787 (1955).
6. Goldman, M.: Spin Temperature and Nuclear Magn. Res. in Solids. London: Oxford Univ. Press 1970.
7. Solomon, I.: Compt. Rend. **248**, 92 (1950).
8. Anderson, A. G., Hartmann, S. R.: Phys. Rev. **128**, 2023 (1960).
9. a) Jeener, J., Broekaert, P.: Phys. Rev. **157**, 232 (1967).
b) Jeener, J., Du Bois, R., Broekaert, P.: Phys. Rev. **139A**, 1959 (1965).
10. McArthur, D. A., Hahn, E. L., Walstedt, R. E.: Phys. Rev. **188**, 609 (1969).
11. Demco, D., Tegenfeldt, J., Waugh, J. S.: Phys. Rev. **B11**, 4133 (1975).
12. Grannell, P. K., Mansfield, P. K., Whittaker, M. A.: Phys. Rev. **B8**, 4149 (1973).
13. Ernst, H.: Wiss. Z. Karl-Marx-Univ. Leipzig Math.-Naturw. **23**, 449 (1974).
14. a) Kanert, O., Mehring, M.: NMR: Basic Principles and Progress, Vol. 3. Berlin-Heidelberg-New York: Springer 1971.

b) Schmid, D.: Springer Tracts in Modern Physics. Berlin-Heidelberg-New York: Springer 1973.
15. Walstedt, R. E., McArthur, D. A., Hahn, E. L.: Phys. Letters **15**, 7, (1965).
16. Bleich, H. E., Redfield, A.: J. Chem. Phys. **55**, 5406 (1971).
17. a) Yannoni, C. S., Bleich, H. E.: J. Chem. Phys. **55**, 5406 (1971).
b) Yannoni, C. S.: J. Chem. Phys. **58**, 1773 (1973).
18. a) Maier, G., Haeberlen, U., Wolf, H. C., Hausser, K. H.: Phys. Letters **25a**, 384 (1967).
b) Maier, G., Wolf, H. C.: Z. Naturforsch. **23a**, 1068 (1968).
c) Lau, P., Stehlik, D., Hausser, K. H.: J. Magn. Res. **15**, 270 (1970).
d) Stehlik, D., Doehring, A., Colpa, J. P., Callaghan, E., Kesmarky, S.: Chem. Phys. **7**, 165 (1975).
e) Hausser, K. H., Wolf, H. C.: Advan. Mag. Res. Vol. 8 (1976)
19. a) Goldman, M., Chapellier, M., Vu Huang, Chau, Abragam, A.: Phys. Rev. **B10**, 226 (1974).
b) Jacquinot, J. F., Wenckenach, W. Th., Goldman, M., Abragam, A.: Phys. Rev. Letters **32**, 1096 (1974).
20. Demco, D., Kaplan, S., Pausak, S., Waugh, J. S.: Chem Phys. Letters **1**, 77 (1975).
21. Goldman, M.: Compt Rend. **246**, 1038 (1958).
22. Goldman, M., Landesmann, A.: a) Compt. Rend. **252**, 263 (1961); b) Phys. Rev. **132**, 610 (1963).
23. Goldman, M.: Phys. Rev. **138**, A1668 (1965).
24. Schwab, M., Hahn, E. L.: J. Chem. Phys. **52**, 3152 (1970).
25. a) Sarles, L. R., Cotts, R. M.: Phys. Rev. **111**, 853 (1958).
b) Bloch, F.: Phys. Rev. **111**, 841 (1958).
c) Ackermann, H., Fujara, F.: presented at the spring meeting of the DPG, Freudenstadt, Germany (1976)
d) Fujara, F.: Diploma work, Heidelberg 1975.
26. Mehring, M., Sinning, G., Pines, A., Shattuck, T. W.: a) XXVth IUPAC, Jerusalem 1975; b) Second Specialized Colloque Ampere, Budapest 1975.
27. Mehring, M., Pines, A., Rhim, W.-K., Waugh, J. S.: J. Chem Phys. **54**, 3239 (1971).
28. Mehring, M., Raber, H., Sinning, G.: 18th Congress Ampere, p. 35. Nottingham, England 1974.
29. Pines, A.: First Specialized Colloque Ampere, Krakow, Poland, August 1973.
30. Pines, A., Shattuck, T. W.: J. Chem. Phys. **61**, 1255 (1974).
31. Strombotne, R. L., Hahn, E. L.: Phys. Rev. **133**, A 1616 (1964).
32. Müller, L., Kumar, A., Baumann, T., Ernst, R. R.: Phys. Rev. Letters **32**, 1902 (1974).
33. Hester, R. K., Ackermann, J. L., Cross, V. R., Waugh, J. S.: Phys. Rev. Letters **34**, 993 (1975).
34. Abragam, A., Winter, J.: Compt. Rend. **249**, 1633 (1959).
35. Walstedt, R. E.: Phys. Rev. **B5**, 41 (1972).
36. a) Mori, H., Kawasaki, K.: Progr. Theor. Phys. (Kyoto) **27**, 529 (1962).
b) Mori, H.: Progr. Theor. Phys. (Kyoto) **33**, 423 (1965).
37. a) Lado, F., Memory, J. D., Parker, G. W.: Phys. Rev. **B4**, 1406 (1971).
b) Parker, G. W., Lado, F.: Phys. Rev. **B9**, 22 (1974).
c) Parker, G. W., Lado, F.: Phys. Rev. **B8**, 3081 (1973).
38. Silberzyc, W.: (private communication).
39. Mehring, M., Sinning, G., Pines, A.: a) Z. Physik B **24**, 73 (1976); b) to be published.
40. Pines, A., Ruben, D. J., Vega, S., Mehring, M.: Phys. Rev. Letters **36**, 110 (1976).
41. a) Snyder, L. C., Meiboom, S.: J. Chem. Phys. **58**, 5096 (1973).
b) Hewitt, R. C., Meiboom, S., Snyder, L. C.: J. Chem. Phys. **58**, 5089 (1973).

Chapter 5

1. Appleman, B. R., Dailey, B. P.: Advan. Magn. Res. **7**, 231 (1974).
2. Emsley, J. W., Feeney, J., Sutcliff, L. H.: High Resolution NMR Spectroscopy. Oxford: Pergamon 1965.
3. Pople, J. A., Scheider, W. G., Bernstein, H. J.: High Resolution Nuclear Magnetic Resonance. New York: McGraw Hill 1959.

4. Ramsey, N. F.: Phys. Rev. **78**, 699 (1950).
5. Snyder, L. C., Parr, R. G.: J. Chem. Phys. **34**, 837 (1961).
6. Karplus, M., Das, T. P.: J. Chem. Phys. **34**, 1683 (1961).
7. Haeberlen, U.: Advan. Magn. Res. supplement (1976)
8. Ditchfield, R., Miller, D. P., Pople, J. A.: J. Chem. Phys. **54**, 4186 (1971).
9. Kolker, H. J., Karplus, M.: J. Chem. Phys. **41**, 1259 (1964).
10. Reid, R. V., May-Chu, A. H.: Phys. Rev. **A9**, 609 (1974).
11. Kempf, J., Spiess, H. W., Haeberlen, U., Zimmermann, H.: Chem. Phys. **4**, 269 (1974).
12. a) Gierke, T. D., Flygare, H. W.: J. Am. Chem. Soc. **94**, 7277 (1972).
b) Gierke, T. D., Tigelaar, H. L., Flygare, H. W.: J. Am. Chem. Soc. **94**, 330 (1972).
13. Malli, G., Fraga, S.: Theoret. Chim. Acta **5**, 284 (1966).
14. Haeberlen, U., Kohlschütter, U., Kempf, J., Spiess, H. W., Zimmermann, H.: Chem. Phys. **3**, 248 (1974).
15. Grosescu, R., Achlama, A. M., Haeberlen, U., Spiess, H. W.: Chem Phys. **5**, 119 (1974).
16. Achlama, A. M., Kohlschütter, U., Haeberlen, U.: Chem Phys. **7**, 287 (1975).
17. Lau, K. F., Vaughan, R. W.: Chem. Phys. Letters (in press).
18. a) Schmiedel, H.: Phys. Stat. Sol. **b67**, K27 (1975).
b) Neubauer, R., Schnabel, B.: First Spec. Colloque Ampere, Krakow, Poland 1973.
19. Haeberlen, U., Kohlschütter, U.: Chem. Phys. **2**, 76 (1973).
20. Raber, H., Brünger, G., Mehring, M.: Chem. Phys. Letters **23**, 400 (1973).
21. Schreiber, L. B., Vaughan, R. W.: Chem. Phys. Letters **28**, 586 (1974).
22. Rhim, W.-K., Elleman, D. D., Vaughan, R. W.: J. Chem. Phys. **59**, 3740 (1973).
23. Haddix, D. C., Lauterbuhr, C. C.: Natl. Bur. Std. (U.S.) Spec. Publ. No. 301, 403 (1969).
24. Haubenreisser, U., Schnabel, B.: Proc. 1st. Spec. Colloque Ampere, Krakow 1973, p. 140.
25. a) Burghoff, U., Scheler, G., Müller, R.: Phys. Stat. Sol. **25a**, K31 (1974).
b) Terao, T., Hashi, T.: J. Phys. Soc. Japan **36**, 989 (1974).
26. Van Hecke, P., Weaver, J. C., Neff, B. L., Waugh, J. S.: J. Chem. Phys. **60**, 1668 (1974).
27. Kohlschütter, U.: Frühjahrstagung der DPG Köln 1975 and Ph. D. Thesis, Heidelberg 1975.
28. a) Spiess, H. W., Achlama, A. M., Haeberlen, U.: Frühjahrstagung der DPG Köln 1975, to be published.
b) Van Hecke, P., Spiess, H. W., Haeberlen, U.: J. Mag. Res. **22**, 103 (1976)
c) Van Hecke, P., Spiess, H. W., Haeberlen, U., Haussuehl, S.: J. Mag. Res. **22**, 93 (1976)
d) Pollak-Stachura, M., Haeberlen, U.: to be published
29. Pines, A., Ruben, D. J., Vega, S., Mehring, M.: Phys. Rev. Lett. **36**, 110 (1976)
30. Waugh, J. S., Huber, L. M., Haeberlen, U.: Phys. Rev. Letters **20**, 180 (1968).
31. a) Vaughan, R. W., Elleman, D. D., Rhim, W.-K., Stacey, L. M.: J. Chem. Phys. **57**, 5383 (1972).
b) Stacey, L. M., Vaughan, R. W., Ellleman, D. D.: Phys. Rev. Letters **26**, 1153 (1971).
32. Sears, R. E.: J. Chem. Phys. **59**, 5213 (1973).
33. a) Phillips, J. C.: Rev. Mod. Phys. **42**, 317 (1970).
b) Gordy, W.: Phys. Rev. **69**, 604 (1946).
c) Gordy, W., Thomas, W. J. O.: J. Chem. Phys. **24**, 439 (1956).
34. Stoeckmann, H. J., Ackermann, H., Dubbers, D., Group, M., Heitjans, P.: Z. Phys. **269**, 47 (1974).
35. Lau, K. F., Vaughan, R. W.: Chem. Phys. Letters (in press).
36. a) Sears, R. E.: J. Chem. Phys. **61**, 4368 (1974).
b) Mehring, M., Pines, A., Rhim, W.-K., Waugh, J. S.: J. Chem. Phys. **54**, 3239 (1971).
37. a) Mehring, M., Griffin, G. R., Waugh, J. S.: J. Am. Chem. Soc. **92**, 7222 (1970).
b) Mehring, M., Griffin, G. R., Waugh, J. S.: J. Chem. Phys. **55**, 746 (1971).
38. Chan, S. I., Dubin, A. S.: J. Chem. Phys. **46**, 1745 (1967).
39. Griffin, R. G., Ellett, J. D., Mehring, M., Bullitt, M., Waugh, J. S.: J. Chem. Phys. **57**, 2147 (1972).
40. Mehring, M., Raber, H., Sinning, G.: Proc. of the 18th. Colloque Ampere, p. 35. Nottingham 1974.
41. a) Nehring, J., Saupe, A.: J. Chem. Phys. **52**, 1307 (1970).
b) Snyder, L.: J. Chem. Phys. **43**, 4041 (1965).

42. Griffin, R. G., Yeung, H. N., LaPrade, M. D., Waugh, J. S.: J. Chem. Phys. **59**, 777 (1973).
43. O'Reilly, D. E., Peterson, E. M., El Saffar, Z. M., Scheie, C. E.: Chem. Phys. Letters **8**, 470 (1971).
44. Carolan, J. L.: Chem Phys. Letters **12**, 389 (1971).
45. a) Hunt, E., Meyer, H.: J. Chem. Phys. **41**, 353 (1964).
b) Brooks Harris, A., Hunt, E., Meyer, H.: J. Chem. Phys. **42**, 2851 (1965).
46. Andrew, E. R., Tunstall, D. P.: Proc. Phys. Soc. **81**, 986 (1963).
47. Yannoni, C. S., Dailey, B. P., Ceasar, G. P.: J. Chem. Phys. **54**, 4020 (1971).
48. Hull, W. E., Sykes, B. D.: J. Mol. Biol. **98**, 121 (1975).
49. a) Raber, H., Mehring, M.: DPG Frühjahrstagung München 1973.
b) Long, R. C., Goldstein, J. H.: J. Chem. Phys. **54**, 1563 (1971).
50. Wilson III, C. W.: J. Polymer. Sci. **61**, 403 (1962).
51. Blinc, R., Zupancic, I., Maricic, S., Veksli, Z.: J. Chem. Phys. **39**, 2109 (1963), **40**, 3739 (1964).
52. Blinc, R., Pirkmayer, E., Slivnik, J., Zupancic, I.: J. Chem. Phys. **45**, 1488 (1966).
53. Hendermann, D. K., Falconer, W. E.: J. Chem. Phys. **50**, 1203 (1969).
54. Van der Hart, D. L., Gutowsky, H. S., Farrar, T. C.: J. Chem. Phys. **50**, 1058 (1969).
55. a) Lauterbur, P. C.: Phys. Rev. Letters **1**, 343 (1958).
b) Yannoni, C. S., Whipple, E. B.: J. Chem. Phys. **47**, 2508 (1967).
56. a) Pines, A., Rhim, W.-K., Waugh, J. S.: J. Chem. Phys. **54**, 5438 (1971).
b) Spiess, H. W., Schweitzer, D., Haeberlen, U., Hausser, K. H.: J. Magn. Res. **5**, 101 (1971).
56. c) Spiess, H. W., Mahnke, H.: Z. Naturforsch. **27a**, 1536 (1972), Ber. Bunsen-Ges. Phys. Chem. **76**, 991 (1972).
57. Kempf, J., Spiess, H. W., Haeberlen, U., Zimmermann, H.: Chem. Phys. Letters **17**, 39 (1972).
58. Pines, A., Gibby, M. G., Waugh, J. S.: a) J. Chem. Phys. **56**, 1776 (1972); b) J. Chem. Phys. **59**, 569 (1973).
59. Pausak, S., Pines, M. G., Waugh, J. S.: J. Chem. Phys. **59**, 591 (1973).
60. a) Pines, A., Gibby, M. G., Waugh, J. S.: Chem. Phys. Letters **15**, 373 (1972).
b) Waugh, J. S., Gibby, M. G., Pines, A., Kaplan, S.: Proc. of the 17th Colloque Ampere, p. 13. Turku, Finland 1972.
61. Kaplan, S., Griffin, R. G., Waugh, J. S.: Chem. Phys. Letters **25**, 78 (1974).
62. Chan, J. J., Griffin, R. G., Pines, A.: a) J. Chem. Phys. **60**, 2561 (1974); b) J. Chem. Phys. **62**, 4923 (1975).
c) Pines, A., Abramson, E.: J. Chem. Phys. **60**, 5130 (1974).
63. Pausak, S., Tegenfeldt, J., Waugh, J. S.: J. Chem. Phys. **61**, 1338 (1974).
64. Pines, A., Chang, J. J., Griffin, R. G.: J. Chem. Phys. **61**, 1021 (1974).
65. Ando, I., Nishioka, A., Kondo, M.: Chem. Phys. Letters **25**, 212 (1974).
66. Stoll, M. E., Vaughan, R. W., Saillant, R. B., Cole, T.: J. Chem. Phys. **61**, 2896 (1974).
67. Spiess, H. W., Grosescu, R., Haeberlen, U.: Chem. Phys. **6**, 226 (1974).
68. Kaplan, S., Pines, A., Griffin, R. G., Waugh, J. S.: Chem. Phys. Letters **25**, 78 (1974).
69. Schweitzer, D., Spiess, H. W.: J. Mag. Res. **16**, 243 (1974).
70. Gibby, M. G., Griffin, R. G., Pines, A., Waugh, J. S.: Chem. Phys. Letters **17**, 80 (1972).
71. Bhattacharyya, P. K., Dailey, B. P.: J. Chem. Phys. **59**, 5820 (1973).
72. Schweitzer, D., Spiess, H. W.: J. Mag. Res. **15**, 529 (1974).
73. Gibby, M. G., Pines, A., Rhim, W.-K., Waugh, J. S.: J. Chem. Phys. **56**, 991 (1972).
74. Gibby, M. G., Pines, A., Waugh, J. S.: J. Am. Chem. Soc. **94**, 6231 (1972).
75. Koma, A., Tanaka, S.: Solid State Commun. **10**, 823 (1972).
76. Koma, A.: Phys. Stat. Sol. **b57**, 299 (1973).
77. Terao, T., Hashi, T.: J. Phys. Soc. Japan **36**, 989 (1974).
78. Zumbulyadis, N., Dailey, B. P.: Chem. Phys. Letters **26**, 273 (1974).
79. Lucken, E. A. C., Williams, D. F.: Mol. Phys. **16**, 17 (1969).
80. Bensoussan, M.: J. Phys. Chem. Sol. **28**, 1533 (1967).

Chapter 6

1. Abragam, A.: The Principles of Nuclear Magn. London: Oxford Univ. Press 1961.
2. Goldman, M.: Spin Temperature and Nuclear Magn. Res. in Solids. London: Oxford Univ. Press 1970.
3. Mansfield, P.: Progr. NMR Spectroscopy **8**, 41 (1971).
4. Look, D. C., Lowe, I. J.: J. Chem. Phys. **44**, 2995 (1966).
5. Jones, G. P.: Phys. Rev. **148**, 332 (1966).
6. Haeberlen, U., Waugh, J. S.: Phys. Rev. **185**, 420 (1969).
7. Gründer, W., Schmiedel, H., Freude, D.: Ann. Phys. (London) **27**, 409 (1971).
8. Gründer, W.: Wiss. Z. Karl-Marx-Univ. Leipzig Math.-Naturw. **23**, 466, (1974).
9. Andrew, E. R., Jasinski, A.: J. Phys. C. Sol. Stat. Phys. **4**, 391 (1971).
10. Ostroff, E. D., Waugh, J. S.: Phys. Rev. Letters **16**, 1097 (1966).
11. Mansfield, P., Ware, D.: Phys. Letters **22**, 133 (1966).
12. Lee, M., Goldberg, W. I.: Phys. Rev. **A410**, 1261 (1965).
13. Roeder, St. B. W., Douglass, D. C.: J. Chem. Phys. **52**, 5525 (1970).
14. Mehring, M., Raber, H., Sinning, G.: Proceedings of the 18th Congress Ampere, p. 35. Nottingham 1974.
15. Mehring, M., Griffin, R. G., Waugh, J. S.: J. Chem. Phys. **55**, 746 (1971).
16. Mehring, M., Raber, H.: J. Chem. Phys. **59**, 1116 (1973).
17. Hilt, R. L., Hubbard, P. S.: Phys. Rev. **134A**, 392 (1964).
18. Wolff, E.: Diploma work, Dortmund 1975.
19. Schütz, J. U. v., Wolf, H. C.: Z. Naturforsch. **27a**, 42 (1972).
20. a) Tse, D., Hartmann, S. R.: Phys. Rev. Letters **21**, 511 (1968).
    b) Lin, N. A., Hartmann, S. R.: Phys. Rev. Letters **21**, 511 (1973).
21. a) Tse, D., Lowe, I. J.: Phys. Rev. **166**, 279 (1968).
    b) Lowe, I. J., Tse, D.: Phys. Rev. **166**, 292 (1968).
22. Wolfe, J. P., Markiewicz, R. S.: Phys. Rev. Letters **28**, 1105 (1973).
23. Kaplan, J. I.: Phys. Rev. **B3**, 604 (1971).
24. Gibby, M. G., Pines, A., Waugh, J. S.: Chem. Phys. Letters **16**, 296 (1972).
25. Pines, A., Gibby, M. G., Waugh, J. S.: J. Chem. Phys. **59**, 569 (1973).
26. a) Willsch, R., Müller, R., Scheler, G.: 2nd Spez. Colloque Ampere, Budapest (Hungary) 1975.
    b) Müller, R., Willsch, R.: J. Mag. Res. **21**, 135 (1976).

Chapter 7

1. a) Rose, M. E., Elementary Theory of Angular Momentum. New York: J. Wiley 1967.
   b) Brink, D. M., Satchler, G. R.: Angular Momentum. London: Oxford Univ. Press 1968.
2. Cook, R. L., De Lucia, F. C.: Am. J. Phys. **39**, 1433 (1971).
3. a) Magnus, W.: Com. Pure Appl. Math. **7**, 649 (1954).
   b) Pechukas, P., Light, F. C.: J. Chem. Phys. **44**, 3897 (1966).
   c) Wilcox, R. M.: J. Math. Phys. **8**, 962 (1967).
   d) Haeberlen, U., Waugh, J. S.: Phys. Rev. **175**, 453 (1968).
4. Haeberlen, U.: Advan. Magn. Res. (in press).
5. Bloch, F., Siegert, A.: Phys. Rev. **57**, 552 (1940).
6. Abragam, A.: Principles of Nuclear Magnetism , Chap. IV, London: Oxford Univ. Press 1961.
7. a) Bosse, J.: DFG-Conference on Magnetic Relaxation Hirschegg, Austria, Sept. 1974.
   b) Mansfield, P.: Progr. Nucl. Magn. Res. **8**, 43 (1971).
8. a) Fano, U.: Rev. Mod. Phys. **29**, 74 (1957).
   b) Shimizu, T.: J. Phys. Soc. Jap. **28**, 790 (1970), **28**, 811 (1970).
9. Demco, D., Tegenfeldt, J., Waugh, J. S.: Phys. Rev. **B11**, 4133 (1975).
10. a) Zwanzig, R.: Lectures in Theoretical Physics, New York: Interscience 1961.
    b) Mori, H.: Progr. Theoret. Phys. Jap. **34**, 399 (1965).
11. a) Lado, F., Memory, J. D., Parker, G. W.: Phys. Rev. **B4**, 1406 (1971).
    b) Parker, G. W., Lado, F.: Phys. Rev. **B8**, 3081 (1973).

Author Index Volumes 1–11

Bergmann, K.: Untersuchung von Beweglichkeiten in Polymeren durch NMR. **4**, 233–246 (1971).
Bovey, F. A.: High Resolution NMR Spectroscopy of Polymers. **4**, 1–9 (1971).

Cantow, H.-J., Elgert, K. F., Seiler, E., Friebolin, H.: NMR-Untersuchungen an Poly-α-Methylstyrol und dessen Copolymeren mit Butadien. **4**, 21–46 (1971).
Connor, T. M.: Magnetic Relaxation in Polymers. The Rotating Frame Method. **4**, 247–270 (1971).

Diehl, P., Kellerhals, H., Lustig, E.: Computer Assistance in the Analysis of High-Resolution NMR Spectra. **6**, 1–96 (1972).
Diehl, P., Khetrapal, C. L.: NMR Studies of Molecules Oriented in the Nematic Phase of Liquid Crystals. **1**, 1–96 (1969).

Fischer, H.: ESR-Untersuchungen an Hochpolymeren. **4**, 301–309 (1971).
Forslind, E.: Nuclear Magnetic Resonance Wide Line Studies of Water Sorption and Hydrogen Bonding in Cellulose. **4**, 145–166 (1971).

Guillot, J.: Penultimate Effects in Radical Copolymerization I – Kinetical Study. **4**, 109–118 (1971).

Harwood, H. J.: Problems of Aromatic Copolymer Structure. **4**, 71–99 (1971).
Hilbers, C. W., MacLean, C.: NMR of Molecules Oriented in Electric Fields. **7**, 1–52 (1972).
Hill, H. A. O.: The Proton Magnetic Resonance Spectroscopy of Proteins. **4**, 167–179 (1971).
Hoffmann, R. A., Forsén, S., Gestblom, B.: Analysis of NMR Spectra. **5**, 1–165 (1971).

Jones, R. G.: The Use of Symmetry in Nuclear Magnetic Resonance. **1**, 97–174 (1969).

Kanert O., Mehring, M.: Static Quadrupole Effects in Disordered Cubic Solids. **3**, 1–81 (1971).
Keller, H. J.: NMR-Untersuchungen an Komplexverbindungen. **2**, 1–88 (1970).
Khetrapal, Kunwar, Tracey, Diehl: Nuclear Magnetic Resonance Studies in Lyotropic Liquid Crystals. **9**, 1–85 (1975).
Klesper, E., Gronski, W., Johnsen, A.: Complete Triad Assignment of Methylmethacrylate-Methacrylic Acid Copolymers. **4**, 47–69 (1971).
Kosfeld, R., Mylius, U. v.: Linienbreiten- und Relaxationsphänomene bei der NMR-Festkörperspektroskopie. **4**, 181–208 (1971).

Mehring, M.: High Resolution NMR Spectroscopy in Solids. **11**, 1–243 (1976).

Noack, F.: Nuclear Magnetic Relaxation Spectroscopy. **3**, 83–144 (1971).

Pfeifer, H.: Nuclear Magnetic Resonance and Relaxation of Molecules Adsorbed on Solids. **7**, 53–153 (1972).
Pham, Q. T.: The Cotacticity of (Acrylonitrile-Methyl-Methacrylate) Copolymer by NMR Spectroscopy. **4**, 119–128 (1971).

Richard, C., Granger, P.: Chemically Induced Dynamic Nuclear and Electron Polarizations-CIDNP and CIDEP. **8**, 1–127 (1974).
Rummens, F. H. A.: Van der Waals Forces in NMR Intermolecular Effects. **10**, 1–118 (1975).

Shimanouchi, T.: Conformations of Polymer Chains as Revealed by Infrared Spectroscopy. **4**, 287–299 (1971).
Slichter, W. P.: NMR Studies of Solid Polymers. **4**, 209–231 (1971).

Tosi, C.: New Concepts in Copolymer Statistics. **4**, 129–144 (1971).

Williams, G., Watts, D. C.: Some Aspects of the Dielectric Relaxation of Amorphous Polymers Including the Effects of a Hydrostatic Pressure. **4**, 271–285 (1971).

Zambelli, A.: Research of Homopolymers and Copolymers of Propylene. **4**, 101–108 (1971).

Related Titles

M. v. Ardenne, K. Steinfelder, R. Tümmler
Elektronenanlagerungs-Massenspektrographie organischer Substanzen
109 Abbildungen. VIII, 403 Seiten. 1971

K. Scheffler, H. B. Stegmann
Elektronenspinresonanz
Grundlagen und Anwendung in der organischen Chemie
145 Abbildungen. VIII, 506 Seiten. 1970
(Organische Chemie in Einzeldarstellungen)

Ch. K. Jørgensen
Oxidation Numbers and Oxidation States
VII, 291 pages. 1969
(Molekülverbindungen und Koordinationsverbindungen in Einzeldarstellungen)

New Theoretical Aspects
60 figs. 7 tables. IV, 186 pages. 1975
(Topics in Current Chemistry, Vol. 58)
Contents: A. Julg, On the Description of Molecules Using Point Charges and Electric Moments. R. S. Butler, CRAMS–An Automatic Chemical Reaction Analysis and Modeling System. R. Geick, IR Fourier Transform Spectroscopy.

Structure of Liquids
88 figs. 38 tables. IV, 205 pages. 1975
(Topics in Current Chemistry, Vol. 60)
Contents: P. Schuster; W. Jakubetz; W. Marius, Molecular Models for the Solvation of Small Ions and Polar Molecules. S. A. Rice, Conjectures on the Structure of Amorphous Solid and Liquid Water.

Theoretical Inorganic Chemistry
22 figs. 18 tables. IV, 159 pages. 1975
(Topics in Current Chemistry, Vol. 56)
Contents: C. K. Jørgensen, Continuum Effects Indicated by Hard and Soft Antibases (Lewis Acids) and Bases. H. Brunner, Stereochemistry of the Reactions of Optically Active Organometallic Transition Metal Compounds. L. H. Pignolet, Dynamics of Intramolecular Metal-Centered Rearrangement Reactions of Iris-Chelate Complexes. S. Vepřek, A Theoretical Approach to Heterogeneous Reactions in Non-Isothermal Low Pressure Plasma.

Springer-Verlag
Berlin Heidelberg New York